Lecture Notes in Computer Science

Commenced Publication in 1973
Founding and Former Series Editors:
Gerhard Goos, Juris Hartmanis, and Jan van Leeuwen

Maarten Keijzer   Andrea Tettamanzi
Pierre Collet   Jano van Hemert
Marco Tomassini (Eds.)

Genetic
Programming

8th European Conference, EuroGP 2005
Lausanne, Switzerland, March 30 – April 1, 2005
Proceedings

Volume Editors

Maarten Keijzer
WL | Delft Hydraulics
Rotterdamseweg 181, Delft, The Netherlands
E-mail: mkeijzer@xs4all.nl

Andrea Tettamanzi
Università degli Studi di Milano, Dipartimento di Tecnologie dell'Informazione
c/o Polo Didattico e di Ricerca di Crema
Via Bramante 65, 26013 Crema , Italy
E-mail: andrea.tettamanzi@unimi.it

Pierre Collet
Université du Littoral Côte d'Opale, Laboratoire d'Informatique du Littoral
B.P. 719, 62100 Calais cedex, France
E-mail: pierre.collet@univ-littoral.fr

Jano van Hemert
Napier University, School of Computing, Centre for Emergent Computing
10 Colinton Road, Edinburgh EH10 5DT, UK
E-mail: jvhemert@cwi.nl

Marco Tomassini
University of Lausanne, HEC, Information Systems Department
1015 Lausanne, Switzerland
E-mail: marco.tomassini@hec.unil.ch

Cover illustration: *Triangular Urchin*, by Chaps (www.cetoine.com). Chaps has obtained an MSc in Physics at the Swiss Federal Institute of Technology. He is the developer of the ArtiE-Fract software that was used to create *Triangular Urchin*. *Triangular Urchin* (an Iterated Functions System of 2 polar functions) emerged from an urchin structure after a few generations using ArtiE-Fract. The evolutionary process was only based on soft mutations, some of them directly induced by the author.

Library of Congress Control Number: 2005922866
CR Subject Classification (1998): D.1, F.1, F.2, I.5, I.2, J.3

ISSN   0302-9743
ISBN-10   3-540-25436-6 Springer Berlin Heidelberg New York
ISBN-13   978-3-540-25436-2 Springer Berlin Heidelberg New York

Springer is a part of Springer Science+Business Media

springeronline.com
© Springer-Verlag Berlin Heidelberg 2005
Printed in Germany

Typesetting: Camera-ready by author, data conversion by Scientific Publishing Services, Chennai, India
Printed on acid-free paper   SPIN: 11410591   06/3142   5 4 3 2 1 0

Preface

In this volume we present the contributions for the 8th European Conference on Genetic Programming (EuroGP 2005). The conference took place from 30 March to 1 April in Lausanne, Switzerland. EuroGP is a well-established conference and the only one exclusively devoted to genetic programming. All previous proceedings were published by Springer in the LNCS series. From the outset, EuroGP has been co-located with the EvoWorkshops focusing on applications of evolutionary computation. Since 2004, EvoCOP, the conference on evolutionary combinatorial optimization, has also been co-located with EuroGP, making this year's combined events one of the largest dedicated to evolutionary computation in Europe.

Genetic programming (GP) is evolutionary computation that solves complex problems or tasks by evolving and adapting a population of computer programs, using Darwinian evolution and Mendelian genetics as its sources of inspiration. Some of the 34 papers included in these proceedings address foundational and theoretical issues and there is also a wide variety of papers dealing with different application areas, such as computer science, engineering, language processing, biology and computational design, demonstrating that GP is a powerful and practical problem-solving paradigm.

A rigorous, double-blind, peer-review selection mechanism was applied to 64 submitted papers. This resulted in the acceptance of 20 plenary talks and 14 poster presentations. Each paper was reviewed by three or four members of the international Programme Committee who were selected as fairly as possible by matching a reviewer's particular interests and special expertise to the topics covered by the paper. The results of this process are reflected in the quality of the contributions published within this volume. This year the overall acceptance rate for talks and poster presentations was 53%.

We would like to express our sincere gratitude to the two internationally renowned invited speakers who gave keynote talks at the conference: Prof. Matteo Fischetti of the University of Padova, Italy and Prof. Alberto Piazza from the University of Torino.

The success of any conference results from the input of many people, to whom we would like to express our appreciation. Firstly, we would like to thank the members of the Programme Committee for their attentiveness, perseverance and willingness to provide high-quality reviews. The local team (Mario Giacobini, Leslie Luthi, Denis Rochat and Leonardo Vanneschi), headed by Prof. Marco Tomassini, must also be thanked: the smooth development of the conference has been their feat. Finally, we would also like to thank

Jennifer Willies for her continuous efforts and support, as well as for her valuable and professional help with all the organizational and logistic aspects of organizing the event.

April 2005

Maarten Keijzer
Andrea Tettamanzi
Pierre Collet
Jano van Hemert
Marco Tomassini

Organization

EuroGP 2005 was organized by EvoGP, the EvoNet Working Group on Genetic Programming.

Organizing Committee

Program Co-chairs: Maarten Keijzer
KiQ Ltd., Amsterdam and WL\Delft Hydraulics, Delft, The Netherlands
Andrea Tettamanzi
Genetica S.r.l. and University of Milan, Italy

Publication Chair: Pierre Collet
Université du Littoral (Calais), France

Local Chair: Marco Tomassini
University of Lausanne, Switzerland

Publicity Chair: Jano van Hemert
Napier University, Edinburgh, UK

Programme Committee

Abbass, Hussein. University of New South Wales. Australia
Araujo, Lourdes. Universidad Complutense de Madrid. Spain
Aydin, Mehmet Emin. London South Bank University. UK
Azad, R. Muhammad Atif. University of Limerick. Ireland
Bailleux, Olivier. Université de Bourgogne. France
Banzhaf, Wolfgang. Memorial University of Newfoundland. Canada
Borovik, Alexandre. University of Manchester. UK
Brabazon, Anthony. University College Dublin. Ireland
Burke, Edmund Kieran. University of Nottingham. UK
Cagnoni, Stefano. University of Parma. Italy
Carbajal, Santiago García. University of Oviedo. Gijon. Spain
Cardoso, F. Amilcar. University of Coimbra. Portugal
Cheang, Sin Man. Hong Kong Institute of Vocational Education. China
Clergue, Manuel. Laboratoire I3S. France
Collard, Philippe. Université de Nice–Sophia-Antipolis. France
Collet, Pierre. Université du Littoral, Calais. France
Costa, Ernesto Jorge. University of Coimbra. Portugal
De Jong, Edwin. Utrecht University. The Netherlands
Ebner, Marc. Universität Würzburg. Germany

Ekart, Aniko. Computer and Automation Research Institute, Hungarian Academy of Sciences. Hungary
Essam, Daryl Leslie. ADFA, UNSW. Australia
Fernandez, Francisco. University of Extremadura. Spain
Folino, Gianluigi. ICAR-CNR. Italy
Fonlupt, Cyril. Université du Littoral, Calais. France
Foster, James A. University of Idaho. USA
Gustafson, Steven. University of Nottingham. UK
Hao, Jin-Kao. University of Angers. France
Hemert, van, Jano. Napier University. UK
Hirsch, Laurence Benjamin. Royal Holloway University of London. UK
Hochreiter, Ronald. University of Vienna. Austria
Howard, Daniel. QinetiQ. UK
Johnson, Colin. C.G. Johnson. UK
Kalganova, Tatiana. Brunel University. UK
Kendall, Graham. University of Nottingham. UK
Kim, DaeEun. Max Planck Institute for Human Cognitive and Brain Sciences. Germany
Kochenderfer, Mykel John. University of Edinburgh. UK
Kubalik, Jiri. Czech Technical University in Prague. Czech Republic
Kuo, Tzu-Wen. AI-ECON Research Center. Taiwan
Langdon, William B. University of Essex. UK
Leung, Kwong Sak. Chinese University of Hong Kong. Hong Kong, China
Levine, John. University of Strathclyde. UK
Lopes, Heitor Silvério. CEFET-PR/BIOINFO. Brazil
Lucas, Simon Mark. Essex University. UK
MacCallum, Bob. Stockholm Bioinformatics Center. Sweden
Machado, Penousal. Centre for Informatics and Systems, University of Coimbra. Portugal
Martin, P.N. Naiad Consulting. UK
McKay, Robert Ian. University of New South Wales. Australia
Mehnen, Jörn. ISF, Department of Machining Technology. Germany
Miller, Julian Francis. University of York. UK
Monsieurs, Patrick. Expertise Center for Digital Media. Belgium
Nicolau, Miguel. University of Limerick. Ireland
Nievola, Julio Cesar. PUCPR. Brazil
O'Neill, Michael. University of Limerick. Ireland
Pizzuti, Clara. ICAR-CNR. Italy
Poli, Riccardo. University of Essex. UK
Robilliard, Denis. Univ. Littoral, Côte d'Opale. France
Rodriguez-Vazquez, Katya. UNAM. Mexico
Rothkrantz, Leon. Delft University of Technology. The Netherlands
Ryan, Conor. University of Limerick. Ireland
Saitou, Kazuhiro. University of Michigan. USA
Sapin, Emmanuel. LERSIA. France

Schoenauer, Marc. INRIA. France
Sebag, Michèle. CNRS, Université Paris-Sud. France
Sekanina, Lukas. Brno University of Technology. Czech Republic
Skourikhine, Alexei. Los Alamos National Laboratory. USA
Spezzano, Giandomenico. Institute of High Performance Computing and
    Networking (ICAR)-CNR. Italy
Streeter, Matthew J. Carnegie Mellon University. USA
Tavares, Jorge. University of Coimbra. Portugal
Tommassini, Marco. University of Lausanne. Switzerland
Vanneschi, Leonardo. University of Milan-Bicocca. Italy
Ványi, Róbert. Institute of Informatics, University of Szeged. Hungary
Wilson, Garnett Carl. Dalhousie University. Canada
Wolff, Krister. Chalmers University of Technology. Sweden
Woodward, John Robert William. Birmingham University. UK

Sponsoring Institutions

EvoNet: The Network of Excellence in Evolutionary Computing, funded by the
European Commission's IST Programme

Table of Contents

Talks

Posters

An Algorithmic Chemistry for Genetic Programming

Christian W.G. Lasarczyk[1] and Wolfgang Banzhaf[2],[*]

[1] Department of Computer Science, University of Dortmund,
D-44221 Dortmund, Germany
christian.lasarczyk@uni-dortmund.de
[2] Department of Computer Science, Memorial University of Newfoundland,
St. John's, NL, A1B 3X5, Canada
banzhaf@cs.mun.ca

Abstract. Genetic Programming has been slow at realizing other programming paradigms than conventional, deterministic, sequential von–Neumann type algorithms. In this contribution we discuss a new method of execution of programs introduced recently: Algorithmic Chemistries. Therein, register machine instructions are executed in a non–deterministic order, following a probability distribution. Program behavior is thus highly dependent on frequency of instructions and connectivity between registers. Here we demonstrate the performance of GP on evolving solutions to a parity problem in a system of this type.

1 Introduction

Representations in genetic programming encode functionality both explicitly by choosing from a set of operations and implicitly by choosing a position within the genome. While it is "easy" to inherit the explicitly encoded portion of functionality, variable genome length leads to difficulties in inheritance of implicitly encoded functionality.

In this contribution we present a different way of looking at transformations from input to output that does not require a prescribed sequence of computational steps and therefore no implicitly coded functionality. Instead, the elements of the transformation, which in our case are single instructions from a multiset $I = \{I_1, I_2, I_3, I_2, I_3, I_1, \ldots\}$ are drawn in a random order to produce a transformation result. In this way we dissolve the explicit sequential order usually associated with an algorithm for our programs.

A program in this sense is thus not a sequence of instructions but rather an assemblage of instructions that can be executed in arbitrary order. By randomly

---

[*] The authors gratefully acknowledge support from a grant of the Deutsche Forschungsgemeinschaft DFG (German Research Foundation) to W.B. under Ba 1042/7–3.

M. Keijzer et al. (Eds.): EuroGP 2005, LNCS 3447, pp. 1–12, 2005.

choosing one instruction at a time, the program proceeds through its transformations until a predetermined number of instructions has been executed. It is therfore more akin to a chemical system with data as educts and products, and operations as reactions than to an "orderly" execution of code.

Programs of this type can be seen as artificial chemistries, where instructions interact with each other (by taking the transformation results from one instruction and feeding them into another). Different multisets can be considered different programs, whereas different passes through a multiset can be considered different behavioral variants of a single program.

Because instructions are drawn randomly in the execution of the program, it is really the concentration of instructions that matters most. It is thus expected that "programming" of such a system requires the proper concentration of instructions, while an explicit sequencing is not required.

At first, this kind of repeated execution of instructions seems to be a waste of computational power. While it is always possible to transform a snapshot[1] of an individual into a linear program, hardware centered improvements of execution speed are imaginable, too. E.g., a huge number of processors could execute the same multiset of instructions in parallel. In the extreme case of the number of processors equal to the number of instructions running time is reduced to a minimum predetermined by depth of data flow. Specialized multiprocessor systems, such as the wavescalar–architecture[1, 2], hold potential to achieve this speed up using less processors.

Due to the stochastic nature of results, it might be advisable to execute a program multiple times before a conclusion is drawn about its "real" output. In this way, it is again the concentration of output results that matters. Therefore, a number of n passes through the program should be taken before any reliable conclusion about its result can be drawn. Reliability in this sense would be in the eye of the beholder. Should results turn out to be not reliable enough, simply increasing n would help to narrow down the uncertainty. Thus the method is perfectly scalable, with more computational power thrown at the problem achieving more accurate results.

We believe that, despite the admitted inefficiency of our approach in the small, it might well beat sequential or synchronized computing at large, if we imagine tens of thousands or millions of processors at work.

Algorithmic Chemistries were considered earlier in the work of Fontana [3]. In our contribution we use the term as an umbrella term for those kinds of artificial chemistries [4] that aim at algorithms. As opposed to terms like randomized or probabilistic algorithms, in which a certain degree of stochasticity is introduced explicitely, our algorithms have an implicit type of stochasticity. Executing the sequence of instructions every time in a different order has the potential of producing highly unpredictable results.

It will turn out, however, that even though the resulting computation is unpredictable in principle, evolution will favor those multisets of instructions that

---

[1] Ambiguousness starts, if different instructions share the same target.

turn out to produce approximately correct results after execution. This feature of approximating the wished-for results is a consequence of the evolutionary forces of mutation, recombination and selection, and will have nothing to do with the actual order in which instructions are being executed. Irrespective of how many processors would work on the multiset, the results of the computation would tend to fall into the same band of approximation. We submit, therefore, that methods like this can be very useful in parallel and distributed environments.

Following previous work on Artificial Chemistries (see, for example [5, 6, 7, 8]), [9] introduces a very general analogy between chemical reaction and algorithmic computation, arguing that concentrations of results would be important. [10] was the first step in this new direction. Here we want to deepen our understanding of the resulting system by studying the GP task of even-parity.

2   Algorithmic Chemistry

On executing a sequence of instructions using linear GP[11], each point in execution time is assigned to exactly one instruction, which is executed at that very moment. This principle is even the same, if instructions are stored in a tree like data structure (e.g. Tree–GP[12]).

Applying Tree–GP, functional dependence of instructions is related to their distance within the tree. Subtrees possess sub–functionality, an edge carries an implicit specification for the subtree it connects to the tree. This specification has to be satisfied during recombination. Using linear GP, functional dependence is determined by both, distance within the genome and source and target registers used by instructions. Therefore successful recombination has to consider both.

2.1   GP to AC — A Gradual Transition

Here we shall use 3–address machine instructions. The genotype of an individual is a list of those instructions. Each instruction consists of an operation, a destination register, and two source registers[2]. Initially, individuals are produced by randomly choosing instructions. As is usual, we employ a set of fitness cases in order to evaluate (and subsequently select) individuals.

A time–dependent probability distribution determines the sequence of instructions. Linear GP uses a discrete distribution:

$$P_t(X = x_i) = \begin{cases} 1, \text{if } i = t \\ 0, \text{else} \end{cases}, \quad t, i \in 1, 2, \ldots, n. \tag{1}$$

Position in memory is denoted by x_i, and the individual consists of n instructions. Starting at $t = 1$ exactly one instruction gets executed at each moment in time, followed by the next instruction in memory until at $t = n$ all instructions got executed in exactly the same order as they appear in memory. This is shown on left side of Fig. 1. Thus, the location in memory space determines

---

[2] Operations, which require only one source register, simply ignore the second register.

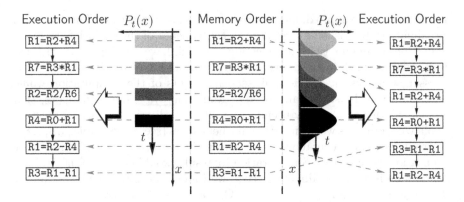

Fig. 1. Execution of an individual. Transition from memory to execution order is determined by a time dependent distribution function. Left side shows transition using distribution function of Eq. 1, emulating linear GP. On right side transition occurs by using a normal distribution. Different gray tones of distribution functions represent different points in time

the particular sequence of instructions. Classically, this is realized by the program counter. Each instruction is executed, with resulting data stored in its destination register.

If we use a distribution to access instructions as described above, we come to a new class of algorithms by changing this distribution. On the right side of Fig. 1 shows a different execution order result from using a Gaussian distribution. If the standard deviation σ is increased the influence of time on instruction selection decreases. In the extreme case $\sigma \to \infty$ a uniform distribution results and all instructions have the same probability to be drawn at any moment. This we call an Algorithmic Chemistry.

Using a uniform distribution, behavior of a program during execution will differ from instance to instance. There is no guarantee that an instruction is executed, nor is it guaranteed that this happens in a definite order or frequency. If, however, an instruction is more frequent in the multi-set, then its execution will be more probable. Similarly, if it should be advantageous to keep independence between data paths, the corresponding registers should be different in such a way that the instructions are not connecting to each other. Both features would be expected to be subject to evolutionary forces.

As shown in Fig. 2, also 1–Point–*Crossover* could be described, using time depended distributions. While the first part of an offspring is formed by instructions drawn from first parent during time interval $1 \leq t \leq c_1 (\leq n_1)$, the second part is drawn from second parent during time interval $(1 \leq) c_2 \leq t \leq n_2$.

Though the instructions inherited from each of the parents are located in contiguous memory locations, the actual sequence of the execution is not determined by that order once we use a distribution to access instructions. The probability that a particular instruction is copied into an offspring again depends on the frequency of that instruction in the parent. Inheritance therefore

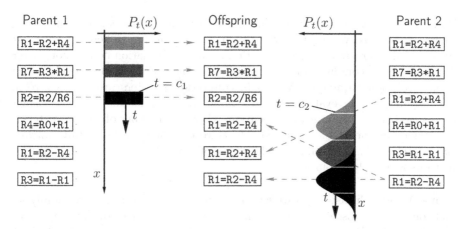

Fig. 2. Recombination accesses parents instruction via the same time dependent distribution function $P_t(x)$ used at evaluation. While we take instruction from the first parent during time span $1 \leq t \leq c_1$, we draw instruction from the second parent during time span $c_2 \leq t \leq n_2$. In contrast to the depict situation $P_t(x)$ is the same for accessing both parents instructions

is inheritance of frequencies of instructions, rather than of particular sequences of instructions.

Estimation of Distribution Algorithms (EDAs). Estimation of Distribution Algorithms[13] are a relatively new class of approaches to evolutionary computation. These population based algorithms generate offsprings by two steps, omitting crossover and mutation. At first, they estimate the probability distribution of a selected subset of the current population, and subsequently they sample a new population from this distribution. We can think of an Algorithmic Chemistries as an implicit description of an instruction distribution by storing a set of samples from this distribution. Recombination is similar to creating a new common distribution based on two of the selected individuals and sampling an offspring from it. While this kind of sampling is not able to create something new, mutation is still needed.

2.2   Algorithmic Chemistry in Detail

Having derived execution, utilizing a 3–address–machine, and crossover of individuals on Algorithmic Chemistry for Genetic Programming(ACGP) from linear GP, we explain further details in this section, including additional information on crossover and evaluation.

Registers. We distinguish between three different kinds of registers:

- connection registers
- input registers
- registers containing constant values

While instructions could read from all of them, and thus can use them as a source register, they can just write to *connection registers*. Therefore these are the only valid targets to store instruction results. Information flows among them in the course of computation. The number of connection registers could be set as an evolution parameter. Values in connection registers are set to zero at the beginning of an evaluation. Each *input register* contains data of a single fitness case at the beginning of execution, where the number of these registers is determined by the problem tackled. The third type of registers contain *constant values* evolved during evolution. The choice of a *result register* out of connection registers is done by evolution.

More About Crossover and Evaluation. Parents are chosen randomly for each offspring. Crossover rate assigns the proportion of offspring created by recombination, the rest of offsprings is created by reproductive cloning. In both cases mutation is applied afterwards. During crossover constant register values will be copied with equal probability from each parent, as is done for the choice of the result register if necessary.

The number of executed instructions on linear GP and Tree–GP is limited by the number of instructions contained in an individual's genome. As described in Sec. 1 an instructions in an Algorithmic Chemistry can be executed successfully – in fitness improving sense – if all required sources contain correct inputs. Therefore, it could be reasonable to increase the upper limit on execution and cycle ($P_{n+t} = P_t$) through individuals more than once. In the case of a constant uniform distribution, as used by the Algorithmic Chemistry presented here, this means that we could execute available instructions multiple times by drawing them randomly. The number of *cycles*, is an additional evolution parameter.

Because evaluating an individual is a stochastic process, it could be useful to evaluate individuals more than once and combine the results to get a single fitness value. We will discuss this in detail later.

Initialization and Mutation. Initialization and mutation of an individual are the same for both the ACGP and usual linear GP. Mutation changes single instructions by changing operation, target register and the source registers according to a prescribed probability. Register values are mutated by using a Gaussian with mean at present value and standard deviation 1.

Selection. We use a (μ, λ)–strategy. In doing so a set of μ parents produce λ offspring first. The λ best individuals of these offspring form the set of next generation's parents.

3   Results and Outlook

Since in [10] we already discussed an approximation and a real-world classification problem, we now evolve a Boolean function using Genetic Programming of Algorithmic Chemistries.

3.1   Even–Parity Problem

Boolean problems are used as popular benchmark problems in GP. The even-parity problem, widely discussed in [12], tries to generate the value of a bit, so that with an input of three external bits, even parity is provided. The individuals can use four logical operations {AND, OR, NAND, NOR}. The cost for random search has been discussed in [14].

ACGP uses real–valued registers. For boolean operations, > 0 values will be mapped to true, ≤ 0 values to false. The fitness function corresponds to the fraction of fitness cases an individual could not generate even parity for. The solution hoped for is to have a fitness of zero.

3.2   ACGP Settings

We do non claim, that we use optimized settings. Nevertheless we think it is important to describe the amount of optimization done so far and describe our settings.

To chose an appropriate setting we create a space filling latin hypercube design with 50 runs on a reduced subspace of our parameter space. Roughly speaking this means, that we divide each parameter into 50 evenly spaced levels[3] and then choose 50 points in parameter space maximizing minimum distance between points considering theses levels.

Table 1. ACGP settings and ranges of our space filling design

parameter	setting	design range min	max
offsprings	450	200	500
crossover rate	0.45	0.0	0.6
mutation rate	0.01	0.0	0.05
initial length	50	10	50
maximal length	150	300	500
cycles	3.5	1	5
connection register	40	30	60
parents	100		
evaluations per ind. (m)	1,2,4,8,16	8	
evolved constants	2		

For each design point we start four runs, executing 10^{10} instructions each. Influenced by those runs showing good average test performance we choose our setting. Table 1 shows ranges of considered parameter subspace and finally selected settings.

---

[3] We even do so for integers and round afterwards.

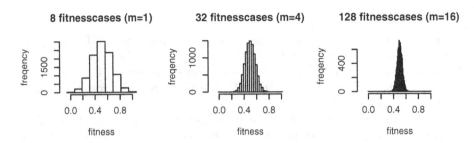

Fig. 3. Fitness distribution of a single initial individual using different training set sizes. These training sets are generated by multiplication of original set containing 8 fitness cases m times. As this is an individual of initial population mean value is expected to be 0.5, standard deviations of noise are $\sigma_{m=1} = 0.163, \sigma_{m=4} = 0.083$ and $\sigma_{m=16} = 0.041$

3.3 Stochastic Noise

During the observation of the even–parity problem a difficulty has occurred, that is unknown for other GP variants: Originating from the small training set of only 8 fitness cases the stochastic noise gains in influence. Through the non–deterministic order of execution of instructions, a multiple execution of a single individual can lead to different results. Especially at the beginning of evolution, while instructions of a chemistry are purely random, and different instructions share the same target register.

To reduce this noise we initially use the concept of repeated evaluation. To do this, we not only execute the algorithmic chemistry of the individual once on the training set of 8 fitness cases, but m times. Accordingly, with $m = 2$ the fitness corresponds to the fraction of 16 fitness cases that did not generate even parity, while there are still just 8 different fitness cases. As a side effect, the resolution of the fitness calculation increases. Instead of 8 different values for $m = 1$ resulting in a 0.125 margin distance, an $m = 4$ approach results in 32 values with a 0.03125 margin distance.

Figure 3 displays the histograms of the fitness values from 10000 analyses of an individual of the initial population using differently sized sets as basis of valuation. In addition to the higher resolution, reduced standard deviation can be observed.

Each increase on size of training set reduces the number of generations for an unchanged number of instructions allowed to be executed. In our system the number of executed instructions serves as a measure of time, each run is allowed to execute 10^{10} instructions.

Figure 4 reveals different fitness development for $m \in \{1, 2, 4, 8, 16\}$, averaged over 100 runs. For evaluation during the training phase, a corresponding number of fitness cases was used. A validation was performed in regular intervals, choosing the best individual based on 128 fitness cases. For testing purposes, fitness is compared to this larger number of fitness cases afterwards.

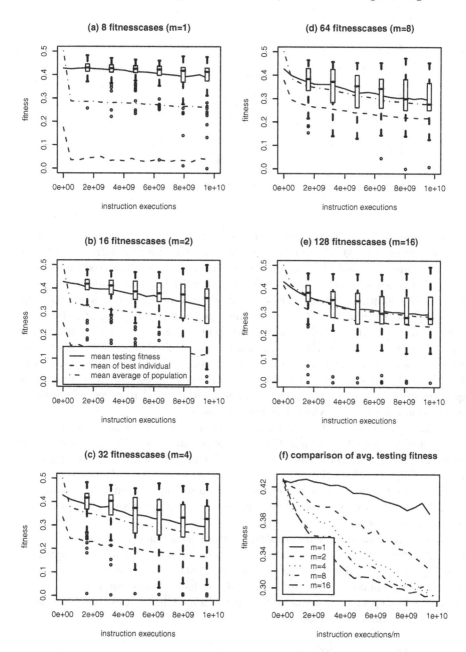

Fig. 4. Using different training set sizes we show fitness of best individual, mean population fitness and fitness on testing set of population's best individual on validation set averaged over 100 runs. The chart in bottom right corner is for direct comparison of achieved testing set fitness

input

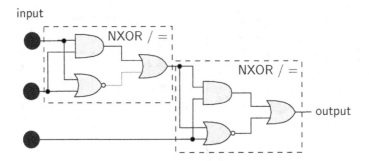

Fig. 5. Circuit of a solution found using Algorithmic Chemistries

The smaller the size of training set, the better the best individual of a population looks on average. Assuming, that the noise for all initial individuals equals that of figure 3, it easily becomes obvious, that among 100 observed individuals one might be chosen whose noise leads to a good fitness. This could prove problematic for evolution, because this selection is not founded and does not hold out against further evaluation. Also, in the next generation, offspring of individuals selected this way might underachieve. This leads to early stagnation among runs with a small training set. For instance, if $m = 1$ (original training set size) most runs do not improve much after initialization and good runs are a very rare event (cf. Fig.4(a) and solid line in Fig.4(f)). Even worse, these runs can pass most generation tests, because using the same individual $m = 1$ requires half as much instruction executions – which are limited – as $m = 2$.

Things get better for $m = 2$, Fig.4(b). While evolutionary improvement take place slowly, these Algorithmic Chemistries continue to evolve. Fitness of the best individual, however, is on average inferior to the case $m = 1$, though more realistic. This trend continues in Fig.4(c-e) with $m \in 4, 8, 16$. Here we can also observe that testing fitness converges to the mean population fitness, which is an indication that effective evolutionary progress is more strongly coupled with population dynamics.

While Algorithmic Chemistries suffer from noise they introduced by non–deterministic instruction execution, this problem can be handled by increasing training set size: By duplicating fitness cases, as done here, noise can be reduced. This reduction in noise, however, comes with an increase in computational power demand. We expect noise reduction in Algorithmic Chemistries to be an important topic for further investigations.

3.4   A Glance at a Solution

Because the data flow of this individual is "nearly unique", it is easy to extract the corresponding circuit, as shown in Fig.5. Here "nearly unique" means, that there are for one connection register two different instructions using it as their target. Because data flow in Algorithmic Chemistries via connection registers, this flow can be symbolized on circuits though conductors. "Nearly unique" conductors are drawn using gray color. Missing uniqueness is caused by a functional intron like this:

```
r27 = r27 AND true
```

The Boolean value **true** resides in a constant register, and it is obvious that register **r27** is not changed through execution. The evolved solution presented here consists of two NXOR–gates. These gates test for equivalence and return 'true' if their inputs are equal.

3.5 Outlook

While using additional cases to evaluate an individual's fitness is able to reduce noise, enlarging a training set is computationally expensive. In fact, applying a training set m times only reduces noise by $\sqrt{m}$. In the future, therefore, sequential sampling techniques [15] shall be used to compare the fitness of individuals. This technique does not fix the number of fitness cases in advance, but ensures a desired level of confidence by treating fitness cases one at a time.

In Sec.3.2 we used a very simple approach to choose settings for ACGP from parameter space. In further work we plan to use methodologies described in [16, 17] useful to analyze and optimize evolutionary algorithms and other search heuristics. Beside an improved system performance we hope for further insights in behavior of algorithmic chemistries.

In this study on our Algorithmic Chemistries for Genetic Programming we considered only uniform distributions of instruction choice. However, other distributions are possible as well. A uniform distribution is, however, the most extreme case, since it ignores order completely when drawing instructions from an equal distribution. By using a normal distribution, we plan to investigate the algorithmic space between linear GP and ACGP.

As mentioned before, there are already some similarities of EDAs and the present approach. By mixing all selected Algorithmic Chemistries (creating a common multi–set of instructions) and drawing new offspring from this "common distribution" we can go one further step in the direction of EDAs.

References

1. Swanson, S., Oskin, M.: Towards a universal building block of molecular and silicon computation. In *1st Workshop on Non-Silicon Computing (NCS-1)* (2002)
2. Swanson, S., Michelson, K., Oskin, M.: Wavescalar. Technical report, University of Washington, Dept. of Computer Science and Engineering (2003)
3. Fontana, W.: Algorithmic chemistry. In Langton, C.G., Taylor, C., Farmer, J.D., Rasmussen, S., eds.: Artificial Life II, Redwood City, CA, Addison-Wesley (1992) 159–210
4. Dittrich, P., Ziegler, J., Banzhaf, W.: Artificial Chemistries - A Review. Artificial Life **7** (2001) 225–275
5. Banzhaf, W.: Self-replicating sequences of binary numbers. Comput. Math. Appl. **26** (1993) 1–8

6. di Fenizio, P.S., Dittrich, P., Banzhaf, W., Ziegler, J.: Towards a Theory of Organizations. In Hauhs, M., Lange, H., eds.: Proceedings of the German 5th Workshop on Artificial Life, Bayreuth, Germany, Bayreuth University Press (2000)
7. Dittrich, P., Banzhaf, W.: Self-Evolution in a Constructive Binary String System. Artificial Life **4** (1998) 203–220
8. Ziegler, J., Banzhaf, W.: Evolving Control Metabolisms for a Robot. Artificial Life **7** (2001) 171–190
9. Banzhaf, W.: Self-organizing Algorithms Derived from RNA Interactions. In Banzhaf, W., Eeckman, F., eds.: Evolution and Biocomputing. Volume 899 of LNCS. Springer, Berlin (1995) 69–103
10. Banzhaf, W., Lasarczyk, C.W.G.: Genetic programming of an algorithmic chemistry. In O'Reilly, U.M., Yu, T., Riolo, R., Worzel, B., eds.: Genetic Programming Theory and Practice II. Volume 8 of Genetic Programming. Kluwer/Springer, Boston MA (2004) 175–190
11. Banzhaf, W., Nordin, P., Keller, R., Francone, F.: Genetic Programming - An Introduction. Morgan Kaufmann, San Francisco, CA (1998)
12. Koza, J.: Genetic Programming. MIT Press, Cambridge, MA (1992)
13. Mühlenbein, H., Paaß, G.: From recombination of genes to the estimation of distributions: I. Binary parameters. In Voigt, H.M., Ebeling, W., Rechenberg, I., Schwefel, H.P., eds.: Parallel Problem Solving from Nature – PPSN IV, Berlin, Springer (1996) 178–187
14. Langdon, W.B., Poli, R.: Boolean functions fitness spaces. In Poli, R., Nordin, P., Langdon, W.B., Fogarty, T.C., eds.: Genetic Programming: Second European Workshop EuroGP'99, Berlin, Springer (1999) 1–14
15. Branke, J., Schmidt, C.: Sequential sampling in noisy environments. In Yao, X., Burke, E., Lozano, J.A., Smith, J., Merelo-Guervós, J.J., Bullinaria, J.A., Rowe, J., Kabán, P.T.A., Schwefel, H.P., eds.: Parallel Problem Solving from Nature - PPSN VIII. Volume 3242 of LNCS., Birmingham, UK, Springer-Verlag (2004) 202–211
16. Bartz-Beielstein, T., Markon, S.: Tuning search algorithms for real-world applications: A regression tree based approach. In Greenwood, G.W., ed.: Proc. 2004 Congress on Evolutionary Computation (CEC'04), Portland. Volume 1., Piscataway NJ, IEEE Press (2004) 1111–1118
17. Bartz-Beielstein, T., Parsopoulos, K.E., Vrahatis, M.N.: Analysis of particle swarm optimization using computational statistics. In Simos, T.E., Tsitouras, C., eds.: Proc. Int'l Conf. Numerical Analysis and Applied Mathematics (ICNAAM), Weinheim, Wiley-VCH (2004) 34–37

Assessing the Effectiveness of Incorporating Knowledge in an Evolutionary Concept Learner

Federico Divina

Computational Linguistics and AI Section
Tilburg University, Tilburg, The Netherlands
`F.Divina@uvt.nl`

Abstract. Classical methods for Inductive Concept Learning (ICL) rely mostly on using specific search strategies, such as hill climbing and inverse resolution. These strategies have a great exploitation power, but run the risk of being incapable of escaping from local optima. An alternative approach to ICL is represented by Evolutionary Algorithms (EAs). EAs have a great exploration power, thus they have the capability of escaping from local optima, but their exploitation power is rather poor. These observations suggest that the two approaches are applicable to partly complementary classes of learning problems. More important, they indicate that a system incorporating features from both approaches could benefit from the complementary qualities of the approaches. In this paper we experimentally validate this statement. To this end, we incorporate different search strategies in a framework based on EAs for ICL. Results of experiments show that incorporating standard search strategies helps the EA in achieving better results.

1 Introduction

Learning concepts from examples can be viewed as a search problem in the space of all possible hypotheses [1]. Given a description language used to express possible hypotheses, a background knowledge, a set of positive examples, and a set of negative examples, one has to find a hypothesis which covers all positive examples and none of the negative ones (cf. [2]). This problem is NP-hard even if the language to represent hypotheses is propositional logic.

We are interested in learning concepts expressed in first–order formulas containing variables. In particular we are interested in knowledge represented by a fragment of first–order logic, called Horn clauses that do not contain negative literals nor function symbols. This knowledge can be directly used in programming languages based on logic programming, e.g., Prolog. When the description language used is first–order logic, we refer to ICL as Inductive Logic Programming (ILP).

The approach used in the majority of first–order based learning systems, e.g., FOIL [3] and Progol [4], is to use specific search strategies, like the general-to-specific (hill climbing) search [3] and the inverse resolution mechanism [5].

M. Keijzer et al. (Eds.): EuroGP 2005, LNCS 3447, pp. 13–24, 2005.

However, the greedy selection strategies adopted for reducing the computational effort render these techniques often incapable of escaping from local optima.

An alternative approach based on evolutionary algorithms (EAs) can be used for ILP. This approach is motivated by two major characteristics of EAs: their good exploration power, that gives them the possibility of escaping from local optima, and their ability to cope well when there is interaction among arguments and when arguments are of different type. However, the use of stochastic variation operators is often responsible for the rather poor performance of EAs on learning tasks which are easy to tackle by algorithms that use specific search strategies. Examples of EAs for ILP are REGAL [6], G-Net [7], DOGMA [8], SIA01 [9] and GLPS [10].

The above observations suggest that the two approaches are applicable to partly complementary classes of learning problems. More important, they indicate that a system incorporating features from both approaches would benefit from the complementary qualities of the approaches. In fact, EAs are characterized by good exploration qualities, but by rather poor exploitation qualities. On the other hand standard ILP techniques have good exploitation quality but have less exploration power.

This motivates us to investigate a framework based on EAs for ILP that incorporates effective search strategies, like those used in FOIL or Progol.

To this aim, we incorporated knowledge in the evolutionary system ECL [11, 12]. In particular knowledge is used in ECL by means of the selection operator, greedy mutation operators and an optimization phase that follows mutation.

In this paper we want to experimentally assess the effectiveness of incorporating knowledge in ECL by means of the mutation operators and of an optimization phase. Results of experiments show that the use of greedy mutation operator and of an optimization phase that follows the mutation is helpful in helping the system achieving better results. This paper is structured in the following way. In section 2 we give a description of ECL. In particular we describe in detail the mutation operators used and the optimization phase. In section 3 we present and discuss the results of the experiments. Finally, in section 4 we give some conclusions.

2   ECL

A detailed description of ECL can be found in [11]. Here we will give a general explanation of the main features used in the algorithm. ECL is a hybrid EA. ECL iteratively builds a Final_population as the union of various Population evolved with an EA. At each execution of the EA, a part of the background knowledge is chosen by means of a simple stochastic sampling mechanism. This partial background knowledge will be used for building and evaluating individuals inside each iteration of the algorithm. A Population is evolved by the repeated application of selection, mutation and optimization. At each generation n individuals are selected. An individual of the population is chosen using a slight modification, introduced in [13] and extensively validated in [14], of the so-called Universal Suffrage (US) selection operator [6]. This operator works in two steps: first, a positive example is selected by a mechanism that favors "harder" examples, that is, covered by few clauses. In order to determine the "hardness" of an example, a weight is assigned to each example, and is adjourned at every generation. The weight depends on the number of individuals covering the example. This is different from the standard US selection operator, where examples are randomly chosen. Next, a roulette wheel is performed on the individuals of the actual `Population` covering that example. If the selected example is not covered by any individual (for instance when the population is empty) then a new clause is created as follows. The example becomes the head of the clause, and suitable elements of the (partial) background knowledge BK having arguments in common with those of the example are added to the body of the emerging clause. The use of the variant of the US operator is motivated by the fact that the use of greedy mutation operators and of the optimization phase may cause the system to converge to some local optima, so maintaining a good diversity in the population becomes an important factor. As in most ILP systems, a maximum number of body atoms is allowed, which is specified by a user defined parameter lc.

Each individual undergoes a mutation and an optimization phase. The mutation consists in the application of one of the four generalization/specialization operators. A clause can be generalized either by deleting a predicate from its body or by turning a constant into a variable. With the dual operations a clause can be specialized. After that an individual has been mutated, the optimization phase is performed. Mutation operators and the optimization phase are described in the next section. The system does not make use of any crossover operator. Experiments with a simple crossover operator, which uniformly swaps atoms of the body of two rules, have been conducted. However the obtained results did not justify its use. The so modified individuals are then inserted in the population. If the population is not full then the individuals are simply inserted. If the population has reached its maximum size, then n tournaments are made among the individuals in the population and the resulting n worst individuals are substituted by the new individuals.

At the end of the run a solution is extracted from the final population. To this end, in [13] a fast heuristic for solving weighted set covering problems was used, but it presented problems with the precision of the extracted solution. For this reason, in the version of ECL used in this paper, another procedure based on precision of individuals is used. The procedure builds a solution from the rules present in the population that is as accurate as possible, by adding each time the most precise rule in the population to the emerging solution, until its accuracy does not decrease. The fitness of an individual x is given by the inverse of its accuracy: $fitness(x) = \frac{1}{Acc(x)} = \frac{P+N}{pos_x+(N-n_x)}$. In the formula P and N are respectively the total number of positive and negative examples in the training set, while p_x and n_x are the number of positive and negative examples covered by the individual x. We take the inverse of the accuracy, because ECL was originally designed to minimize a fitness function.

2.1   Mutation and Optimization

For moving in the hypothesis space ECL makes use of four mutation operators, and takes advantage of the general-to-specific order of the hypothesis space. In fact, two mutation operators are used for generalizing clauses, and two for specializing. Here we use the concept of generality normally used in ICL: a rule r_1 is more general of another rule r_2 if the examples covered by r_2 are also covered by r_1. A clause can be then generalized by either deleting an atom from its body (the atom is deactivated) or by turning a constant into a variable. With the inverse operations a clause can be specialized.

These mutation operators do not act completely at random, but consider a number of mutation possibilities and among these possibilities the best one is applied. The number of mutation possibilities considered are determined by the values of four parameters $N_1, \ldots, N_4$.

In order to determine which mutation possibility is the best, a gain function is used. When applied to clause C and mutation operator τ, the gain function yields the difference between the clause fitness before and after the application of τ: $gain(C, \tau) = f(C) - f(\tau(C))$. The four operators are defined below.

Atom Deletion. Consider the set Atm of N_1 atoms of the body of C randomly chosen. If the number of atoms in the body of C is less than N_1 then Atm contains all the atoms in the body of C. For each A in Atm, compute $gain(C, -A)$, the gain of C when A is deleted from C.

Choose an atom A yielding the highest gain $gain(C, -A)$ (ties are randomly broken), and generalize C by deleting A from its body.

Insert the deleted atom A in a list D_C associated with C containing atoms which are deactivated. Atoms from this list may be added to the clause (activated) during the evolutionary process by means of a specialization operator.

Constant into Variable. Consider the set Var of variables present in C plus a new variable. Consider also the set Con consisting of N_2 constants of C randomly chosen. If the number of constants of C is less than N_2 then Con consists of all the constants of C.

For each a in Con and each X in Var, compute $gain(C, \{a/X\})$, the gain of C when all occurrences of a are replaced by X.

Choose a substitution $\{a/X\}$ yielding the highest gain (ties are randomly broken), and generalize C by applying $\{a/X\}$.

Atom Addition. Consider the set Atm consisting of N_3 atoms of B_C (list built at initialization time) and of N_3 atoms of D_C, all randomly chosen. If B_C contains less than N_3 atoms, then Atm contains all the atoms of B_C. The same holds for D_C.

For each A in Atm compute $gain(C, +A)$, the gain of C when A is added to the body of C.

Choose an atom A yielding the highest gain $gain(C, +A)$ (ties are randomly broken), and specialize C by adding A to its body.

Remove A from its original list (B_C or D_C).

Variable into Constant. Consider the set Con consisting of N_4 constants (of the problem language) randomly chosen, and a variable X of C randomly chosen. If the constants of the problem language are less than N_4, then Con contains all the available constants.

For each a in Con, compute $gain(C, \{X/a\})$, the gain of C when all occurrences of X are replaced by a.

Choose a substitution $\{X/a\}$ yielding the highest gain (ties are randomly broken), and specialize C by replacing all occurrences of X with a.

```
Optimization(φ, max_opt_steps)
    opt_steps = 1, cont = true
    while (opt_steps < max_opt_steps ) ∧ (cont)
        φ' = mutate(φ)
        evaluate(φ')
        opt_steps = opt_steps + 1
        if fitness(φ) < fitness(φ')
            cont = false
        else
            cont = true
            φ = φ'
    return φ
```

Fig. 1. Optimization function used after mutation. The procedure takes as input an individual φ and a maximum number of optimization steps, max_opt_steps

The optimization phase, performed after an individual has been mutated, is shown in figure 1. Optimization consists of a repeated application of the greedy operators to the selected individual, until either its fitness does not increase or a maximum number of iterations (in figure denoted as max_opt_steps) is reached. The default value of max_opt_steps is 10, and can be modified by setting a relative parameter. If the procedure ends because an application of a mutation operator negatively affected the fitness of the individual being optimized then the last mutation applied is retracted. This control, and the relative action, is performed in the **if** statement of the procedure.

3 Experiments

In this section we experimentally assess the effectiveness of both incorporating the optimization phase that follows the mutation and using the non–random mutation operators described in section 2.1. The list of datasets and the relative parameter setting used in the experiments are shown in table 1. The last column of the table indicates whether a dataset is relational or propositional. The propositional datasets were taken from the UCI Machine Learning repository [15]. The Accidents and the Congestions originates from the Traffic dataset [16, 17], while the Mutagenesis from [18]. Originally the Accidents and the Congestions datasets represented two classes in the Traffic dataset. Here we consider the two classes as different datasets.

Table 1. Parameter settings used in the experiments: *gen* is the number of generations performed by the GA, *sel* is the number of individuals selected per generation, N_i, $i \in [1, 4]$, are the greediness parameters of the mutation operators, *lc* is the maximum length of a clause, and *pbk* is the probability of selecting a BK fact. The last column indicate whether the dataset is a relational dataset (R) or a propositional dataset (P)

Dataset	pop size	gen	sel	max_iter	Ni	lc	pbk	Type
Accidents	30	10	10	1	(10,2,2,2)	8	1.0	R
Australian	50	10	15	1	(4,4,4,4)	6	0.4	P
Breast	50	5	5	1	(3,3,3,3)	5	0.6	P
Congestions	30	10	10	1	(10,2,2,2)	8	1.0	R
Crx	80	20	15	1	(4,4,4,4)	7	1.0	P
Ecochardiogram	40	8	10	10	(4,4,4,4)	4	0.7	P
German	200	10	10	2	(3,4,3,4)	6	0.4	P
Glass2	150	15	20	3	(2,8,2,9)	5	0.8	P
Heart	50	10	15	1	(4,4,4,4)	6	1.0	P
Hepatitis	50	10	10	5	(4,4,4,4)	7	0.2	P
Ionosphere	50	10	15	6	(4,8,4,8)	6	0.2	P
Mutagenesis	50	10	15	2	(4,8,2,8)	3	0.8	R
Pima-Indians	60	10	7	5	(2,5,3,5)	4	0.2	P

The parameter settings were obtained after few, in the order of 10, runs of the system on the relative dataset. We emphasize that the parameter settings chosen was the one which led to the best classification accuracy in the training set, i.e., the test set was never accessed during the runs allocated for parameter setting. We use 10-fold cross validation. Each dataset is divided in ten disjoint sets of similar size; one of these sets is used as test set, and the union of the remaining nine forms the training set. Then ECL is run on the training set and the accuracy of solution found is assessed on the test set. Three runs with different random seed are performed on each dataset.

In order to assess the effectiveness of incorporating knowledge in ECL, we perform experiments with ECL in three settings:

ECL-GA. In this setting ECL is run with all the values of N_i set to 1 and with no optimization phase. In this way all the mutation operators act randomly, as in standard GA operators;

ECL-NOT. In this setting ECL is run with the values of N_i set as reported in table 1. The optimization phase is not performed.

ECL-Opt. In this setting ECL is run with the parameters N_i set as reported in table 1. The optimization phase is performed after mutation, with a maximum of 10 optimization steps.

In order to perform a fair comparison, we increased the values of the parameter *sel* in ECL-GA and in ECL-NOT so that the three settings perform about the same number of evaluations. Table 2 reports the average results obtained by ECL in the three different settings.

Table 2. Experiments with various setting of greediness of ECL. In ECL-GA ECL runs as a standard GA. In ECL-NOT greedy mutation operators are used and in ECL-Opt both greedy mutation operators and the optimization phase are used. Standard deviation is reported between brackets

Dataset	Setting	Accuracy	Time (s)	Simplicity
	ECL-GA	0.82 (0.01)	**2752.16 (82.69)**	35.3 (10.92)
Accidents	ECL-NOT	0.88 (0.01)	3092.62 (103.72)	23.37 (9.69)
	ECL-Opt	**0.95 (0.02)**	3395.01 (136.82)	**3.55 (0.49)**
	ECL-GA	0.82 (0.03)	**1353.02 (7.72)**	14.4 (2.77)
Australian	ECL-NOT	0.83 (0.03)	1444.05 (16.54)	12.30 (2.58)
	ECL-Opt	**0.85 (0.01)**	1686.38 (144.07)	**6.10 (2.18)**
	ECL-GA	0.92 (0.01)	**173.67 (2.33)**	11.30 (2.20)
Breast	ECL-NOT	0.93 (0.02)	238.72 (11.52)	11.50 (2.01)
	ECL-Opt	**0.96 (0.02)**	286.13 (37.00)	**8.60 (0.41)**
	ECL-GA	0.91 (0.02)	**2532.98 (98.43)**	5.70 (1.25)
Congestions	ECL-NOT	0.92 (0.02)	2983.15 (38.65)	5.46 (1.46)
	ECL-Opt	**0.94 (0.02)**	3246.30 (138.73)	**3.95 (0.35)**
	ECL-GA	**0.84 (0.04)**	**1707.82 (66.47)**	13.30 (3.34)
Crx	ECL-NOT	0.83 (0.03)	1852.97 (54.47)	11.70 (3.13)
	ECL-Opt	**0.84 (0.01)**	2668.00 (176.45)	**4.80 (0.05)**
	ECL-GA	0.70 (0.03)	**1245.21 (6.21)**	**2.50 (0.53)**
Echocardiogram	ECL-NOT	0.73 (0.03)	1311.95 (10.31)	**2.50 (0.53)**
	ECL-Opt	**0.74 (0.01)**	1443.63 (36.62)	2.60 (0.70)
	ECL-GA	**0.74 (0.02)**	**1041.74 (19.83)**	14.2 (2.20)
German	ECL-NOT	0.73 (0.03)	1153.52 (11.32)	16.70 (3.09)
	ECL-Opt	**0.74 (0.01)**	1605.75 (144.34)	**11.70 (0.24)**
	ECL-GA	0.82 (0.04)	**956.17 (2.74)**	**3.90 (0.99)**
Glass2	ECL-NOT	**0.85 (0.03)**	1143.05 (21.49)	4.40 (1.51)
	ECL-Opt	**0.85 (0.01)**	1246.00 (55.94)	4.20 (1.23)
	ECL-GA	0.77 (0.03)	**345.34 (8.40)**	9.20 (1.93)
Heart	ECL-NOT	0.78 (0.02)	474.51 (13.41)	7.40 (2.12)
	ECL-Opt	**0.80 (0.03)**	436.38 (57.59)	**4.20 (1.32)**
	ECL-GA	0.82 (0.02)	**878.202 (6.31)**	13.00 (1.63)
Hepatitis	ECL-NOT	**0.83 (0.03)**	954.61 (10.34)	13.40 (1.51)
	ECL-Opt	**0.83 (0.02)**	1056.73 (63.84)	**7.60 (0.95)**
	ECL-GA	0.87 (0.04)	**4364.59 (15.96)**	25.90 (3.96)
Ionosphere	ECL-NOT	**0.89 (0.03)**	4498.72 (13.21)	25.10 (2.72)
	ECL-Opt	**0.89 (0.02)**	5276.83 (138.93)	**12.50 (1.48)**
	ECL-GA	0.85 (0.03)	**407.88 (4.56)**	**4.56 (0.73)**
Mutagenesis	ECL-NOT	0.87 (0.02)	470.32 (5.14)	4.70 (0.95)
	ECL-Opt	**0.90 (0.01)**	542.88 (27.88)	7.92 (1.51)
	ECL-GA	0.73 (0.03)	**1031.71 (10.95)**	9.70 (3.06)
Pima-Indians	ECL-NOT	0.74 (0.02)	1157.64 (14.47)	9.20 (1.87)
	ECL-Opt	**0.76 (0.01)**	1214.75 (31.86)	**8.40 (1.84)**

A first aim of these experiments was to assess how the incorporation of the optimization phase and the use of greedy mutation operators affect the computational time required by the evolutionary process. From the experiments, it emerges that the computational time required by ECL-GA and by ECL-NOT is smaller than the time required by ECL-Opt. In particular ECL-GA is the fastest setting. This result was expected, since in ECL-GA mutations are done randomly and no optimization phase takes place.

The second, and main, aim of these experiments was to establish if the incorporation of the optimization phase and of the greedy mutation operators was beneficial in order to improve the accuracy of the found solutions. It can be noticed that ECL-Opt generally obtained the best accuracies. Only on the Crx dataset ECL-GA obtained the same accuracy as the one obtained by ECL-Opt, but with a higher standard deviation. It can also be noticed that generally ECL-NOT obtained better results than ECL-GA. In some cases the solution found by ECL-NOT is equal to the solution found by ECL-Opt. However neither ECL-GA nor ECL-NOT were capable of finding solutions with higher accuracy than the solutions found by ECL-Opt. This is evident especially for the relational datasets. This is due to the fact that for the relational datasets it is important to find good relations among the arguments of a clause. It is more likely to find good relations with greedy operators and an optimization phase than using only random operators.

It is interesting to notice that generally the solutions induced by the three settings become simpler with the use of greedy mutation operators and with the inclusion of the optimization phase. This is mostly due to the optimization phase. In fact during this phase individuals are rapidly refined so that less clauses are needed for obtaining a logic program with good accuracy. It can be seen that only on three datasets, namely on the Echocardiogram, the Glass2 and the Mutagenesis dataset, ECL-Opt did not find the simplest solution. Only on the Mutagenesis dataset the difference in the simplicity of the solution is significant, while in the other two cases the simplicity of the solutions is comparable. On all the other datasets ECL-Opt found the simplest solution. This can be noticed especially for the Accidents dataset where the solution found by ECL-Opt is almost 10 times simpler than the one obtained by ECL-GA and almost 7 times simpler than the one obtained by ECL-NOT.

In order to summarize the performance of the three settings of ECL and the significance of the results with respect to the accuracy, we compute the statistical paired two-tailed t-test, with confidence level of 1% and 5%. The t-test is performed on the 30 results obtained from the 10 folds and the 3 random seeds. For the Echocardiogram, the Glass2, the Heart and the Hepatitis datasets, the results are not normally distributed and so the t-test cannot be performed on these datasets. Table 3 reports the results of the t-test for the other datasets used in this section. Using a confidence level of 1%, we can see that ECL-Opt outperformed once ECL-GA, namely it obtained significantly better results on the Accidents dataset. If we extend the confidence level to 5%, we have that ECL-Opt outperforms ECL-GA on three datasets, namely the Accidents, the Congestion and the Mutagenesis datasets. Using a 5% confidence level we have that ECL-Opt outperforms ECL-NOT on the Accidents dataset.

Table 3. Results of the two-tailed paired t-test for the used datasets with 1% confidence level: each entry contains the number of datasets on which the algorithm in the row is significantly better than the one in the column. The results of the test using 5% confidence level are reported between brackets when they differ from those using 1% confidence level

	ECL-GA	ECL-NOT	ECL-Opt	Total
ECL-GA	–	0	0	0
ECL-NOT	0	–	0	0
ECL-Opt	1 (3)	0 (1)	–	1 (4)
Total	1 (3)	0 (1)	0	

The difference of performance between ECL-Opt and the other two settings is evident on the Accidents dataset. For this reason we want to analyze the dynamics of the three settings on this dataset. In graphs 2(a), 3(a) and 4(a), we show the best and average fitness of the population at every generation, computed over 10 runs of the various settings. In graphs 2(b), 3(b) and 4(b) we report the average number of positive and negative examples covered by an

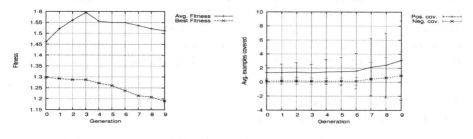

(a) Average and best fitness   (b) Average coverage of individuals

Fig. 2. Graphs relative to fitness and coverage for 10 runs of ECL-GA on the Accidents dataset

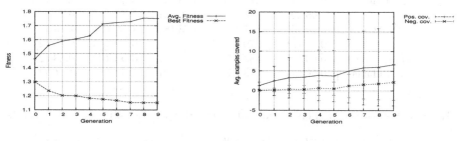

(a) Average and best fitness   (b) Average coverage of individuals

Fig. 3. Graphs relative to fitness and coverage for 10 runs of ECL-NOT on the Accidents dataset

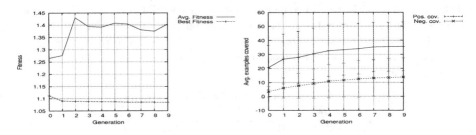

(a) Average and best fitness   (b) Average coverage of individuals

Fig. 4. Graphs relative to fitness and coverage for 10 runs of ECL-Opt on the Accidents dataset

individual in the population evolved with the three settings, computed over 10 runs. From the graphs, it can be seen that the fitness of the best individual in the population is better if more knowledge is incorporated in ECL. It is interesting to notice that the same does not hold for what concerns the average fitness. In fact the average fitness is better in the population evolved with ECL-GA than in the population evolved with ECL-NOT, and is comparable with the average fitness of the population evolved with ECL-Opt. By looking at the graphs relative to the coverage of each individual evolved in the three settings, it can be noticed that individuals evolved by ECL-GA are more specific than those evolved by ECL-Opt and by ECL-NOT. In particular individuals evolved

Table 4. Average accuracies obtained by various systems for ICL on the propositional datasets. Standard deviation between brackets, where (x) stands for $(0.0x)$

Dataset	ECL-Opt	C4.5	HIDER*	GAssist	SMO
Australian	**0.85 (1)**	0.85 (4)	**0.85 (3)**	**0.85 (5)**	**0.85 (1)**
Breast	**0.96 (2)**	0.94 (2)	**0.96 (2)**	**0.96 (2)**	**0.96 (1)**
Crx	0.84 (1)	0.85 (4)	0.83 (5)	**0.86 (5)**	0.85 (2)
Echocardiogram	0.74 (1)	0.71 (1)	**0.79 (13)**	0.72 (2)	0.75 (3)
German	0.74 (1)	0.72 (4)	0.73 (4)	0.72 (2)	**0.76 (1)**
Glass2	**0.85 (1)**	0.78 (4)	0.79 (3)	0.82 (8)	0.65 (2)
Heart	0.80 (3)	0.77 (4)	0.78 (8)	0.80 (7)	**0.83 (1)**
Hepatitis	0.83 (2)	0.79 (4)	0.83 (2)	**0.89 (8)**	0.85 (2)
Ionosphere	0.89 (2)	0.89 (7)	0.89 (6)	**0.93 (4)**	0.88 (2)
Pima-Indians	0.76 (1)	0.73 (3)	0.74 (2)	0.74 (2)	**0.76 (1)**

Table 5. Average accuracy on relational datasets. Standard deviation between brackets

Dataset	ECL-Opt	ICL	Tilde	Progol
Mutagenesis	**0.88 (0.01)**	**0.88 (0.08)**	0.86 (0.03)	**0.88 (0.02)**
Traffic	0.93 (0.02)	0.93 (0.04)	**0.94 (0.04)**	0.94 (0.03)

by ECL-Opt are much more general than those evolved by the other two settings. This fact explains the difference in the simplicity of the solution found, which is much higher in ECL-GA and ECL-NOT than in ECL-Opt. So, on the Accidents dataset, incorporating more knowledge has the effect of allowing ECL-Opt to evolve more general individuals, that have on average not better performances on the training sets, but when used on the test sets yield better results.

Even if the focus of this paper was not to globally assess the performance of ECL it is nevertheless interesting to briefly compare the results obtained by ECL-Opt with those obtained by other popular propositional and ILP learners.

Tables 4 and 5 shows the results obtained by ECL-Opt and by others learner, on the propositional and on relational datasets, respectively. In table 5 we consider the whole Traffic dataset, consisting of three classes. From these tables, it can be noticed that ECL-Opt obtains better or comparable performances with those obtained by other state of the art learners.

4   Conclusions

In this paper we have experimentally validated the effectiveness of incorporating knowledge by means of greedy mutation operators and an optimization phase in the evolutionary systems ECL. The results of the experiments confirm that including the optimization phase that follows the mutation phase and a degree of greediness in the mutation operators is beneficial for improving the accuracy of the found solutions, especially for the relational datasets. The drawback of incorporating knowledge in the evolutionary system, is represented by the computational time, that increases with the inclusion of greediness and of the optimization phase. The fact that including knowledge in ECL helps the system in achieving better results, is an important result, since it confirms what theorized in our motivations for this paper: a system incorporating features of both standard EAs and of standard ICL systems can benefit from the complementary qualities of the two approaches.

References

1. Mitchell, T.: Generalization as search. Artificial Intelligence **18** (1982) 203–226
2. Kubat, M., Bratko, I., Michalski, R.: A review of Machine Learning Methods. In Michalski, R., Bratko, I., Kubat, M., eds.: Machine Learning and Data Mining. John Wiley and Sons Ltd., Chichester (1998)
3. Quinlan, J.: Learning logical definition from relations. Machine Learning **5** (1990) 239–266
4. Muggleton, S.: Inverse entailment and Progol. New Generation Computing, Special issue on Inductive Logic Progra mming **13** (1995) 245–286
5. Muggleton, S., Buntine, W.: Machine invention of first order predicates by inverting resolution. In: Proceedings of the Fifth International Machine Learning Conference, Morgan Kaufmann (1988) 339–351
6. Giordana, A., Neri, F.: Search-intensive concept induction. Evolutionary Computation **3** (1996) 375–416

7. Anglano, C., Giordana, A., Bello, G.L., Saitta, L.: An experimental evaluation of coevolutive concept learning. In: Proc. 15th International Conf. on Machine Learning, Morgan Kaufmann, San Francisco, CA (1998) 19–27

8. Hekanaho, J.: Background knowledge in GA-based concept learning. In: International Conference on Machine Learning. (1996) 234–242

9. Augier, S., Venturini, G., Kodratoff, Y.: Learning first order logic rules with a genetic algorithm. In Fayyad, U.M., Uthurusamy, R., eds.: The First International Conference on Knowledge Discovery and Data Mining, Montreal, Canada, AAAI Press (1995) 21–26

10. Wong, M.L., Leung, K.S.: Inducing logic programs with genetic algorithms: The genetic logic programming system. In: IEEE Expert 10(5). (1995) 68–76

11. Divina, F.: Hybrid Genetic Relational Search for Inductive Learning. PhD thesis, Department of Computer Science, Vrije Universiteit, Amsterdam, the Netherlands (2004)

12. Divina, F., Marchiori, E.: Handling continuous attributes in an evolutionary inductive learner. IEEE Transactions on Evolutionary Computation **to appear** (2004)

13. Divina, F., Keijzer, M., Marchiori, E.: Non-universal suffrage selection operators favor population diversity in genetic algorithms. In et al., E.C.P., ed.: Genetic and Evolutionary Computation – GECCO-2003. Volume 2724 of LNCS., Chicago, Springer-Verlag (2003) 1571–1573

14. Divina, F., Marchiori, E.: Knowledge–based evolutionary search for inductive concept learning. In: Knowledge Incorporation in Evolutionary Computation. Springer-Verlag (2004) 237–254

15. Blake, C., Merz, C.: UCI repository of machine learning databases (1998)

16. Džeroski, S., Jacobs, N., Molina, M., Moure, C.: ILP experiments in detecting traffic problems. In: European Conference on Machine Learning. (1998) 61–66

17. Džeroski, S., Jacobs, N., Molina, M., Moure, C., Muggleton, S., Laer, W.V.: Detecting traffic problems with ILP. In: International Workshop on Inductive Logic Programming. (1998) 281–290

18. Debnath, A., de Compadre, R.L., Debnath, G., Schusterman, A., Hansch, C.: Structure-Activity Relationship of Mutagenic Aromatic and Heteroaromatic Nitro Compounds. Correlation with molecular orbital energies and hydrophobicity. Journal of Medical Chemistry **34(2)** (1991) 786–797

Automated Re-invention of a Previously Patented Optical Lens System Using Genetic Programming

Sameer H. Al-Sakran[1], John R. Koza[2], and Lee W. Jones[3]

[1] Genetic Programming Inc., Mountain View, California USA
[2] Stanford University, Stanford, California USA
koza@stanford.edu
[3] Genetic Programming Inc., Mountain View, California USA
{sameer, lee}@genetic-programming.com

Abstract. The three dozen or so known instances of human-competitive designs of antennas, mechanical systems, circuits, and controllers produced by genetic programming suggest the question of whether genetic programming can be extended to the design of complex structures from other fields. This paper describes how genetic programming can be used to automatically create a complete design for an optical lens system "from scratch"—without starting from a pre-existing good design and without pre-specifying the number of lenses, the layout of lenses, or the numerical parameters of the lenses. More particularly, genetic programming created an optical system that infringed a previous patent (the Konig patent) and improved upon another previous patent (the Tackaberry-Muller patent). The genetically evolved design is an example of a human-competitive result produced by genetic programming in the field of optical design.

1 Introduction

There are now over three dozen known instances where genetic programming has produced "human-competitive" designs (as defined in [7]), including the automatic synthesis of an X-Band Antenna for NASA's Space Technology 5 Mission [11], quantum computing circuits [14], a previously patented mechanical system [10], project schedules in civil engineering [5], previously patented controllers [8], analog electrical circuits patented in the 20th-century [7] and 21st-century [8], and new controller designs that have been granted patents [4].

These past human-competitive results raise the question of whether genetic programming can be extended to the design of complex structures from other fields.

This paper describes how genetic programming can be used to automatically create a complete design for an optical lens system "from scratch"—without starting from a pre-existing good design and without pre-specifying the number of lenses, the layout of lenses, or the numerical parameters of the lenses. Specifically, genetic programming created an optical system that infringed a patent (the Konig patent [6]) and improved upon another (the Tackaberry-Muller patent [15]). The genetically evolved design in this paper is, therefore, an example of a human-competitive result (as defined in [7]) produced by genetic programming in the field of optical design.

M. Keijzer et al. (Eds.): EuroGP 2005, LNCS 3447, pp. 25–37, 2005.
© Springer-Verlag Berlin Heidelberg 2005

Section 2 provides background on the design of optical lens systems, using the patented Tackaberry-Muller system [15] as an example. Section 3 discusses the issues involved in applying genetic programming to the design of optical systems. Section 4 presents the results. Section 5 compares the techniques in this paper with previous work on automatic synthesis of complex structures. Section 6 is the conclusion.

2 Background on the Design of Optical Lens Systems

An optical lens system is an arrangement of refractive or reflective materials that manipulate light [3, 12, 13]. Optical design is more of an art than a science. As Warren J. Smith states in *Modern Optical Engineering* [13, page 393]:

> "There is no 'direct' method of optical design for original systems; that is, there is no sure procedure that will lead (without foreknowledge) from a set of performance specifications to a suitable design."

> "Optical design is in great measure a systematic application of the cut-and-try process."

An existing design is frequently the starting point of optical design by humans (and by conventional optimization software). As Smith [13, page 393] states:

> "[O]ne part of the art in optical design ... consists of the choice of the point at which the designer begins."

2.1 The Konig and Tackaberry-Muller Lens System

Figure 1 shows a four-lens system intended for use as a telescope eyepiece that was patented in 1958 by Tackaberry and Muller [15]. The Tackaberry-Muller lens system is an improved design within a subclass of the Konig lens systems patented in 1940 [6]. The object is shown at the far left and the image is shown at the far right. Spherical surfaces 1 and 2 (and the glass between them) define the first lens (and surfaces 3 and 4, the second). Spherical surfaces 5, 6, and 7 define a *doublet* (i.e., two lenses fitting together). Light rays a, b, and c (so-called *axial rays*) enter the system in parallel at the entry pupil and converge to a single point (the *focal point* f) on the image surface at the far right. Light ray d (the so-called *chief ray*) runs from the point farthest from the axis of the object plane that is visible through the lens system (the *field of view*), to the center of the aperture stop (*entry pupil* here), and to the image surface (at the so-called *image height*). This system appears in Smith's *Modern Optical Engineering* [13, page 508].

2.2 Elements of Design of an Optical System

A complete design for a classical optical lens system entails numerous decisions, including the number of lenses in the system, the layout of the lenses, choices for numerical parameters, and choices for non-numerical parameters.

The layout decisions include the sequential arrangement of lenses between the object and the image, decisions as to whether consecutive lenses touch or are separated by air (or other material), the nature of the mathematical expressions defining the curvature of each surface (often spherical, but sometimes aspherical), and the location and size of the field and aperture stops, which determine the visible field and the maximum illumination of the image, respectively.

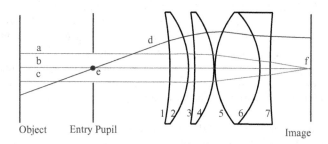

Fig. 1. The Tackaberry-Muller lens system

Table 1. Lens file for the Tackaberry-Muller lens system

Surface	Distance	Radius	Material	Aperture
OBJ	10^{10}	10^{10}	air	
EP	0.88	10^{10}	air	0.18
1	0.21900	−3.52361	BK7	0.62
2	0.07280	−1.05274	air	0.62
3	0.22500	−4.40723	BK7	0.62
4	0.01360	−1.07043	air	0.62
5	0.52100	1.02491	BK7	0.62
6	0.11800	−0.93493	SF61	0.62
7	0.47485	7.94281	air	0.62
IMS		10^{10}		

The numerical choices required to define a lens system include the physical distance between each surface of each lens, the numerical coefficients for the mathematical expressions defining the curvature of each surface of each lens (which, in turn, implies whether each surface is concave, convex, or flat), and the apertures (semi-diameter) of each surface.

The non-numerical choices include the type of glass (or other material) for each lens. Each type of glass has various properties of interest to optical designers, notably including the index of refraction, n (which varies by wavelength); the Abbe number, V; and the cost. Choices of glass are typically drawn from a standard glass catalog.

2.3   Prescription (Lens File) for the Tackaberry-Muller Lens System

A classical lens system is conventionally specified by a table called a *prescription* (or, if the system is being analyzed by modern-day optical simulation software, a *lens file*).

Table 1 shows a prescription (slightly modified for tutorial purposes) for the Tackaberry-Muller lens system [15] of figure 1. Each row in the table represents a surface. The object appears in the first row and is labeled "OBJ." The entry pupil appears in the second row of the table and is labeled "EP." The image appears in the last row and is labeled "IMS". The other surfaces are consecutively numbered (from 1 to 7). All entries are normalized so that the system has a focal length of 1. The system has a *half field of view* of 19.8°. The object is at infinite distance (10^{10}).

The distance shown in column 2 of each row of table 1 refers to the distance between the surface represented by that row and the surface represented by the next row (which, in figure 1, is the surface to the right). Similarly, the material in column 4 refers to the material (glass, air, or other) between the surface represented by that row and the surface represented by the next row. Columns 3 (radius of curvature of the surface) and 5 (aperture) apply to the surface itself.

Surface 1 in table 1 is located at a distance (column 2) of 0.88 from the previous surface (i.e., entry pupil EP) and defines the *eye relief* of the eyepiece. The entry in column 3 of −3.52361 for the radius of curvature of surface 1 indicates that surface 1 is a sector of a sphere with a radius 3.52361. The negative sign indicates that the sphere's center is located to the surface's left. The material (column 4) located to the right of surface 1 is BK7 (a commercially available crown glass). The apertures for surfaces 1 through 7 (found in column 5) have a uniform value of 0.62 because this particular system is a telescope eyepiece that is encased entirely inside a cylinder.

Surface 2 is located at distance 0.21900 from surface 1. The material to the right of surface 2 is air. Together, surfaces 1 and 2 define a lens of thickness 0.21900 composed of BK7 glass with a concave left surface and a convex right surface.

Surfaces 5, 6, and 7 together define a *doublet* lens. The material to the right of surface 5 is BK7 glass and the material to the right of surface 6 is SF61 (a commercially available flint glass).

The material between surface 7 and the image surface (IMS) is air and its distance is 0.47485 (the *back focal length*).

2.4   Analysis of an Optical System

Once a classical optical system is specified by means of its prescription (lens file), many of its optical properties can be calculated by mathematically tracing the path of light rays of various wavelengths through the system. Ray-tracing analysis is extremely tedious and, nowadays, is typically performed by optical simulation software (e.g., OSLO, Zemax, V-Code, KOJAC). Figure 2 shows conventional curves for distortion (figure 2a), astigmatism (figure 2b), chromatic aberration (figure 2c), and spherical aberration (figure 2d) for the Tackaberry-Muller system.

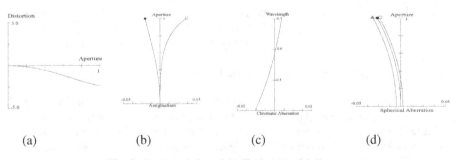

(a)   (b)   (c)   (d)

Fig. 2. Characteristics of the Tackaberry-Muller system

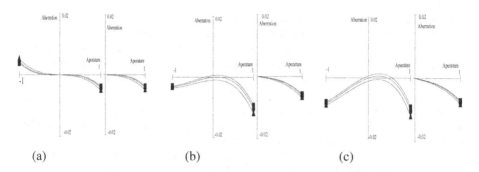

(a)   (b)   (c)

Fig. 3. Additional characteristics of the Tackaberry-Muller system

Figure 3 shows the on-axis ray intercept diagram (figure 3a), the partial field ray intercept diagram (figure 3b), and the full field ray intercept diagram (figure 3c). The diagram for the meridional plane is on the left, with the sagittal on the right.

2.5   Previous Work in Optics in Genetic and Evolutionary Computation

Genetic algorithms have been extensively used for optimizing the choices of parameters of optical systems having a pre-specified number of lenses and a pre-specified layout, as listed in Jarmo Alander's voluminous *An Indexed Bibliography of Genetic Algorithms in Optics and Image Processing* [1]. In a noteworthy paper, Beaulieu, Gagné, and Parizeau [2] used genetic programming to "re-engineer" the design of a four-lens system (produced by genetic algorithm) and thereby create an improvement over the best design produced by 11 human teams in a design competition held at the 1990 International Lens Design Conference. Their approach used functions that incrementally adjusted (additively or multiplicatively) the distance between lens surfaces, radius of curvature of lens surfaces, and stop location values.

3   Applying Genetic Programming to Optical Design

This section briefly describes the most important issues in applying genetic programming to the design of an optical lens system. A technical report with a fuller explanation of the preparatory steps and other details is available on the web at www.genetic-programming.com/techreports.html.

3.1   Representation Scheme (Functions and Terminals) for Optical Systems

The widely-used and well-established format for optical prescriptions (and lens files for optical analysis software) suggests a developmental representation for use in conjunction with genetic programming. In this representation, a turtle (similar to the turtle used in Lindenmayer systems and used to synthesize antennas using genetic programming [8]) starts at a specified point on the entry pupil surface (point e in figure 1). On each move, the turtle does three things. First, it inserts a spherical surface with a specified radius of curvature. Second, it inserts a specified type of material (e.g., air or a type of glass) to the right of the surface. Third, it moves to the right by a specified distance along the system's main *axis* (line b in figure 1).

The three-argument SS ("spherical surface") function causes the turtle to insert a spherical surface with specified radius and thickness and to move a specified distance.

The two-argument PROGN2 function is a connective function that first executes its first argument and then executes its second argument.

Radius and distance values are each established by a value-setting subtree containing a single perturbable numerical value. The type of material is established by a value-setting subtree containing a type of material (e.g., air or a type of glass).

A constrained syntactic structure specifies how the functions and terminals may be combined in a program tree. The constrained syntactic structure enforces the use of one function set and terminal set for each value-setting subtree that establishes the numerical value for thickness and radius of curvature; another terminal set for establishing the type of material; and another function set for all other parts of the program tree.

The object surface (OBJ), image surface (IMS) and entry pupil (EP) constitute the *test fixture* when it comes time to evaluate the behavior and characteristics of the lens system (analogous to the test fixtures described in [7] and [8]).

3.2   Special Toroidal Mutation Operation for Radius of Curvature

A flat surface can be viewed as a spherical surface with a very large radius of curvature. Given our aperture of 0.62 and effective focal length of 1, a radius of curvature of +10 or –10 both approximately represent a flat surface. That is, the space of curvature values is toroidal, so that seemingly distant values such as +10 and –10 are, in fact, nearby. Therefore, our mutation operation operates in a toroidal way when it is applied to a terminal representing a radius of curvature.

3.3   Special Glass Mutation

In real-world situations, the optical designer usually does not freely choose the index of refraction, n, and the Abbe number, V, for a glass, but, instead, chooses one of a relatively small number of commercially available types of glass (such as those found in the Schott catalog or other standard glass catalogs). Moreover, the 199 commercially available glasses in the Schott catalog reside in a relatively small and compact crescent-shaped area occupying only about 27% of the area of the rectangle bounded by the extreme values of n and V for the catalog, namely $1.46 < n < 2.02$ and $20 < V < 77$ for the Schott catalog. Independently perturbing n and V would usually yield a glass lying outside the crescent-shaped area and, even if the offspring were inside, it would almost never correspond to a commercially available type of glass. Thus, it is advisable to employ a domain-specific mutation operation that permits mutation only to a "nearby" type of glass in the chosen catalog.

3.4   Operator Probabilities

The function set here is unusual in that there is only one active function (SS) and all of its arguments are terminals. Moreover, the glass mutation operator is unusually important in that a change in material simultaneously affects the refraction of light rays as well as the chromatic corrective (dispersive). These considerations suggest performing crossover, numerical mutation, and glass mutation at 30% probability each, with reproduction at 9% and tree mutation at 1%. The population size was 346,000 (500 individuals at 692 nodes of a cluster computer).

3.5   Fitness Measure

The fitness measure used in optical design (whether by evolutionary search, other types of search, or human design) is multi-objective.

We extensively modified and augmented KOJAC (public-domain educational software for optical ray tracing) to create an optical simulator that we could embed inside our genetic programming system operating on our Beowulf cluster computer. KOJAC was originally authored by Olivier Scherler and is currently maintained by Olivier Ripoll. We also used a commercially available software package (Lambda Research's OSLO) for post-run checking of final results on a single computer.

The fitness measure begins by analyzing an axial and chief ray trace in order to derive aberration and paraxial coefficients (and assigns a fatally high penalty value if either ray trace fails). Fitness is incremented by the weighted sum of the deviations between the behavior of the candidate individual and the target values of various performance measures. In particular, fitness is incremented by the sum of 1,000 times each of the following aberration deviations: spherical aberration, astigmatism, distortion, coma, axial chromatic, lateral chromatic; 100 times the absolute deviation of effective focal length (EFL) from target; 100 times the absolute deviation of back focal length (BFL) from target (but only if it is less than target); and 25 times the absolute deviation of Petzval radius from target (but only if it is less than target). The weights were chosen to approximately equalize the influence of each of the above types of deviations in a manner similar to our recent work with automatic circuit systems involving multi-objective fitness measures [9].

Then, a 17×17 grid is overlaid on the entry pupil (figure 4a) and a ray is shot through the corner defining each grid position contained inside the entry pupil.

Figures 4b, 4c, and 4d are gray scale versions of a three-color spot diagram for the Tackaberry-Muller system. The rays from figure 4a are traced for each of three wavelengths (486, 588 and 656 nm) and projected through the system on the object plane. Figure 4b shows the trace from the axial point. Figure 4c shows the trace from the 70% of the field of view (FOV). Figure 4d shows the full FOV performance.

Figure 4e shows the modulation transfer function performance of the Tackaberry-Muller system in the tangential and sagittal planes of each of the FOV positions. The spot diagram measures the deviation resulting from the compound error of the chosen lens aberration contributions. An ideal diffraction limit spot size (corresponding to the minimum spot size that can be discernable when diffraction effects are taken into account) is determined for the system and the root mean square (RMS) error for the ray intercept deviations is calculated. Fitness is incremented by 200, 340 and 400 times the difference between the target RMS error for the axial, 70% FOV and full FOV, respectively. The increasing penalty multiplier reflects the increasing difficulty in attaining the desired performance. The modulation transfer functions are each sampled at 30 increments of 10 cycles/mm across the target system and that modulation efficiency is defined as the target to meet.

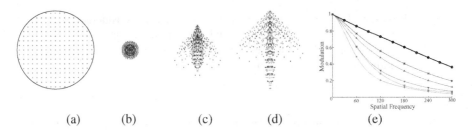

(a)   (b)   (c)   (d)   (e)

Fig. 4. Fitness measure

After an individual reaches a specified satisfactory level for all of the foregoing requirements, individuals are evaluated for parsimony. The parsimony penalty is the sum of 100 times the number of lens, 100 times the number of different types of glass used, the width of the lens system (its footprint), and (optionally, but not used for the work in this paper) the cost of the glass (found in the glass catalog).

For the specific problem discussed herein, each candidate lens system is first examined to see if it is "all air" or whether a lens has a spherical radius smaller than the problem's uniform aperture of 0.62. If so, the individual receives a fatally high penalty value of fitness. In addition, a significant (but not fatal) infeasibility penalty is applied if two lenses overlap (i.e., occupy the same space), if the back focal length (BFL) is negative (meaning the image is inside the lens system, instead of being to the right), or if the final surface is not air (meaning that a flat glass surface touches the image surface).

4   Results

The randomly created individuals in generation 0 exhibit a wide variety of pathological characteristics. For some individuals, none of the light rays from the object may reach the image surface. Randomly created multi-lens systems are also likely to refract some rays outside the eyepiece's housing, causing *vignetting*.

Many of the better individuals in generation 0 have a single positive lens with above-average diffraction. These individuals keep the incoming light rays within the system. These one-lens systems provide a toehold for later evolutionary progress.

The best individual of generation 0 (figure 5) has a fitness of 820.4 and corresponds to a lens system with one positive lens (biconvex in this case) of BASF57

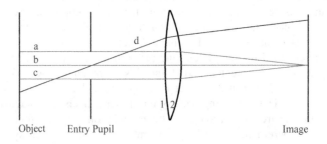

Fig. 5. Best-of-generation individual from generation 0

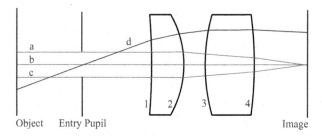

Fig. 6. Best-of-generation individual from generation 11

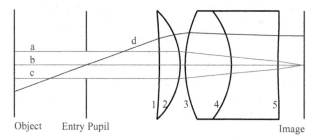

Fig. 7. Best-of-generation individual from generation 18

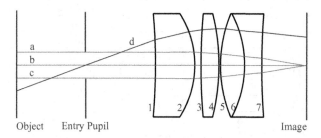

Fig. 8. Best-of-run individual from generation 490

glass (a flint glass). The axial rays (a, b, and c) coming in through the entry pupil are focused in approximately the desired range on the image surface; however, the chief ray, d (determining image height) lands outside the desired range.

In generation 11, the best individual (figure 6) has two positive focusing lenses (using two similar materials, namely LAK31 with an n of 1.696732 and a V of 56.42 for yellow light and LAKN22 with an n of 1.651131 and V of 55.89. The axial rays are focused in approximately the desired range and the chief ray, d, is inside the desired range on the image surface. This individual has a fitness of 483.2.

The best individual from generation 18 (figure 7) has two new features of interest. First, a doublet has emerged (defined by surfaces 3, 4, and 5). Second, two distinctly different types of glasses are employed (LAK8 crown glass for surfaces 3 and 4 and LAF11A flint glass for surfaces 4 and 5). The crown glass elements are primarily responsible for ray shaping while the flint glass corrects aberrations. The front end resembles earlier individuals in that it is a positive lens (in this case, a meniscus lens composed of PSK3 crown glass). This individual has a fitness of 154.43.

Many different systems consisting of widely varying numbers of lenses and layouts appear over the next sixty or so generations. In generation 86, a five-lens system with one positive focusing meniscus lens and two doublets had sufficiently high performance to permit the remainder of the run to begin considering parsimony.

A 4-lens system emerged in generation 207 with the same number of lenses (four) as the Tackaberry-Muller system, with two positive lenses followed by a doublet, and with the doublet consisting of a flint and crown class. The remainder of the run concentrated on optimizing various aspects of the system, including the effective focal length (EFL) to a value of 0.9979, the back focal length (BFL) to a value of 0.570, and the evolved footprint (length) of the system to 1.353.

This optimizing process reached a plateau by generation 490, yielding the best-of-run individual (figure 8). Table 2 shows the lens file for the best-of-run individual.

As can be seen in table 3, the best-of-run individual is a slight improvement over the design in the Tackaberry-Muller patent [15], using our fitness measure employing an exact ray trace of the marginal and chief rays for aberration calculations.

Table 4 shows that the best-of-run lens system from generation 490 infringes claim 1 of the Konig patent [6].

Table 2. Lens file for the best-of-run individual from generation 490

Surface	Distance	Radius	Material	Aperture
OBJ	10^{10}	10^{10}	air	
EP	0.88	10^{10}	air	0.18
1	0.500217794	-7.260474245	SSK3	0.62
2	0.079530961	-1.164063819	air	0.62
3	0.229844668	7.54377332	SSK3	0.62
4	0.02	-2.580718992	air	0.62
5	0.305914688	1.632125441	LAK16A	0.62
6	0.217637521	-1.348604772	SF58	0.62
7	0.5706017465291	8.735622978	air	0.62
IMS		10^{10}		

Table 3. Characteristics of the best-of-run individual from generation 490

	Tackaberry-Muller	Genetically evolved
Spherical aberration	-0.003999242	-0.003102831
Coma	-0.002828716	-0.002496014
Astigmatism	0.002817410	0.0027877324
Petzval	-0.006505427	-0.006353730
Distortion	-0.009906606	-0.009244075
Distortion percentage	2.4166	1.8344
Maximum distortion percentage	2.4166	1.8344
Axial chromatic	-0.001121585	-0.000737052
Lateral chromatic	-0.002213006	-0.001904380

Table 4. Comparison of best-of-run lens system to Konig patent

Claim 1 of Konig patent	Genetically evolved optical system
"An optical system for telescope eyepieces, comprising a front, a medial and a rear element, said elements being convergent and axially spaced by air,"	The evolved solution contains three convergent elements (two single lenses and one doublet lens) and they are separated by air.
"the sum of the distances apart of said elements being at most one-third of the focal length of said system,"	The focal length of the evolved system is 0.9958 and the sum of the distances is 0.099531 (approximately 1/10).
"said rear element being a single lens, the numerical value of the curvature of the rear surface of said lens being smaller than the numerical value of the refractive power of said lens,"	The curvature of the rear element (the lens defined by surfaces 1 and 2) is 0.1377 (i.e., 1/7.260474245) and its refractive power is 0.443518 (computed by the standard textbook formula).
"said medial element consisting of at least one lens and at most two lenses,"	The medial element (the lens defined by surfaces 3 and 4) is a single lens.
"said front element consisting of at least one lens,"	The front element (surfaces 5, 6, and 7) consists of two lenses.
"the front lens of said medial element and that lens of said front element which faces this front lens of said medial element being convergent,"	These two lenses (namely the medial lens defined by surfaces 3 and 4 and the portion of the doublet defined by surfaces 5 and 6) are both convergent.
"at least one optically effective surface of one of said two convergent lenses being a cemented surface,"	The doublet lens defined by surfaces 5, 6, and 7 has a common surface 6.
"the refractive power of one cemented surface of said two convergent lenses amounting to at least eleven twentieths of the algebraic sum of the refractive powers of all cemented surfaces of said convergent lenses,"	The refractive power of the specified surface is 0.13652. It is also the only common surface and hence amounts to the entire sum described.
"the numerical value of last said sum being greater than one twelfth of the sum of the numerical values of the curvatures of those surfaces of said convergent lenses which face each other."	It is equal to 0.136492 which is greater than one twelfth.

Table 5 compares the features of the best-of-run lens system from generation 490 with claim 1 of the Tackaberry-Muller patent [15]. The only difference (a slightly out-of-range radius of curvature) is apparently due to the improved performance of the genetically evolved design compared to the 1958 design. The Tackaberry-Muller patent [15] cites the 1940 Konig patent [6] and is a special case of it.

Table 5. Comparison of best-of-run lens system to Tackaberry-Muller patent

Claim 1 of Tackaberry-Muller patent	Genetically evolved solution
"A telescopic eyepiece adequately corrected for color"	The genetically evolved solution is a slight improvement over the patented design.
"consisting of three convergent components"	There are three convergent components.
"formed of four lenses"	There is a total of four lenses.
"the front component being a doublet"	The front component is a doublet.
"comprising a divergent lens of relatively low dispersion glass"	The first material of the doublet is SF58, a flint (low dispersion glass).
"and a convergent lens of crown glass"	The second material of the doublet is LAK16A, a crown glass (convergent lens)
"the dispersive indices of the two lenses of the doublet having a ratio lying between .415 and .445"	The dispersive indices are 21.51 and 51.78, with a ratio of .415.
"and the radius of curvature of the internal contact surfaces lying between .86 F and 1.01 F,"	The radius of curvature is 1.3.
"the other convergent components being single lenses of crown glass."	The material for the other two single lens is SSK3, a crown glass.

5   Comparison with Circuits, Controllers, and Antenna

We observed several similarities between the work described in this paper and previous work on the automatic synthesis of circuits [7], controllers [8], and antennas [9]. First, a straight-forward representation based on elementary principles of the field was sufficient to enable genetic programming to produce human-competitive results in all four fields. Second, a high percentage of the randomly created individuals in generation 0 were unsimulatable, but the percentage of unsimulatable individuals dropped quickly (to single-digit levels) after only a few generations. Third, the best individuals in generation 0 and the early generations were structures with one (or just a few components) that partially satisfied a single prominent element of the fitness measure (apparently because of the low probability of numerous randomly created components being appropriately combined and harmoniously parameterized).

6   Conclusion

The paper described how genetic programming created the complete design for the Konig optical lens system patented in 1940 and the Tackaberry-Muller system patented in 1958. The entire lens system was automatically synthesized "from scratch"—without starting from a pre-existing good design and without pre-specifying the number of lenses, the layout of lenses, or the numerical parameters of the lenses. The genetically evolved design in this paper is an example, of a human-competitive result produced by genetic programming in the field of optical design.

References

1. Alander, Jarmo T. 2000. *An Indexed Bibliography of Genetic Algorithms in Optics and Image Processing - Draft August 16, 2000*. Report 94-1-OPTICS. Department of Information Technology and Production Economics. Vassa, Finland: University of Vaasa.
2. Beaulieu, Julie, Gagné, Christian, and Parizeau, Marc. 2002. Lens system design and re-engineering experimentations with genetic algorithms and genetic programming. In Langdon, W. B., Cantu-Paz, E., Mathias, K., Roy, R., Davis, D., Poli, R., Balakrishnan, K., Honavar, V., Rudolph, G., Wegener, J., Bull, L., Potter, M. A., Schultz, A. C., Miller, J. F., Burke, E., and Jonoska, N. (editors). *Proceedings of the 2002 Genetic and Evolutionary Computation Conference*. San Francisco, CA: Morgan Kaufmann. Pages 155–162.
3. Fischer, Robert E. and Tadic-Galeb, Biljana. 2000. *Optical System Design*. New York, NY: McGraw-Hill.
4. Keane, Martin A., Koza, John R., and Streeter, Matthew J. 2002. *Improved General-Purpose Controllers*. U.S. patent application filed July 12, 2002.
5. KHosraviani, Bijan, Levitt, Raymond E. and Koza, John. R. 2004. Organization design optimization using genetic programming. In Keijzer, Maarten (editor*). Late-Breaking Papers at the 2004 Genetic and Evolutionary Computation Conference*. Seattle, WA: International Society of Genetic and Evolutionary Computation.
6. Konig, Albert. 1940. *Telescope Eyepiece*. U. S. Patent 2,206,195. Filed in Germany December 24, 1937. Filed in U. S. December 14, 1938. Issued July 2, 1940.
7. Koza, John R., Bennett III, Forrest H, Andre, David, and Keane, Martin A. 1999. *Genetic Programming III: Darwinian Invention and Problem Solving*. San Francisco, CA: Morgan Kaufmann.
8. Koza, John R., Keane, Martin A., Streeter, Matthew J., Mydlowec, William, Yu, Jessen, and Lanza, Guido. 2003. *Genetic Programming IV: Routine Human-Competitive Machine Intelligence*. Kluwer Academic Publishers.
9. Koza, John R., Jones, Lee W., Keane, Martin A., Streeter, Matthew J., and Al-Sakran, Sameer H. 2004. Toward automated design of industrial-strength analog circuits by means of genetic programming. In O'Reilly, Una-May, Riolo, Rick L., Yu, Gwoing, and Worzel, William (editors). *Genetic Programming Theory and Practice II*. Boston: Kluwer Academic Publishers. Chapter 8. Pages 121–142.
10. Lipson, Hod. 2004. How to draw a straight line using a GP: Benchmarking evolutionary design against 19th century kinematic synthesis. In Keijzer, Maarten (editor). *Genetic and Evolutionary Conference 2005 Late-Breaking Papers*. CD ROM. Seattle, WA: International Society for Genetic and Evolutionary Computation.
11. Lohn, Jason D., Hornby, Greg S., and Linden, Derek S. 2004. An evolved antenna for deployment on NASA's Space Technology 5 Mission. In O'Reilly, Una-May, Riolo, Rick L., Yu, Gwoing, and Worzel, William (editors). *Genetic Programming Theory and Practice II*. Boston: Kluwer Academic Publishers. Chapter 18.
12. Smith, Warren J. 1992. *Modern Lens Design: A Resource Manual*. Boston, MA: McGraw-Hill.
13. Smith, Warren J. 2000. *Modern Optical Engineering*. 3rd edition. New York: McGraw-Hill.
14. Spector, Lee. 2004. *Automatic Quantum Computer Programming: A Genetic Programming Approach*. Boston: Kluwer Academic Publishers.
15. Tackaberry, Robert B. and Muller, Robert M. 1958. *Telescope Eyepiece System*. U. S. Patent 2,829,560. Filed October 15, 1956. Issued April 8, 1958.

Bayesian Automatic Programming

Evandro Nunes Regolin and Aurora Trindad Ramirez Pozo

Computer Science Department,
Federal University of Paraná, Brazil
{evandro, aurora}@inf.ufpr.br

Abstract. In this work a new approach, named Bayesian Automatic Programming (BAP), to inducing programs is presented. BAP integrates the power of grammar evolution and probabilistic models to evolve programs. We explore the use of BAP in two domains: a regression problem and the artificial ant problem. Its results are compared with traditional Genetic Programming (GP). The experimental results found encourage further investigation, especially to explore BAP in other domains and to improve the proposed approach to incorporating new mechanisms.

1   Introduction

Evolutionary Algorithms (EAs) have been successfully used to solve a wide variety of optimization and search problems. Two of them are the Genetic Algorithm [1] and the Genetic Programming [2]. Both are based on the principle of selection and variation. While selection tries to collect individuals that better solve a given problem, variation modifies individuals or combines individuals in order to exploit new regions of the search space. Although EAs have been used with success in many contexts, they have many problems such as the linkage problem [3]. The linkage problem occurs when the variation process causes the disruption of building blocks.

Recently, a new type of EA, based on probabilistic models to describe the characteristic of the best chromosomes, has been proposed. It is known as Estimation of Distribution Algorithm (EDA). In EDA, the genetic pool is coded as a distribution of the search space; in each generation, the population is generated from the current distribution; and the distribution is updated from the best (and possibly the worst) individuals in the current population [4]. Many tools that follow this paradigm have been developed, mainly for optimization. A limited effort has been applied in the development of EDAs with the objective of automatic programming. Some algorithms proposed with the objective of automatic programming are the Probabilistic Incremental Program Evolution (PIPE) [5], the Stochastic Grammar-based Genetic Programming (SG-GP) [4] and the Extended Compact Genetic Programming (eCGP) [6].

This work presents a new scheme, called Bayesian Automatic Programming (BAP), that extends the multivariate probability distribution used in Bayesian networks, such as that used in BOA [7] to Context Free Grammar based Genetic

M. Keijzer et al. (Eds.): EuroGP 2005, LNCS 3447, pp. 38–49, 2005.

Programming (CFG-GP) [4][8]. Combining these two techniques, we intend to use the space search restriction of CFG-GP and the Bayesian network population description. This study allows to analyze the effectiveness of exploring the correlation among concrete syntax tree hierarchy to find good solutions. Our initiative was motivated by the good results provided by BOA in the optimization field and its easy adaptation to the genetic programming context through the use of CFG-GP approach.

In order to extend BOA to genetic programming context, a linear representation of a derivation tree similar to the one introduced in Grammar Evolution [8] is used. The linear representation easily adapts the genetic program representation, usually a tree, to the requirements of the Bayesian learning algorithm used in BOA. The approach was applied to traditional problems of genetic programming to evaluate its performance.

This paper is organized as follows. The next section briefly reviews the main genetic programming system based on EDAs. Section 3 summarizes Bayesian network, context free grammar and CFG based genetic programming. The approach BAP is presented in Section 4, its main algorithm, chromosome representation, the chromosome generation algorithm and the mutation operator. Section 5 describes the experiments and its results. Finally, the conclusions of the work are presented in Section 6.

2   Literature Review

Recently, a number of evolutionary algorithms that guide the exploration of the search space by building probabilistic models of promising solutions found so far have been proposed. These algorithms have shown to perform very well on a wide variety of problems [9]. Most of them solve problems in the domain of Genetic Algorithm. Among this kind of algorithm, we can mention PBIL [10], BOA [7] and eCGA [11]. Just few explore the use of stochastic models for Automatic Programming. Among these, we can cite PIPE [5], eCGP [6] and SG-GP [4].

BOA is an optimization system which uses an estimation of a probability distribution of promising solutions in order to generate new candidate solutions. To estimate the distribution, techniques for modeling multivariate data by Bayesian networks are used [7].

The Probabilistic Incremental Program Evolution (PIPE) [5] is an automatic programming algorithm based on PBIL [10] which uses a univariate probabilistic model to evolve programs. The distribution on the GP search space is represented as a Probabilistic Prototype Tree (PPT); in each PPT node stand the probabilities for selecting any variable and operator in this node. After the current individuals have been constructed and evaluated, the PPT is biased toward the current best and the best-so-far individuals. One feature of the PIPE system is that the PPT grows deeper and wider along evolution, depending on the size of the best trees, since the probabilities of each variable/operator have to be defined for each possible position in the tree [4].

SG-GP [4] is a distribution based evolution system that allows the use of context free grammar (CFG) to introduce prior knowledge to the search of best programs. The distribution on the GP search space is represented as a stochastic grammar, where each derivation d_i in a production rule is attached a weight w_i, and the changes for selecting derivation d_i are proportional to w_i [4]. Structures that compose the best (worst) programs have their weight increased (reduced). New programs are generated from the probabilistic model by selecting grammar derivations proportionally to their weight.

The extended compact genetic programming (eCGP) is an algorithm based on the extended compact genetic algorithm (eCGA) and PIPE. It combines the capability of handling multivariate interactions among variables of eCGA with the variable-size program tree representation of PIPE. The probability distribution is modeled using a marginal product model (MPM). The optimal probabilistic model is found using a greedy search algorithm that minimizes the model complexity. The structure of the model found is composed by partitions aggregating nodes. In each of this partition were recognized relevant conditional relationships. Additionally, nodes parameters are calculated as well.

In their work [6], Sastry and Goldberg propose the use of Bayesian networks-based models for automatic programming. Moreover, Blanco and Lozano [12] presented empirical results suggesting that algorithms which use Bayesian network to model the probabilistic distribution have better performance in discrete domains (in relation with convergence velocity, convergence reliability and scalability) for complex function than univariate models. These principal points support our study, that is, to explore the Bayesian network as model in EDAs to induce programs. Beside that, we propose the use of a compact solution representation as alternative to the tree used in PIPE and eCGA.

3   Background

In this section, the main techniques used to construct our approach are presented. First, the Bayesian network, a probabilistic model used to represent the component dependences of the individual evolved. After, we introduce the grammar evolution approach.

3.1   Bayesian Network

A Bayesian network is a graphical model for probabilistic relationships among a set of variables [13]. It is composed by a vector of random variables $\mathbf{X} = (X_1, ..., X_n)$ and an acyclic graph $G = (\mathbf{X}, E)$ whose edges $E \subseteq [\mathbf{X}]$ represent a conditional relationship. Random variables are functions from events in a sample space to a domain representing the qualities or distinctions of interest [14]. If there is an edge between the random variables from X_i to X_j, X_i is said to be parent of X_j. A Bayesian network gives a complete description of the domain, being able to describe the full joint probability distribution from the pair $(\mathbf{X}, G)$.

Learning the network structure can be done using an algorithm with two components: a scoring metric and a search procedure [15]. A scoring metric is a measure of how well the network models the data. Prior knowledge can be incorporated into the metric to reduce the computational time of the learning algorithm. The search procedure explores the space of all possible networks in order to find the network (or a set of networks) with the value of a scoring metric as high as possible. The space of networks can be reduced by constraint operators [7]. A well known algorithm for this task is K2 [16]. K2 uses a model selection approach where a scoring metric guides a simple greedy algorithm in the search for the best network. It has $O(m\ u^2\ n^2\ r)$ time complexity for learning the network, where m is the number of cases, u is the maximum number of parents that any node is permitted to have, n is the number of nodes and r is the number of possibles events [16].

Using the description of domain through a Bayesian network, it is possible to generate new instances of the variables with similar properties as those of the training data [7].

K2 returns the network structure and a conditional probability table (CPT) for each $X_i \in \mathbf{X}$. A conditional probability table stores the chance of the occurrence of an event given that a second event has already occurred.

Each row in a CPT contains the conditional probability of each node value for a possible combination of values for the parent nodes [17].

3.2   Context Free Grammar, Genetic Programming and the Grammar Evolution Approach

Let $G = \{N, T, P, S\}$ be a grammar, where N is the set of non-terminals, T is the set of terminals, P is a set of production rules that maps the elements of N to $T \cup N$, and S is a start symbol that is member of N [4]. An example of a grammar is shown in Figure 1.

$$N=\{\text{expr, boper}\},\ T=\{\text{X, +, -, \%, *}\},\ S=\{\text{expr}\}$$
$$P = \begin{cases} < expr > ::= < expr >< boper >< expr > |\ \text{X} \\ < boper > ::= +\ |\ -\ |\ *\ |\ \% \end{cases}$$

Fig. 1. An example of a free context grammar

To generate a string derivable from G, it is necessary to start with S, choose a formation rule $S \longrightarrow \alpha$ and successively expanding the non-terminals in α until only terminals are left. A tree which represents the steps taken in the production process is denoted concrete syntax trees.

A straightforward way to compose a string is, starting from the start symbol, randomly select a formation rule from a grammar when necessary to translate a non-terminal. This "obvious solution" has two drawbacks. The first one is that the process can generate extremely long sentences. The second one is that not all possible sentences of a maximum size have the same probability of being generated. Given the grammar $S \longrightarrow a \mid A; A \longrightarrow b \mid c$, it will generate a with probability $\frac{1}{2}$, while b and c will be generated with probability $\frac{1}{4}$. The structure of the grammar affects the probability as well. Two distinct grammars can generate the same sentence with different probabilities [18].

CFG can be used to express problem-specific constraints on the GP search space by dictating how a chromosome should be constructed. A GP that uses a CFG in this scheme is called Context Free Grammar based Genetic Programming [4].

As described above, a sentence, which could be a program, can be constructed using a grammar. In order to do that, it is necessary to decide which formation rule should be selected every time a non-terminal is translated. In Grammatical Evolution [8], a chromosome created randomly and structured as a string composed of 0 and 1 is used to make this decision. Initially, the grammar formation rules have an integer assigned to each of them. Then, fixed sized substrings of the chromosome are composed and, after converted to integers, they are used to select a formation rule each time a non-terminal has to be translated.

4   Bayesian Automatic Programming

4.1   General Algorithm

This section describes the BAP general algorithm. Like BOA, BAP uses a Bayesian network to estimate the joint distribution of promising solutions. This estimation is used to generate new candidate solutions. BAP uses the flexibility of the genetic programming paradigm to try to extend the power of description of a Bayesian network introduced in BOA to the automatic program creation.

The pseudocode of the BAP is shown in Algorithm 1.The algorithm initially generates an initial population $P(0)$ using a recursive uniform random generation method. At each generation, the population $P(t)$ is evaluated by a fitness function and selected from the current population using the elitism selection method, but other selection methods can be used as well. A Bayesian network that fits the selected set of solutions $S(t)$ is learned using the K2 algorithm and constraints. New solutions $O(t)$ are generated using the joint distribution encoded by the learned network and can undergo mutation. The set of selected solutions is placed together with the newly generated solutions to replace the entire old population. The process is repeated until the termination criteria is met.

4.2   BAP Individual Codification

Candidate solutions are represented by a fixed size vector of integers. This vector will be called chromosome, and each position in the vector will be called gene. Each gene stores one number associated with a formation rule from the grammar used to model the problem. The formation rules of a grammar are enumerated sequentially starting from zero. Each time a grammar formation rule should be chosen to translate a non-terminal using right reduction, as introduced in

Algorithm 1 General Algorithm

1: $t = 0$
2: Generate the initial population P(0) using a uniform random generation algorithm
3: **while** termination criteria is not met **do**
4:     Select a set of promising solutions $S(t)$ from $P(t)$
5:     Construct the Bayesian network B using K2 and constraints
6:     Generate a set of new solutions $O(t)$ according to the joint distribution encoded by B and a grammar G
7:     Apply mutation on $O(t)$
8:     Create a new population $P(t+1)$ by replacing all solutions from $P(t)$ with $O(t) \cup S(t)$
9:     $t = t + 1$
10: **end while**

Section 3.2, the number assigned to the chosen formation rule is concatenated on the right side of the chromosome. Therefore, a chromosome stores the number of the rule rather than the program itself. This chromosome representation keeps the implicit hierarchy of the derivation tree and can be used to restore this tree through the grammar. Figure 2 shows an example of this representation for a simple arithmetic grammar: the enumerated formation rules; a possible chromosome and its production tree.

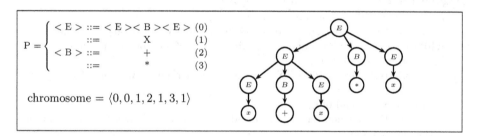

Fig. 2. An example of enumerated formation rules, a chromosome and the concrete syntax tree

Because the K2 algorithm learns the network structure from a set of tuples, the vector representation of the chromosomes is recommended.

As defined previously, a Bayesian network is represented by a graph and a set of random variables. In BAP, there is a random variable associated with each gene of a chromosome. The i-th random variable represents the integers observed in the i-th gene of a chromosome.

The solution representation used in BAP allows the implicit hierarchy of the concrete syntax tree to be used as prior information to the K2 algorithm. Moreover, the use of a grammar to guide the construction of a solution allows the application of prior knowledge about the problem to reduce the search space.

4.3   Individual Generation

BAP applies two different methods for generating new chromosomes: one to the initial population and other to the following generation. To get an initial population with high diversity, a uniform random generation algorithm [18] is used. This algorithm guarantees that all chromosomes of a fixed size, given a grammar, have the same probability of being chosen. This allows a good distribution of the chromosomes through the search space.

All chromosomes created after the initial population are built using the mapping process. The mapping process basically follows the procedures introduced in Section 3.2 and 4.2. However, this mapping process differs in the way that it uses the Bayesian Network to influence the formation rule selection. Algorithm 2 shows a pseudocode of the mapping process to one chromosome C. It begins creating an auxiliary vector I . This vector initially stores only the start symbol S of the grammar G and at the end of the process it will contain the program encoded by C. At each iteration, the algorithm searches for a non-terminal in I and a roulette wheel algorithm [19] consults the CPT of the random variable related to this non-terminal in order to select a formation rule. The number of this formation rule is attributed to c and inserted at the end of C , the non-terminal is replaced in I by the right side of the formation rule chosen.

Algorithm 2 Individual Creation Algorithm

1: $i = 1$. Index of the chromosome vector.
2: $I{=}S \in G$. I is an auxiliary vector which stores non-terminals and terminals.
3: **while** I has non-terminal and $i \leq$ maximum size of a chromosome **do**
4:     r=normalized random number
5:     c=rouletteWheel(r,CPT of X_i). X_i is the i-th random variable of a Bayesian network
6:     Insert c at the end of C.
7:     Remove the first non-terminal in I and replace it by the right side of the formation rule chosen.
8:     $i = i + 1$
9: **end while**
10: **if** I has a non-terminal **then**
11:     Return C
12: **else**
13:     Return FALSE
14: **end. if**

The composition process of a chromosome using a grammar as applied in the mapping has the tendency to generate shorter or very long chromosomes, as exposed in Section 3.2. To avoid that, it was necessary to use a restriction to control their size. This control consists of comparing the size of the produced chromosome with the average size of the chromosomes set used to learn the Bayesian network. A tolerance of δ genes in the size is used to permit the evolution of the model. Chromosomes that does not comply with this restriction or are bigger than the maximum size are discarded.

4.4   Mutation

BAP applies a mutation operator to avoid local optimum and evolution stagnation.

The mutation operation replaces a subtree of the derivation tree mapped in a chromosome. The portion removed is replaced by a new one generated through the uniform random generation algorithm. A mutation constant μ is defined by the user to determine the probability of a gene undergoing mutation. A gene is mutated if the inequality (1) is satisfied.

$$\frac{\mu}{\mid C \mid} \geq r \tag{1}$$

5   Experiments

We applied BAP to two traditional genetic programming problems the symbolic regression and the artificial ant problem, to evaluate its performance. In all tests, BAP is compared with the Lilgp tool set version 1.1 [20], a traditional GP tool. The methods were compared by their cumulative success frequency during 50 generations in 100 runs and by the average of the fitness during 50 generations also in 100 runs.

5.1   Configurations

Symbolic Regression. In this problem, a set of records with input and output values is provided. The objective is to find a function which produces the output value from the input. In our experiments, the objective function is:

$$f(x, y) = x^4 + 2xy + y^4 \tag{2}$$

The input data is composed by two sets of twenty real numbers produced randomly in the range $[-10...10]$. The fitness function is the sum of the error taken over the 20 fitness cases. The Figure 3 shows the grammar used by BAP. Table 1 contains the configuration of BAP and GP parameters for the symbolic regression problem.

$$N = \{expr, boper, var\}, \ T = \{X, Y, +, -, \%, *\}, \ S = \{expr\}$$

$$P = \begin{cases} < expr >::=< expr >< boper >< expr > \\ \qquad \mid (< expr >< boper >< expr >) \mid < var > \\ < boper >::= + \mid - \mid * \mid \% \\ < var >::= X \mid Y \end{cases}$$

Fig. 3. Grammar used by BAP in the symbolic regression problem

Table 1. Configuration of the symbolic regression parameters

BAP		GP	
Parameter	Value	Parameter	Value
Population size	500	Population size	500
Generation number	50	Generation number	50
μ	0.3	Mutation rate	0.01
δ	10	Crossover	0.7
Chromosome size	60	Tournament size	5
		Terminal Operators	$\%, +, *, -$
		Terminal Operands	X, Y

Artificial Ant. The objective in the artificial ant problem is to find a program that controls the movements of an ant which should get as food as possible with a fixed maximum number of actions. The artificial ant specification is the same as defined in [21] . The ant is inserted on a 32 by 32 toroidal with 91 pieces of food composing a discontinuous trail. The fitness function is the amount of food pieces missed on the trail. Figure 4 shows the grammar used by BAP. Table 2 contains the configuration of BAP and GP parameters for the artificial ant problem.

N={expr, line}, T={left(), right(), move()}, S={expr}

$$P = \begin{cases} < expr >::=< line > \mid < expr > < line > \\ < line >::= \text{if food\_ahead() } \{< expr >\} \text{ else } \{< expr >\} \mid < op > \\ < op >::= right() \mid left() \mid move() \end{cases}$$

Fig. 4. Grammar used by BAP in the symbolic regression problem

Table 2. Configuration of the artificial ant parameters

BAP		GP	
Parameter	Value	Parameter	Value
Population size	4000	Population size	4000
Generation number	50	Generation number	50
μ	0.3	Mutation rate	0.01
Chromosome size	70	Crossover	0.7
		Tournament size	5
		Terminal Operators	move(), left(), right(), food_ahead()

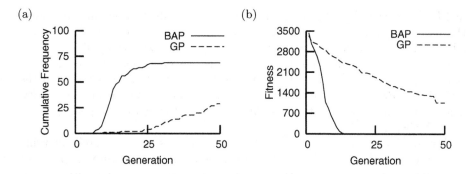

Fig. 5. Cumulative frequency of success (a) and the average (b) of the fitness for symbolic regression

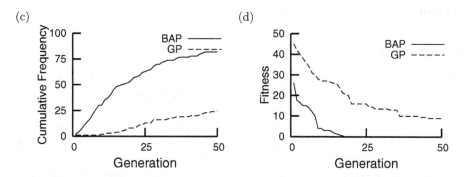

Fig. 6. Cumulative frequency of success (a) and the average (b) of the fitness for the artificial ant

5.2   Experimental Results

Figure 5.a shows the cumulative frequency and Figure 5.b presents the average of the fitness for BAP and GP in the symbolic regression problem. Figure 6.a and Figure 6.b present these same measures now for the artificial ant problem.

As observed in all graphs, BAP outperforms the traditional GP in both problems. Although the time performance of the two systems was not properly compared, it was possible to note that BAP is much slower than GP, when using long chromosomes. This happens due to the high complexity of the K2 algorithm, even when using the node order as prior knowledge. As the chromosome size has influence in this complexity it was limited to a low value.

One important point to note in Figures 5.b and 6.b is the convergence of BAP. The BAP algorithm converges very fast to an optimum, not always the global, this means a lost of diversity in the population. In BAP, the probabilistic model evolves through the generations and becomes more restrictive, the chromosomes generated more similar. This happens because the Bayesian network describes only the chromosomes used as learning data. These chromosomes represents only a reduced part of the whole population. In contrast, the model used in PIPE and eCGP keeps all possibles terminals and functions in every node through the generations. Furthermore, the BAP repopulation process discards all chromosomes of the predecessor generation. In this way, the building blocks belonging to chromosomes not used to train the model are lost. Figure 5.a shows that besides BAP has a better overall result, its evolution does not keep the rate of the beginning of the evolution, when the population has more diversity. The starvation of the evolution also indicates that the mutation operator has a limited efficiency.

We also applied BAP to another experiment of symbolic regression using the expression $x^4 + x^3 + x^2 + x$. To solve this problem, it was used a grammar with more operators (sin, cos, tag, ln, exp) and a training data with values near to zero. In this configuration, BAP presents worse performance compared to GP. One possible explanation is that GP outperform BAP in harder problems. However, further experiments have to be performed on harder and different types of problems to learn true usefulness of BAP.

6 Conclusion

This work presents a novel approach called Bayesian Automatic Programming (BAP), which integrates the power of GP and EDA technologies to induce programs.

BAP has some limitations, such as the prohibitive use of long chromosomes, which is caused by the time complexity of K2 algorithm. In addition, BAP has the tendency to generate very short or very long chromosomes due to the irregular probability distribution that the grammar based creation method undergoes. However, it shows the ability to solve the problems of symbolic regression and artificial ant with advantage over classical GP, when comparing the number of best solutions found and the convergence speed to the best solution.

The good performance observed in the two studied problems encourages its application to new problem domains. In special, we intend to evaluate the performance of the algorithm in the deceptive trap problem [22], and compare its results with PIPE and eCGP.

To avoid local optimal caused by the lost of building blocks during the evolution, new selection and population replacement mechanisms can be applied. The production of chromosome based on grammar can also be improved. Our overall conclusion is that the scheme proposed was successful enough to deserve better studies.

References

1. Goldberg, D.: Genetic Algorithms in Search, Optimization and Machine Learning. Addison-Wesley (1989)
2. Koza, J.R.: Genetic Programming: On the Programming of Computers by Means of Natural Selection. MIT Press, Cambridge, MA, USA (1992)
3. Harik, G.R., Goldberg, D.E.: Learning linkage. In: Foundations of Genetic Algorithms 4. (1996) 247–262

4. Ratle, A., Sebag, M.: Avoiding the bloat with probabilistic grammar-guided genetic programming. In: Artificial Evolution 5th International Conference,Evolution Artificielle, EA 2001. Volume 2310. (2001) 255–266
5. Salustowicz, R., Schmidhuber, J.: Evolving structured programs with hierarchical instructions and skip nodes. In: Proceedings of the 15th International Conference on Machine Learning, Morgan Kaufmann (1998) 488–496
6. Sastry, K., Goldberg, D.: Probabilistic model building and competent genetic programming. Technical Report 2003013, University of Illinois (2003)
7. Pelikan, M., Goldberg, D., Cant-Paz, E.: BOA: The bayesian optimization algorithm. In: GECCO '99: Proceedings of Genetic and Evolutionary Computation Conference. Volume 1. (1999) 525–532
8. O'Neill, M., Ryan, C.: Grammatical Evolution. IEEE Transactions on Evolutionary Computation **5** (2001) 349–357
9. Pelikan, M., Goldberg, D.E., Lobo, F.: A survey of optimization by building and using probabilistic models. Technical Report 99018, Illinois Genetic Algorithms Laboratory, University of Illinois at Urbana-Champaign, Urbana (1999)
10. Baluja, S.: Population-based incremental learning: A method for integrating genetic search based function optimization and competitive learning. Technical Report CMU-CS-94-163, Carnegie Mellon University (1994)
11. Harik, G.: Linkage learning via probabilistic modeling in the ECGA. Technical Report 99010, Illinois Genetic Algorithms Laboratory, University of Illinois, Urbana (1999)
12. Larranaga, P., Lozano, J.A.: Estimation of Distribution Algorithms, A New Tool for Evolutionary Computation. Kluwer Academic Publishers (2001)
13. Heckerman, D.: A tutorial on learning with bayesian networks. Technical Report MSR-TR-95-06, Microsoft Corporation (1995)
14. Krause, P.: Learning probabilistic networks. The Knowledge Engineering Review **13** (1998) 321–351
15. Heckerman, D., Geiger, D., Chickering, M.: Learning bayesian networks: The combination of knowledge and statistical data. Technical Report MSR-TR-94-09, Microsoft Research (1994)
16. Cooper, G., Herskovits, E.: A bayesian method for the induction of probabilistic networks from data. Machine Learning **9** (1992) 309–345
17. Russell, S., Norvig, P.: Artificial Intelligence, A Modern Approach. Second edition edn. Prentice Hall (2003)
18. McKenzie, B.: Generating strings at random from a context free grammar. Technical Report TR-COSC 10/97, University of Canterbury (1997)
19. Rudnick, E.M., Patel, J.H., Greenstein, G.S., Niermann, T.M.: Sequential circuit test generation in a genetic algorithm framework. In: Design Automation Conference. (1994) 698–704
20. Research, G.A., GARAGe, A.G.: Lilgp. http://garage.cps.msu.edu/software/lilgp/lilgp-index.html (ver 1.1)
21. Koza, J.R.: 12. In: Genetic Programming II. MIT Press (1998) 349–353
22. Deb, K., Goldberg, D.E.: Analyzing deception in trap functions. Technical Report 2002007, Illinois Genetic Algorithms Laboratory, University of Illinois, Urbana (1993)

Dynamic Size Populations in Distributed Genetic Programming

Denis Rochat[1], Marco Tomassini[1], and Leonardo Vanneschi[2]

[1] Information Systems Department, University of Lausanne, Switzerland
{marco.tomassini, denis.rochat}@unil.ch
[2] Dipartimento di Informatica Sistemistica e Comunicazione (DISCo),
University of Milano-Bicocca, Italy
vanneschi@disco.unimib.it

Abstract. This paper proposes the association of two approaches of GP which improve efficiency and reduce bloat. The first approach is to use a multi-population version of GP and the second one is to employ populations that can change size dynamically and adaptively. The latter approach consists in deleting or adding individuals in the population as a function of the current fitness and two other parameters. We test this approach on three well-known problems in GP, artificial ant, even parity 5 and one instance of the symbolic regression. We find that the combination of these two methods improves the quality of the individuals in the populations while keeping their size as small as possible and decreases the amount of resources required.

1 Introduction

A well-known drawback of Evolutionary Algorithms is the amount of computational effort that is sometimes required for solving difficult applications. This problem is more marked when algorithms use variable size chromosomes, which has been traditionally the case in Genetic Programming (GP) [1]. In fact, it has been observed that GP individuals tend to steadily grow in size as generations are computed ([2, 3]). The phenomenon goes under the name of *bloat* and there have been several proposals aimed at preventing such an inordinate growth (see [4] for a rather complete survey). For instance, the fitness function may embody a penalty associated to the size of the individuals, so that short individuals are favored. Another proposal is to set a limit for the maximum size. In the last few years some researchers have tried to apply multiobjective techniques to GP, in such a way that both size of individuals and fitness are considered as tasks to be optimized ([5]). Most of these techniques have some drawbacks: either they change the way in which fitness is computed, thus influencing the structure of the fitness landscape and the main characteristics of the problem, or they require additional computational costs to be implemented, often thwarting their positive effects on size. Moreover, most proposed solutions focus on individual size and do not take into account the global problem of population growth as a whole.

M. Keijzer et al. (Eds.): EuroGP 2005, LNCS 3447, pp. 50–61, 2005.

In some studies [6, 7, 8], the bloat and reduction of computational effort problems have been approached employing another perspective: given that the increase in size of individuals produces an overall growth in the population, the population size has been dynamically changed, trying to control the computational effort required for evaluating individuals. The technique could later be combined with any of the traditional ones for controlling individuals length. Variable-size populations have seldom been used: two studies pertaining to GAs are [9] and [10]. Structured models using variable-size populations appears in [11] and [12].

In this study, we further improve the semi-automatic sizing of the population at run time, and we structure the population as an ensemble of multi-populations between which migration of selected individuals is allowed. This approach, called *island model*, has been empirically found very useful for improving the numerical characteristics of the search, as well as to decrease computing time when it is executed on distributed hardware (see [13, 6] and references therein for more details).

The paper is structured as follows: in section 2 we motivate the use of populations of dynamic size in GP and we describe two algorithms allowing to add or suppress individuals from a population "on the run", according to some particular events occurring during the evolution. Section 3 briefly introduces the set of GP problems used to test the suitability of these algorithms and the GP parameters used in our experiments. Section 4 presents and discusses experimental results in terms of efficiency and bloat reduction. Section 5 offers our conclusions and hints for future work.

2   Dynamic Size Populations in Multi-population GP

Dynamic size population means that the number of individuals in the population can be modified during the run. The origin of this idea is found in [14] and is called *plague*. In this early version, some individuals were suppressed from the population at each generation, which is helpful to reduce the effort but can also diminish population diversity to a point were further evolution is impossible. The following step has been to delete individuals from the population while the best individual found keeps improving along generations, and to add individuals to the population if there is no improvement. The suitability of this idea has been empirically shown in [8, 6]. Indeed, with this method, the individuals found are fitter on the average, and the computational effort is considerably reduced. Here we further improve the self-adaptive mechanism by which individuals are added or deleted from/to the population. In addition, we modify the source of new individuals by using a multi-population model instead of a standard panmictic GP population.

2.1   Add or Delete Individuals?

To answer this question, we have to calculate a value called *pivot* (p), which is obtained by dividing the best fitness at the first generation by the maximum allowed number of generations g_{max} in the run for a given problem. The *pivot* calculated in this way is fixed and won't be modified during the run. Furthermore, a value called *period* (t) must also be calculated; it is the number of consecutive generations after which we add or delete individuals. So, after *period* generations, we compute the difference in best fitness at generation g with respect to the best fitness at generation $g - t$. This quantity is divided by t. If the result is larger than the pivot p then we delete some individuals, otherwise we add individuals.

Now that we have described *when* individuals have to be deleted or added from/to the population, we have to specify *which* ones must be deleted and how the new individuals to be added are generated.

2.2   The Removal of Individuals

We use a new method for suppressing N_{del} individuals which is based on the following equation:

$$N_{del} = Pg - t * \frac{f_{g-t} - f_g}{f_{g-t}}.$$

where P is the initial population size, f_g is the best fitness of the population at generation g, and f_{g-t} is the best fitness at generation $g - t$. Thus, the expression $\frac{f_{g-t} - f_g}{f_{g-t}}$ is the fractional *gain* in fitness from generation $g - t$ to generation g.

To suppress N_{del} individuals, we first determine the $2 \times N_{del}$ worst individuals in the population and, among them, we suppress the N_{del} having the largest size, in order to give a contribution for fighting bloat. Furthermore, a lower limit on the size of the population must be fixed in order to insure sufficient diversity in the population. We use two values for the lower limit of the population size: two individuals or half of the initial population size.

2.3   The Addition of Individuals

In single-population varying-size GP, as described in [8, 6], the new individuals are created by applying mutation on the best N_{add} individuals, where N_{add} is the number of individuals to be added. This solution presents a potential problem, since mutation does not guarantee that the newly generated individuals will have a good fitness value. In the approach presented in this paper, the individuals that are added to one subpopulation are (copies of) the best ones in another coevolving subpopulation.

Now the way population size will increase has to be defined. We propose four methods that we call, respectively, M1, M2, M3, M4. M1 increases the population size in order to have the final size equal to the initial size if the run does not find a satisfying solution. In other words, calling P_g the population size at generation g before adding individuals, we add a number of individuals given by $(P_0 - P_g)/(g_{max} - g)$. This is graphically depicted in figure 1 (a).

The second addition method (M2) suggests refilling the population proportionally to a certain coefficient calculated on the first best fitness and the current best fitness. More precisely, if f_g and f_0 are the best fitness at generation g and generation 0 respectively, then the coefficient is: $c = \sqrt{f_g/f_0}$, and the number

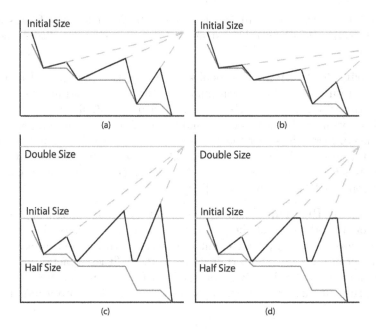

Fig. 1. Graphic representation of the four methods used to increase the population size. (a): M1, (b): M2, (c): M3, (d): M4. The idealized curves show the correlation between the population size (black curve) and the evolution of the best fitness (gray curve) (see text)

of individuals added is $(c\, P_0 - P_g)/(g_{max} - g)$. Figure 1 (b) displays the idea behind the method.

The third method (M3) is similar to the first one with a projected final population size which is twice the initial size.

The last method (M4) is the same as the third one but the population size is never larger than the initial size. The two last methods can be visualized in figure 1(c) and 1(d).

3   Test Problems and GP Parameters

We decided to address a set of problems that have been classically used for testing GP: the symbolic regression problem, the artificial ant problem on the Santa Fe trail and the even parity problem. The following is a brief description of these problems (details can be found in [1]).

Symbolic Regression Problem. The problem aims to find a program that matches a given equation. We employ the classic polynomial equation $f(x) = x^4 + x^3 + x^2 + x$, and the input set is composed of 1000 equidistant points in the range [0,2] (1000 fitness cases). For this problem, the set of functions used for GP individuals is the following: $\mathcal{F} = \{*, //, +, -\}$, where $//$ is like $/$ but returns

1 instead of *error* when the divisor is equal to 0. We define the fitness as the sum of the square errors at each test point. Consequently, lower fitness means a better solution.

Even Parity 5 Problem. The boolean even parity k function of k boolean arguments returns *true* if an even number of its boolean arguments evaluates to true, otherwise it returns *false*. If $k = 5$, then $2^5 = 32$ fitness cases must be checked to evaluate the fitness of an individual. The fitness is computed as 32 minus the number of hits over the 32 cases. Thus a perfect individual has fitness 0, while the worst individual has fitness 32. The set of functions used to code individuals is the following: $\mathcal{F} = \{NAND, NOR\}$. The terminal set is composed of 5 different boolean variables $\mathcal{T} = \{a, b, c, d, e\}$.

Artificial Ant Problem. In this problem, an artificial ant is placed on a 32 × 32 toroidal grid. Some of the grid cells contain food pellets. The goal is to find a navigation strategy for the ant that maximizes its food intake. We use the same set of functions and terminals as in [1]. As fitness function, we use the total number of food pellets lying on the trail (89) minus the amount of food eaten by the ant during his path. This turns the problem into a minimisation one.

GP Parameters. In all the experiments we use the following set of parameters: generational GP, crossover rate : 95%, mutation rate: 0.1%, tournament selection of size: 10, ramped half and half initialization, maximum depth of individuals for the creation phase: 6, maximum depth of individuals for crossover: 10, elitism (i.e. survival of the best individual into the newly created population at each generation). Furthermore, to avoid complicating the issue, we refrained from using advanced techniques such as ADFs, tailored function sets and so on.

Population Structure. We have used an island model with five populations connected according to a random communication topology. Individuals are exchanged every ten generations and the number of migrating individuals, which are the best in their island, is 10% of the current size of the given sub-population. All the populations send their boats of migrants to a master process which accumulates the individuals and then sends an appropriate amount to each population, choosing the destination island in a random way. In the destination island, the migrants replace an equal number of worst individuals. Communication is synchronous i.e., all the populations send and receive individuals before computing the next generation.

Computational Effort. In GP, comparing fitness values during generations or, as it is usually done in GAs, comparing in terms of the number of fitness evaluations performed is inadequate. While this is often acceptable in EAs with fixed length representations, it can be misleading in GP, where individuals change their size dynamically. We thus analyzed the data as in [13] by means of the *effort of computation* at generation g which is the total number of nodes that have been evaluated before generation $g+1$ takes place. Note that, strictly speaking, one should take into account that evaluation times are different for different operators, but our simplification is still useful as a first approximation. Clearly, this measure is problem-specific but it is useful to compare different solutions of the same problem.

4   Experimentals Results

The aim of the simulations is to empirically give evidence of the fact that dynamic size population associated with multi-population GP produces better solutions and saves computational effort than in the case of standard GP. In section 2 we defined a method for deleting individuals from the population, and four methods, called M1, M2, M3, and M4 for the addition of individuals. Thus, in the experiments presented in this section we have tested four different combinations of addition and suppression of individuals in the population for each problem. In section 2.1 we defined a parameter called *period* which was the number of generations elapsed between a suppression or addition operation in the populations. We did different runs for each problem with period values between 1 and 5. Results are very similar. For reasons of space, here we report only the results for a single period value for each problem.

Fitness Evolution. Figures 2 (a), (b), and (c) depict the average fitness of the ensemble of the islands for the three benchmark problems as a function of the computational effort. The values are the average over 100 independent executions for each problem and for each method. The thick black curve refers to the *standard* multi-population model (i.e. the model with fixed size subpopulations), while the other four curves each represent one the four models of variable multipopulation size.

In the ant problem – figure 2 (a) – we see that all four methods using dynamical population size give better results with respect to the standard fixed-size island model. For instance, to reach an average fitness of about 15, the standard distributed model needs a computational effort of about 5.5×10^6, while the same average fitness level is reached by the M1, M2, M3, and M4 variable-island size methods with an effort that is approximately comprised between 2×10^6 and 3×10^6.

For the even 5 parity and symbolic regression problems – figures 2 (b) and (c) – we observe the same general trend i.e., the variable-size methods are all superior to the standard method but the differences are smaller. Indeed, standard deviations at the end of the evolution for each problem (not shown in the figures to avoid cluttering them) indicate that, while differences are significant between the standard model and the ensemble of variyng-population size models for the ant problem, they are not for the other two problems, differences among the various curves being of the same order as the standard deviation.

Success Rate. The success rate is a good performance indicator when the problem solution is known, which is the case here (see for instance [6] for a discussion about this issue). Success rates for each problem are reported in table 1 with their standard deviations. These figures confirm that, for the ant problem,

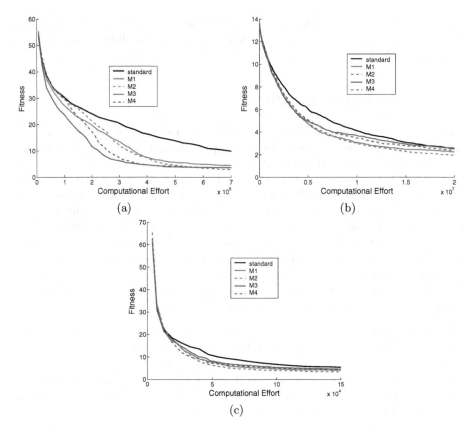

Fig. 2. Average fitness results vs computational effort. Each curve is an average over 100 runs. (a) Artificial Ant, (b) Even Parity 5, (c) Symbolic Regression

variable-size islands, whatever the method used for suppression and addition of individuals, are better than the standard model. For the even 5 parity and symbolic regression problems, the results are still favorable to the dynamical island models since there is always a method among the four that is better than the standard island model. However, for the remaining methods the differences are within the standard deviation and thus the statistical significance is doubtful.

Program Size Evolution. Figures 3 (a), (b), and (c) show the evolution of the size of the programs in the populations with time. It is easily seen that, in general, methods M1, M2, M3 and M4 that automatically adjust the population size in the islands offer an easy and implicit means for limiting bloat. This had already been found to be the case for the usual constant-size island model with respect to standard panmictic genetic programming [15]. Therefore, variable-size populations are a really effective and transparent way for limiting bloat. Now, in the figures, it is apparent that method M3 is less effective than M1, M2, and M4 in this respect. Keeping in mind that method M3 is allowed to add individuals

Table 1. Success rates at three different computational effort values for the three test problems and for the five structured population models. (a): artificial ant, (b): even 5 parity, (c): symbolic regression. Standard deviations of each value is included between parenthesis

	EC 1	EC2	EC 3
Standard	0.20(0.04)	0.28(0.04)	0.31(0.05)
M1	0.51(0.05)	0.56(0.05)	0.59(0.05)
M2	0.55(0.05)	0.60(0.05)	0.63(0.05)
M3	0.59(0.05)	0.61(0.05)	0.63(0.05)
M4	0.61(0.05)	0.65(0.05)	0.65(0.05)

(a)

	EC 1	EC2	EC 3
Standard	0.04(0.02)	0.07(0.03)	0.08(0.03)
M1	0.05(0.02)	0.06(0.02)	0.07(0.03)
M2	0.12(0.03)	0.13(0.03)	0.16(0.04)
M3	0.07(0.03)	0.07(0.03)	0.08(0.03)
M4	0.04(0.02)	0.04(0.02)	0.08(0.03)

(b)

	EC 1	EC2	EC 3
Standard	0.41(0.05)	0.44(0.05)	0.46(0.05)
M1	0.52(0.05)	0.53(0.05)	0.53(0.05)
M2	0.60(0.05)	0.61(0.05)	0.62(0.05)
M3	0.48(0.05)	0.49(0.05)	0.49(0.05)
M4	0.51(0.05)	0.52(0.05)	0.52(0.05)

(c)

up to twice the original population size (see section 2.2), it is clear that this has a negative influence on bloat. However, the other three methods offer performances that are equivalent to M3, and thus program growth can be best controlled by using one of them. It is clear that, since bloat is not controlled explicitly in our models, it would be possible to add other bloat-reducing techniques to the system. A good candidate could be Poli's *Tarpeian* method, which is theoretically justified and extremely easy to implement [16].

Population Size. In figures 4 (a), (b), and (c) we report the evolution of the total population size for the five cases. The first remark is that, except in two cases that will be discussed below, the population size in the adaptive-size models always remains smaller than the standard model (horizontal line). This explains why the computational effort required is lower for the same solution accuracy. We also see that the evolution of the total population size for the four variable-size methods is clearly correlated with the previous curves of program size. This empirically shows that one of our original goals, that of limiting program size by

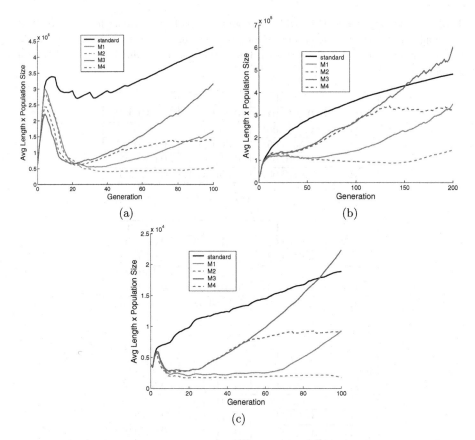

Fig. 3. The average population length times the total population size is represented as a function of generation number. Averages are over 100 independent runs for each problem and for each model. The thick curve refers to the standard island model, while the other curves are for the varying-size population models (see box). (a) Artificial Ant, (b) Even 5 Parity, (c) Symbolic Regression

adjusting the size of the populations has been reached. Again, the only variable-size method that is less effective in this respect, especially for the even 5 parity problem, is M3, for the reasons mentioned above.

Diversity Evolution. With smaller subpopulations one might fear that diversity would tend to be lost, which could have adverse effects on the evolution of good solutions: a too homogeneous population could loose its evolvability. We have measured global population diversity for the test problems and we observe that this is not the case. As an example, in figure 5 we report phenotypic and genotypic entropy against generation number for the even-parity 5 problem (these measures are standard and their description can be found for instance in [17]). We see that diversity is maintained in the variable-population size models

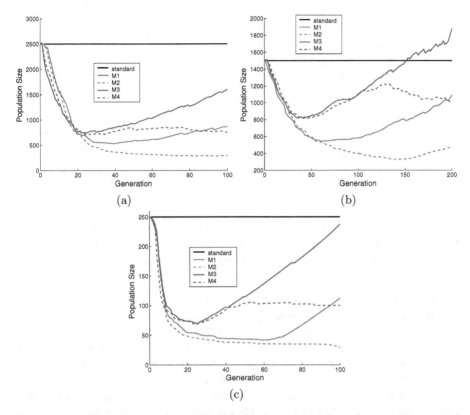

Fig. 4. Population size as a function of the generation number. Curves are averaged over 100 runs for each problem and for each model. (a) Artificial Ant, (b) Even 5 Parity, (c) Symbolic Regression

at approximately the same level as for the standard fixed-size multi-population system. The jumps are due to synchronous individual migration.

5   Conclusions

Following some recent studies [7, 8, 6], we have used a new methodology in genetic programming that consists in subdividing the total population into several subpopulations and making each subpopulation size a dynamical, self-adjusting parameter. It is already empirically well-known that multi-population models are in general beneficial in evolutionary computation (see for instance [6]). In fact, it has been reported that they are more efficient in problem solving, and that they also have an indirect positive effect on program size, thus limiting bloat. In the present work, we have found the same trends, but the results are even better both in terms of success rates, and also from the point of view of the effort expenditure and program size. This shows that the little-studied field of

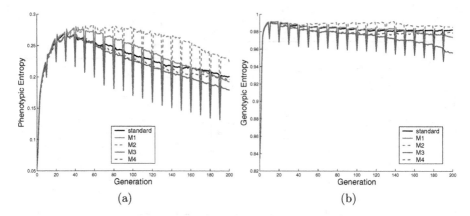

Fig. 5. Diversity as a function of the generation number for the standard island model and the variable-size population models for the even-parity 5 problem. Curves are averaged over 100 runs. (a) phenotypic entropy; (b) genotypic entropy

variable-size populations can offer new insights and more powerful evolutionary algorithms. Our findings are empirical; therefore, we cannot extrapolate them beyond the simple benchmark problems studied here. Indeed, already the theory of standard panmictic GP is rather difficult [4]. Variable-size populations will be even harder to model and analyze. However, the trend seems to be consistent and thus the methodology could be a promising one. To confirm these preliminary results we intend to test our models on a more complete benchmark suite, including some difficulties not present here, such as more difficult regression problems containing constants. A more detailed treatment of the statistical significance of the results is also called for.

References

1. J. R. Koza. *Genetic Programming*. The MIT Press, Cambridge, Massachusetts, 1992.
2. W. Banzhaf and W. B. Langdon. Some considerations on the reason of bloat. *Genetic Programming and Evolvable Machines*, 3:81–91, 2002.
3. W. B. Langdon and R. Poli. Fitness causes bloat. In R. Roy P. K. Chawdhry and R. K. Pant, editors, *Soft Computing in Engineering Design and Manufacturing*, pages 13–22. Springer-Verlag London, 1997.
4. W. B. Langdon and R. Poli. *Foundations of Genetic Programming*. Springer, Berlin, 2002.
5. E.D. De Jong and J.B. Pollak. Multi-objective methods for tree size control. *Genetic Programming and Evolvable Machines*, 4:211–233, 2003.
6. L. Vanneschi. *Theory and Practice for Efficient Genetic Programming*. PhD thesis, University of Lausanne, Switzerland, 2004.
7. F. Fernández, M. Tomassini, and L. Vanneschi. Saving computational effort in genetic programming by means of plagues. In *Congress on Evolutionary Computation (CEC'2003)*. IEEE Press, Piscataway, NJ, 2003.

8. F. Fernández, M. Tomassini, and L. Vanneschi. A new technique for dynamic size populations in genetic programming. In *Congress on Evolutionary Computation (CEC'2004)*. IEEE Press, Piscataway, NJ, 2004.
9. J. Arabas, Z. Michalewicz, and J. Mulawka. A genetic algorithm with varying population size. In *Proceedings of the 1994 IEEE Conference on Evolutionary Computation*, pages 73–78. IEEE Press, Piscataway, NJ, 1994.
10. A. E. Eiben, E. Marchiori, and V. A. Valkó. Evolutionary algorithms with on-the-fly population size adjustment. In X. Yao et al., editor, *Parallel Problem Solving from Nature - PPSN VIII*, volume 3242 of *Lecture Notes in Computer Science*, pages 41–50. Springer-Verlag, Heidelberg, 2004.
11. T. Krink, B. H. Mayoh, and Z. Michalewicz. A patchwork model for evolutionary algorithms with structured and variable size populations. In W. Banzhaf et al., editor, *Genetic and evolutionary conference, GECCO99*, volume 2, pages 1321–1328. Morgan Kaufmann, San Francisco, CA, 1999.
12. B. Olsson. Optimization using a host-parasite model with variable-size distributed populations. In *Third International Conference on Evolutionary Computation*, pages 295–299. IEEE Press, Piscataway, NJ, 1996.
13. F. Fernández, M. Tomassini, and L. Vanneschi. An empirical study of multi-population genetic programming. *Genetic Programming and Evolvable Machines*, 4(1):21–52, 2003.
14. F. Fernández, L. Vanneschi, and M. Tomassini. The effect of plagues in genetic programming: a study of variable-size populations. In C. Ryan et al., editor, *Genetic Programming, Proceedings of the 6th European Conference, EuroGP 2003*, volume 2610 of *LNCS*, pages 317–326. Springer-Verlag, 2003.
15. G. Galeano, F. Fernández, M. Tomassini, and L. Vanneschi. Studying the influence of synchronous and asynchronous parallel GP on programs length evolution. In *Congress on Evolutionary Computation (CEC'02)*, pages 1727–1732. IEEE Press, Piscataway, NJ, 2002.
16. R. Poli. A simple but theoretically motivated method to control bloat in genetic programming. In C. Ryan et al., editor, *Genetic Programming, Proceedings of the 6th European Conference, EuroGP 2003*, volume 2610 of *LNCS*, pages 204–217. Springer-Verlag, 2003.
17. E. K. Burke, S. Gustafson, and G. Kendall. Diversity in genetic programming: An analysis of measures and correlation with fitness. *IEEE Transactions on Evolutionary Computation*, 8(1):47–62, 2004.

Evolution of Robot Controller Using Cartesian Genetic Programming

Simon Harding and Julian F. Miller

Department Of Electronics
University of York, UK YO10 5DD
{slh, jfm}@evolutioninmaterio.com

Abstract. Cartesian Genetic Programming is a graph based representation that has many benefits over traditional tree based methods, including bloat free evolution and faster evolution through neutral search. Here, an integer based version of the representation is applied to a traditional problem in the field: evolving an obstacle avoiding robot controller. The technique is used to rapidly evolve controllers that work in a complex environment and with a challenging robot design. The generalisation of the robot controllers in different environments is also demonstrated. A novel fitness function based on chemical gradients is presented as a means of improving evolvability in such tasks.

1 Introduction

Cartesian Genetic Programming is a graph based representation that has many benefits over traditional tree based methods, including bloat free evolution and faster evolution through neutral search. In this paper we apply this representation to a robot control task. In this section the representation is discussed - including a comprehensive survey of existing work on Cartesian Genetic Programming(CGP) and we look at previous work in evolving robot controllers using genetic programming. We describe the benefits of this representation over traditional techniques. The algorithm and novel fitness function are discussed in section 2. In section 3 we apply apply CGP to various robot tasks and demonstrate generality in evolved solutions.

1.1 Cartesian Genetic Programming

Cartesian Genetic Programming [13] is a graph based form of Genetic Programming that was developed from a representation for evolving digital circuits [7, 8]. In essence, it is characterized by its encoding of a graph as a string of integers that represent the functions and connections between graph nodes, and program inputs and outputs. This gives it great generality so that it can represent neural networks, programs, circuits, and many other computational structures. Although, in general it is capable of representing directed multigraphs, it has so far only been used to represent directed acyclic graphs. It has a number of features that are distinctive compared with other forms of Genetic Programming.

M. Keijzer et al. (Eds.): EuroGP 2005, LNCS 3447, pp. 62–73, 2005.

Foremost among these is that the genotype can encode a non-connected graph (one in which it is not possible to walk between all pairs of nodes by following directed links). This means that it uses a many-to-one genotype-phenotype mapping to produce the graph (or program) that is evaluated. The genetic material that is not utilised in the phenotype is analogous to junk DNA. As we will see, mutations will allow the activation of this redundant code or de-activation of it. Another feature is the ease with which it is able to handle problems involving multiple outputs. Graphs are attractive representations for programs as they are more compact than the more usual tree representation since subgraphs can be used more than once.

CGP has been applied to a growing number of domains and problems: digital circuit design [11, 12], digital filter design [8], image processing [21], artificial life [20], bio-inspired developmental models [9, 14, 10], evolutionary art [1], molecular docking [4] and has been adopted within new evolutionary techniques cell-based Optimization [19] and Social Programming [24]. In addition a more powerful form of CGP with the equivalent of Automatically Defined Functions is also being developed [25].

Figure 1 shows the general form of Cartesian Program for an n input m-output function. There are three user-defined parameters: number of rows (r), number of columns (c) and levels-back (which defined how many columns back a node in a particular column can connect to). Each node has a set of C_i connection genes (according to the arity of the function) and a function gene f_i which defines the nodes's function from a look-up table of available functions. On the far left are seen the program inputs or terminals and on the far right the program output connections O_i

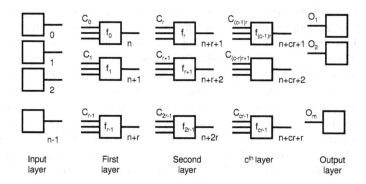

Fig. 1. General form of Cartesian Program

If the graphs encoded by the Cartesian genotype are directed (as in this work) then the range of allowed alleles for C_i are restricted so that nodes can only have their inputs connected to either program inputs or nodes from a previous (left) column. Function values are chosen from the set of available functions. Point mutation consists of choosing genes at random and altering the allele to another value provided it conforms to the above restrictions. Although the use of crossover is not ruled out, most implementations of CGP (including this one) only use point mutation.

We emphasize that there is no requirement in CGP that all nodes defined in the genotype are actually used (i.e. have their output used in the path from program output to input). Although a particular genotype may have a number of such redundant nodes they cannot be regarded as non-coding genes, since mutation may alter genes "downstream" of their position that causes them to be activated and code for something in the phenotype, similarly, formerly active genes can be deactivated by mutation. We refer genotypes with the same fitness as being neutral with respect to each other. A number of studies (mainly on Boolean problems) have shown that the constant genetic change that happens while the best population fitness remains fixed is very advantageous for search ([13, 23, 26]).

1.2   Obstacle Avoiding Robots

A common test problem in genetic programming involves generating a program that can control a robot. Typically, the task is to travel around a closed environment avoiding the walls and any obstacles [6, 2, 15, 3, 16, 17, 22, 18]. The task may be extended to getting the robot to cover as much floor space as possible or to follow the wall. In this scenario the robot has to navigate around an unknown environment avoiding contact with the walls. The control system is able to use the distance sensors on the robot, perform some form of signal processing and in turn control the motion of the robot. Two common robotic platforms are the Khepera miniature robot or a gantry style robot, such as those from iRobot. Both of these types of robots have an array of distance sensors. The Khepera has 8 short range infra red sensors, the gantry robot because of its larger size can accommodate 24 sonar sensors. Generally evolution is performed in simulation. Solutions based on genetic programming and neural network architectures can be run in faster than real time in simulation, as they can ignore (to a degree) the physical properties of the robot and its hardware. A simple obstacle avoiding robot was evolved by Koza in [5]. In this task, a robot had to move around an 8 by 8 grid and avoid obstacles - in order to mop the floor of a room. The obstacles were cells in the grid that the robot was not allowed to enter. The robot is able to move forward one cell, leap forward several cells, turn left, check to see if an obstacle is directly in front and addition (with modulo 8 integer arithmetic). The fitness was calculated as the number of squares visited in a fixed period of time. Without the use of automatically defined functions (ADFs), it took fewer than 50 generations to evolve a successful behaviour, with ADFs this result was lowered to 29 generations. However, Koza uses relatively large populations - in this example each population contained 500 individuals. The function set available to the system was quite basic. Lazarus and Hu also used a grid based system to evolve robot controllers. In [6] Lazarus and Hu evolved wall following robots, that could find, then move around the edge of a room. The room was a square room, with a number of extrusions in the walls. The most complex map had four extrusions - where each wall was interrupted by a section projecting into the centre of the room. The program could move the robot to adjoining cells and sense the state of those cells. To process the sensor information, the function set included IF, AND, OR and NOT. Using a population of size 1000, solutions for the most complex room were evolved in 90 generations. In [16] Reynolds uses genetic programming (GP) to evolve a program that can take visual information and control a simulated robot that avoids obstacles. The input to the program was a simple sensor that measures how strongly an object is seen in a particular direction. This information is then processed using basic mathematical operations and conditionals. The program output allowed the robot to drive forward and turn. In this example, the robot was simulated, however, in [15] this style of control was used to control a real robot. The input terminals were readings from the distance sensors of the Khepera robot. In [3], Ebner evolves a controller for a gantry style robot using genetic programming. The distance readings from 24 sensors were mapped into 6 virtual sensors and these were used as terminals in the GP program. Output terminals for movement are restricted to go forward, stop and turn. To allow for the evolution of a hierarchical control mechanism the program could make use of conditional statements, which allows for greater program complexity. The task was to navigate around a short stretch of a straight corridor without colliding with the walls. With the robot running in simulation, evolution was performed for 50 generations on a population of 75 individuals. When the process was moved into a physical robot, evolution took a long time (197 hours) and produced similar results to the simulated work. In [15], Nordin and Banzhaf apply genetic programming to evolving robot controllers in a real environment. The program was encoded as a binary string and evolved using a standard genetic algorithm The function set comprised of ADD, SUB, MUL, SHL, SHR, AND, OR, XOR and integer terminals. A population size of 30 was used, and successful individuals were found within 200 to 300 generations. A pleasure-pain fitness function was used. The robot received pleasure from going straight and fast, and pain from coming close to the obstacles. The scores from this pleasure/pain reward are then weighted and summed to produce an overall fitness. In this work, we present a robot, with fine grained control, several environments of greater complexity to those described above. We show that CGP is suitable for controlling the robot, and that the results are competitive to previous techniques.

2   Algorithm

2.1   The Robot

The simulated robot used in this set of experiments is similar to a Kephera robot. There are two wheels driven by motors, these motors can turn in either direction and are variable speed. Driving both motors in the same direction (and speed) moves the robot forward or backward in a straight line. By using different speeds the robot can be made to turn, turning motors in opposite directions increases the turning speed. The robot is equipped with two distance sensors mounted on the front, which are separated by an angle of 20 degrees. The sensors return the distance to the nearest wall in the direction they are pointed. There is no grid representation used in this simulation (other than for the fitness function), and the robot can move anywhere to within the resolution of double precision floating point numbers. The robot's orientation is also defined by a real number.

2.2   Function Set

For these experiments we use an integer version of CGP, in which the nodes operate on signed integer values and output signed integer values. The levels back parameter has been set for the whole width of the graph, and multiple rows are used. The nodes in the CGP graph can use the following functions: Add, Subtract, Multiply, Divide, Compare, Min, Max, Fixed integer and Input node. Add, subtract, multiply and divide are all two input nodes that perform integer arithmetic. The divide function performs integer division and is safe: dividing by 0 returns 0. Min and max respectively output the minimum and maximum of the two inputs to that node. Compare returns -1 if the first input is less than the second, 0 if the inputs are equal and +1 if the first input is larger than the second. Some nodes can store a fixed integer in the range -100 to +100. These are terminal nodes, i.e. nodes with no inputs. The first column in the CGP graph is made of input nodes. The speed used to drive the motors was taken from 2 nodes on the last column in the graph. The integer value for each node was found, truncated to fall between -100 and 100 and then scaled to a value between -1 and +1.

2.3   Operators, Parameters and Fitness Function

The evolutionary algorithm used for these experiments is very simple. The population size was set to 40 individuals. Elitism was used, with the best 5 individuals retained for the next generation. Tournament selection was used, with a tournament size of 5. No crossover was used when generating subsequent populations. Mutation was set to 5 percent of the node count, with entire nodes being mutated in each operation. Evolutionary runs were limited to 1000 generations, with each run being aborted when a solution was found i.e. when an individual's fitness was greater than 9995. For these experiments, the CGP graph was set to 20 nodes wide by 2 nodes tall. The first column of the graph is used for input nodes, leaving 38 nodes to perform processing. Typically for obstacle avoiding robots, fitness values are computed based on factors such as time spent moving forward, total path length and Euclidian distance travelled. For example, Thompson[22] uses the following calculation:

$$fitness = \frac{1}{T} \int_0^T \left(e^{-k_x c_x(t)^2} + e^{-k_y c_y(t)^2} - s(t) \right) \; where \; s(t) =_0^1 \; {}^{when \; stationary}_{otherwise}$$

where the distance of the robot from the centre of the room in the x and y directions at time t was c_x and c_y, for an evaluation for T seconds. However, during initial experiments it appeared that this method for calculating fitness had many drawbacks including local minima which resulted in poor evolutionary characteristics (e.g. mean fitness did not increase smoothly). In environments with obstacles this style of fitness function fails to capture the difficulty of getting to hard to reach locations in the map. To address this a fitness map of the environment was calculated, where each area in the map had an absolute measure of the difficulty in reaching that point. The fitness for the robot was calculated as the highest fitness measure seen during the robot's movement around the map. The calculation for this fitness map was performed by modeling chemical diffusion within the environment. If we imagine the room to be filled with a fluid such as water, and then add some coloured ink in a particular location, the colour would diffuse through the water. Near the point where the dye was added, the concentration would be greatest, the further away from the source the lower the concentration. The actual concentration (after a period of time) at a point is related to the shortest possible path to the point where the dye was added. Using a model of diffusion, we can easily approximate the shortest path required to reach any point in the map. This model automatically takes into account any shape of environment and the obstacles within it. By adding a "dye" at the starting position of the robot, and allowing it to diffuse until the chemical level at all points in the map is above 0 percent, an absolute fitness can be calculated for the map.

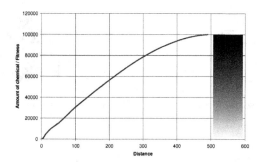

Fig. 2. Plot of fitness against distance from start position. The gradient on the right can be used a key for figures 4 and 6

The diffusion algorithm works by breaking the map area into a grid (in this experiment of size 500 x 500) which stores the amount of chemical in each part of the map, initially all cells are set to 0. The diffusion is calculated using a simple cellular automata style technique. For each cell, if the average amount of chemical in its eight neighbouring cells is greater than the amount in that cell then the amount of chemical in the cell is increased. When all the cells contain some chemical, the map is normalised so that the fitness score falls between 0 and 10000, with 10000 being the maximum fitness. Figure 2 shows the relationship between fitness and distance from the starting position. The gray-scale gradient on the right shows the colour corresponding to the fitness value, and is used in the fitness maps throughout this paper. Figures 4 and 6 show the fitness maps for the two environments used. The darker colours show the areas of greatest difficulty to reach from the robots starting position in the top left corner. Robots that can successfully navigate around the obstacles, and explore large amounts of the map will pass through areas marked as having high fitness. Solutions where the robot does not move far or travels in a circle will obtain low fitness. In these experiments the robots were allowed to travel until they collided with a wall, or a timeout situation occurs i.e. the simulation has been updated 10000 times.

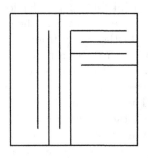

Fig. 3. Map 1

Fig. 4. Fitness values for map 1

Fig. 5. Escape Map

Fig. 6. Fitness values for the escape map

3 Experiments

3.1 Escaping a Room

The first problem involves the robot escaping from a small room. Figure 5 shows the layout of the room, with the robot starting in the centre of the map. It should be noted that all the map images are drawn to the same scale.

Results. 140 runs of the algorithm were performed, with 81% of evolutionary runs providing a solution within 1000 generations. We found the average number of evaluations required for a solution to be 8515, however the standard deviation was high (8091) - the minimum time to discover a solution was 8 generations. On reviewing some of the paths taken by the robot, it was seen that some became stuck in the bottom part of the maze. As the population converged, it would have become harder to escape this local optimum. With the neutrality in CGP, the represenation is more robust to this type of situation - even in converged populations, small mutations can produce large changes in phenotypic behaviour.

In this problem scenario, it is expected that there are many local optima, and with the high success rate it is clear that CGP is capable of escaping these.

The following is a sample program for a good solution. Motor1 and motor2 are the motor speeds of the left and right motors. INPUT_0 and INPUT_1 are the two distance sensor readings. The remaining functionality is described previously. An interesting observation from this is the reuse of the node -25. A non-graph based representation would not easily allow for this and would have to replicate the node. But here GCP can reuse nodes, and share subgraphs that are integral to the behaviour of both motor controllers. In the motor1 control program, it appears evolution has used this node to manufacture a 0 rather than evolve a integer node with a value of 0.

```
motor1= ADD(MINUS(-25 , -25), MULT(ADD(INPUT_0, -25),
        MIN(INPUT_1, INPUT_0)))
motor2=ADD(MINUS(ADD(INPUT_0, MAX(MIN(MIN(INPUT_1, INPUT_0),
 MIN(INPUT_1, INPUT_0)), -25 ), 18), MINUS(MIN(INPUT_1, INPUT_0),
 MIN(ADD(MINUS(INPUT_0, INPUT_0), MULT(MAX(MIN(MIN(INPUT_1, INPUT_0),
 MIN(INPUT_1, INPUT_0)), -25), MAX(MIN(MIN(INPUT_1, INPUT_0),
 MIN(INPUT_1, INPUT_0)), -25))), ADD(INPUT_0, -25))))
```

If we take the above programs and turn them into a more human readable form, we find the following rules have been evolved.

```
IF INPUT_0 <= INPUT_1 THEN
     motor1 := INPUT_0 ( INPUT_0 - 25 )
     ELSE
     motor1 := INPUT_1 ( INPUT_0 - 25 )
ENDIF

IF INPUT_0 <= INPUT_1 THEN
   motor2 := (2 * INPUT_0) - 43
ELSE
   IF (INPUT_1 ^ 2) <= (INPUT_0 - 25) THEN
      motor2 := INPUT_0 + (2 * INPUT_1) - (INPUT_1 ^ 2) - 18
   ELSE
      motor2 := (2 * INPUT_1) - 43
   ENDIF
ENDIF
```

The program for motor1 is very simple. If INPUT_0 detects a wall, then the speed of the motor is negative - and the robot will start to turn. Otherwise the motor is on in a forward direction. For motor2 the program is slightly more complex. However, it still has the same basic functionality - the motor speed is dependent on the value of sensor that is nearest the wall. However, motor2 appears to have it's speed regulated to ensure that in general it is going forward, and to slow down when INPUT_1 detects a wall. This code demonstrates a sophistication beyond the binary control of Braitenberg type vehicles - in this result the robot speeds up and slows down depending on the current state of its sensors.

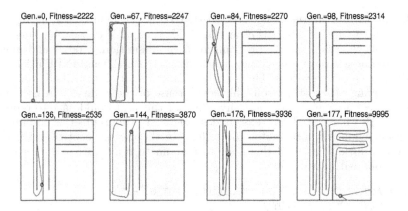

Fig. 7. Evolutionary history of maze solving behaviour

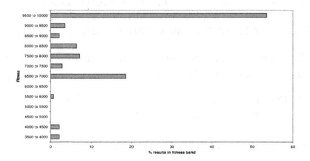

Fig. 8. Distribution of peak fitness scores during evolution, for the initial maze problem (figure 3)

3.2   Solving a Maze

The challenge in this scenario is for robot to solve a complicated maze. The maze, shown in figure 3 has many tight u-bends which are placed at different orientations. Figure 4 shows the fitness scores throughout the map.

Out of 140 runs, 51% of the runs produced individuals with perfect fitness. For each run the maximum fitness obtained and the number of evaluations required to reach that fitness were logged. The average fitness evolved was 8722 (with a standard deviation of 1635). Figure 8 shows the distribution of fitness scores, and from these results we can see that the least successful robots all managed to make their way around the first bend before crashing or looping back on themselves. The range 6500-7000 contains 26 individuals. Based on their average fitness of 6932, we can see that the robots fail to navigate into the second half of the map - where the walls change from vertical to horizontal. The ability to see where the local optima are is a useful feature of the fitness function. Without having to plot all the runs onto a map and observing where the robot becomes stuck, we can use the absolute fitness values on the fitness map to locate any trouble spots in the environment.

3.3   Generalisation of Evolved Behaviour

It is important to demonstrate that the solutions evolved generalise to different starting conditions. In this experiment we perform two tests for generalisation. In the first, evolution is allowed to solve the maze problem and on success the evolved program is tried from a different starting position. The second starting position is on the right hand side of the map, half way down. The fitness map for this starting condition is shown in figure 9 - the point of highest fitness is now in the top left of the map.

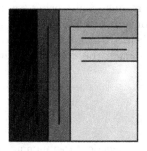

Fig. 9. Fitness values for the maze - starting at different location

Fig. 10. Solution to the reverse maze problem

Fig. 11. Fitness values for the unseen environment

Fig. 12. Example behaviour of robot in unseen enivronment

In the second test the robot is evolved to solve the original maze, and then is put into a different environment, the map with fitness values is shown in figure 11. The maze incorporates features not seen in the previous maps - e.g. walls at angles. In both these scenarios the second test is performed as soon as a solution for the first map has been evolved. We do not allow further evolution to occur.

For the first test of generalisation, from 70 evolutionary runs it was found that 97% of programs that evolved to solve the first map successfully completed the maze when started from a different location. With the new environment, 91% of programs that can complete the first maze can fully explore the second maze. This shows that the evolved programs have achieved a high degree of generalisation.

4   Conclusions

This work has demonstrated the suitability of CGP for control applications. Although it is most often misleading to directly compare results (because of different simulators, environments and methodologies), the results indicate that CGP is highly competitive when compared to previous results using traditional GP. In future work we hope to use CGP to control a physical robot, and to perform the entire evolutionary algorithm in hardware. Currently, we are working on an FPGA implementation of the algorithm, which could be used with the Kephera robots.

References

1. L. Ashmore. An investigation into cartesian genetic programming within the field of evolutionary art. Technical report, Final year project, Department of Computer Science, University of Birmingham, 2000. http://www.gaga.demon.co.uk/evoart.htm.
2. R.A. Dain. Developing mobile robot wall-following algorithms using genetic programming. In *Applied Intelligence*, volume 8, pages 33–41. Kluwer Academic Publishers, 1998.
3. Marc Ebner. Evolution of a control architecture for a mobile robot. In *Proc. of the 2nd International Conference on Evolvable Systems*, pages 303–310. Springer-Verlag, 1998.
4. A. Beatriz Garmendia-Doval, Julian Miller, and S. David Morley. Cartesian genetic programming and the post docking filtering problem. In Una-May O'Reilly and Tina Yu et al., editors, *Genetic Programming Theory and Practice II*, chapter 14. Kluwer, 2004.
5. John R. Koza. *Genetic Programming II: Automatic Discovery of Reusable Programs*. MIT Press, Cambridge Massachusetts, 1994.
6. C. Lazarus and H. Hu. Using genetic programming to evolve robot behaviours. In *Proc. of the 3rd British Workshop on Towards Intelligent Mobile Robots*, 2001.
7. J.F. Miller, P. Thomson, and T. Fogarty. Designing electronic circuits using evolutionary algorithms arithmetic circuits: A case study. In D. Quagliarella and J. Périaux et al., editors, *Genetic Algorithms and Evolution Strategies in Engineering and Computer Science*. John Wiley and Sons, West Sussex, England, 1997.
8. Julian F. Miller. An empirical study of the efficiency of learning boolean functions using a cartesian genetic programming approach. In Wolfgang Banzhaf and Jason Daida et al., editors, *Proc. of GECCO*, volume 2, pages 1135–1142. Morgan Kaufmann, 1999.

9. Julian F. Miller. Evolving developmental programs for adaptation, morphogenesis, and self-repair. In Wolfgang Banzhaf and Thomas Christaller et al, editors, *Proc. of ECAL*, volume 2801 of *LNAI*, pages 256–265. Springer, 2003.

10. Julian F. Miller and Wolfgang Banzhaf. Evolving the program for a cell: from french flags to boolean circuits. In Sanjeev Kumar and Peter J. Bentley, editors, *On Growth, Form and Computers*, pages 278–301. Academic Press, 2003.

11. Julian F. Miller, Dominic Job, and Vesselin K. Vassilev. Principles in the evolutionary design of digital circuits-part I. *Genetic Programming and Evolvable Machines*, 1(1/2):7–35, 2000.

12. Julian F. Miller, Dominic Job, and Vesselin K. Vassilev. Principles in the evolutionary design of digital circuits-part II. *Genetic Programming and Evolvable Machines*, 1(3):259–288, 2000.

13. Julian F. Miller and Peter Thomson. Cartesian genetic programming. In Riccardo Poli and Wolfgang Banzhaf et al., editors, *Proc. of EuroGP 2000*, volume 1802 of *LNCS*, pages 121–132. Springer-Verlag, 2000.

14. Julian Francis Miller. Evolving a self-repairing, self-regulating, french flag organism. In Kalyanmoy Deb and Riccardo Poli et al., editors, *Proc. of GECCO*, volume 3102 of *LNCS*, pages 129–139. Springer-Verlag, 2004.

15. Peter Nordin and Wolfgang Banzhaf. Genetic programming controlling a miniature robot. In E. V. Siegel and J. R. Koza, editors, *Working Notes for the AAAI Symposium on Genetic Programming*, pages 61–67, MIT, Cambridge, MA, USA, 1995. AAAI.

16. Craig W. Reynolds. An evolved, vision-based behavioral model of obstacle avoidance behaviour. In Christopher G. Langton, editor, *Artificial Life III*, volume 16 of *SFI Studies in the Sciences of Complexity*. Addison-Wesley, 1993.

17. Craig W. Reynolds. Evolution of corridor following behavior in a noisy world. In *Simulation of Adaptive Behaviour (SAB-94)*, 1994.

18. Craig W. Reynolds. Evolution of obstacle avoidance behavior: using noise to promote robust solutions. pages 221–241, 1994.

19. J. Rothermich, F. Wang, and J. F. Miller. Adaptivity in cell based optimization for information ecosystems. In IEEE Press, editor, *Proc. of the CEC2003*, pages 490–497, 2003.

20. Joseph A. Rothermich and Julian F. Miller. Studying the emergence of multicellularity with cartesian genetic programming in artificial life. In Erick Cantú-Paz, editor, *GECCO Late Breaking Papers*, pages 397–403. AAAI, 2002.

21. Lukas Sekanina. *Evolvable Components: From Theory to Hardware Implementations*. SpringerVerlag, 2004.

22. A. Thompson. Evolving electronic robot controllers that exploit hardware resources. In F. Morán and A. Moreno et al., editors, *Proc. 3rd Eur. Conf. on Artificial Life*, volume 929 of *LNAI*, pages 640–656. Springer-Verlag, 1995.

23. Vesselin K. Vassilev and Julian F. Miller. The advantages of landscape neutrality in digital circuit evolution. In *ICES*, pages 252–263, 2000.

24. M. S. Voss and J. Holland. Financial modelling using social programming. In *FEA 2003: Financial Engineering and Applications*, 2003.

25. James Alfred Walker and Julian Francis Miller. Evolution and acquisition of modules in cartesian genetic programming. In Maarten Keijzer and Una-May O'Reilly et al., editors, *Genetic Programming 7th European Conference, EuroGP 2004, Proceedings*, volume 3003 of *LNCS*, pages 187–197. Springer-Verlag, 2004.

26. Tina Yu and Julian Miller. Neutrality and the evolvability of boolean function landscape. In Julian F. Miller and Marco Tomassini et al., editors, *Proc. of EuroGP*, volume 2038 of *LNCS*, pages 204–217. Springer-Verlag, 2001.

Evolving L-Systems to Capture Protein Structure Native Conformations

Gabi Escuela[1], Gabriela Ochoa[2], and Natalio Krasnogor[3]

[1,2] Department of Computer Science, Simon Bolivar University, Caracas, Venezuela
gabiescuela@netuno.net.ve, gabro@ldc.usb.ve
[3] School of Computer Science and I.T., University of Nottingham
Natalio.Krasnogor@nottingham.ac.uk

Abstract. A protein is a linear chain of amino acids that folds into a unique functional structure, called its native state. In this state, proteins show repeated substructures like alpha helices and beta sheets. This suggests that native structures may be captured by the formalism known as Lindenmayer systems (L-systems). In this paper an evolutionary approach is used as the inference procedure for folded structures on simple lattice models. The algorithm searches the space of L-systems which are then executed to obtain the phenotype, thus our approach is close to Grammatical Evolution. The problem is to find a set of rewriting rules that represents a target native structure on the lattice model. The proposed approach has produced promising results for short sequences. Thus the foundations are set for a novel encoding based on L-systems for evolutionary approaches to both the Protein Structure Prediction and Inverse Folding Problems.

1 Introduction

The Protein Structure Prediction Problem (PSP) is among the most outstanding open problems in Biochemistry. A successful approach for efficient and accurate prediction would hasten a new era for biotechnology. A protein is as a linear sequence of units, called amino acids, that under certain physical conditions folds into a unique functional structure known as the native state or tertiary structure. This native state is the key to understanding a proteins' functionality in a living organism as an enzyme, a storage, transport, messenger, antibody, or regulation molecule. The simplest models for studying the properties of protein folding and structure prediction are based on lattices (of 2 or 3 dimensions), these models capture the essential aspects of the folding process while keeping low computational costs. The on-lattice hydrophobic-hydrophilic (HP) model, assumes the hydrophobic effect of amino acids as the main force governing folding.

The correspondence between amino acids and positions within a lattice is called *embedding* of the protein. It was shown that finding the embedding of a protein is NP-hard even for very simple lattice models [7,33]. Thus, the use of heuristics and approximation algorithms became the most promising approach for the PSP. In particular, several evolutionary algorithms have been suggested for solving this problem [12,18,19,27,34]. All these approaches employ a direct encoding of the folded chain (See Section 2).

M. Keijzer et al. (Eds.): EuroGP 2005, LNCS 3447, 2005, pp. 74–84, 2005.

This paper suggests a *novel encoding* scheme for the PSP based on Lindenmayer systems. The rationale for this is twofold:

1. A protein structure often exhibits a high degree of regularity, with a wealth of secondary structures, preferred motifs, and tertiary symmetries [8]; also, proteins have been compared to fractals [32]. This is consistent with the recursive nature of L-systems where rewriting rules lead to modular, auto-similar structures.
2. It is not clear that the encoding currently used in evolutionary algorithms for HP models, namely, internal coordinates (see section 2.1) are suitable for crossover and building block transfer between individuals [4,15,16].

An evolutionary algorithm is proposed as the inference procedure for folded structures under the HP model in 2D lattices. *The problem is to find a set of rewriting rules (an L-system) that captures a target folded structure (which represents the native state for a given protein) on the selected lattice model.*

Evolutionary algorithms have been successfully applied to a variety of design problems, but it is not clear whether evolutionary techniques can scale to the complexities of real world designs. It has been argued that a generative or grammatical encoding scheme, (i.e. an encoding that specifies how to construct the phenotype, instead of a direct encoding of the phenotype) can achieve greater scalability through self-similar and hierarchical structure [1,2,10]. Moreover, by reusing parts of the genotype while generating the phenotype, a generative encoding is a more compact encoding of a solution. These approaches to encoding have had an enormous success; we point the reader to [25,30,31] for a general overview of grammatical evolution and to [26] for an application of grammatical evolution to a problem related to the one we focus on here.

L-systems as a generative encoding have been used in previous applications of evolutionary algorithms to problems in biology, medicine, engineering, and computer graphics. The production of plant structures [3,6,13,24,28] has been the most studied case; where results have shown the usefulness of this encoding, both to obtain structures resembling natural organisms, and in the generation of artificial designs with novel features. Furthermore, L-systems grammars have proved to be a powerful genotype encoding to represent blood circulation of the human retina [14], physical design of tables, robots, and virtual creatures [9,11], and in the design of transmission towers [29].

We proceed as follows: Section 2 provides the theoretical basis of the PSP; section 3 describes the mathematical formalism of L-systems. The proposed approach is presented in Section 4. Section 5 describes the experiments and results; and finally section 6 concludes and comments on future work.

2   The Protein Structure Prediction Problem Simplified

Proteins are the building blocks and functional units of all biological systems. There are 20 naturally occurring amino acids that make up protein chains. The amino acid's chain of a protein is known as its primary structure and usually contains about 30 to 400 acids. The primary structure folds in space and forms secondary structures. These secondary structures present specific signatures like α-helices and β-sheets. In turn, secondary motifs fold yet again and aggregate in space giving raise to a 3D conformation, the tertiary structure. The tertiary structure conforms a very specific geometric pattern (the native state).

2.1 The HP Model

In the HP model [5], only two types of monomers are distinguished: *hydrophobic* (H), and *polar* or *hydrophilic* (P). The hydrophobic monomers tend to occupy the center of the protein, staying close to each other to avoid surrounding water, whereas the polar residues are attracted to water and are frequently found on the convex hull of the native state. The set of valid protein structure conformations is the space of all self-avoiding paths (on a selected lattice, e.g., square 2D, triangular, cubic, diamond, etc.), with each amino acid located on a lattice bead. Hydrophobic units that are adjacent in the lattice but non-adjacent in the sequence (also called non-local H-H contacts) add a constant negative factor (generally $\varepsilon=-1$) and all other interactions are ignored. The native state is thought to be the global energy minimum.

In the HP model, the structures can be represented by Cartesian coordinates, internal coordinates or distance geometry. We concentrate here on internal coordinates, which can be defined as absolute or relative. Under the absolute encoding, the structures are represented by a list of absolute moves. In a 2D square lattice, for example, a structure s is encoded as a string $s = \{$**U**p, **D**own, **L**eft, **R**ight$\}^+$. When using a relative coordinates, each move is interpreted in terms of the previous one, like in LOGO turtle graphics; a structure s is encoded as a string $s = \{$**F**orward, **T**urn**L**eft, **T**urn**R**ight$\}^+$. Designing with black the hydrophobic residues and white the polar ones, the structure of Figure 1 is coded either as $s =RDDLULDLDLUURULURRD$ (absolute encoding) or $s = RFRRLLRLRRFRLLRRFR$ (relative encoding), with 9 non-local H-H contacts.

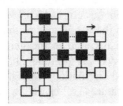

Fig. 1. Native structure in the square 2D lattice for the primary sequence HPHPPHHPHPPHPHHPPHPH. The arrow indicates the starting point, and the dotted lines the non-local H-H contacts

3 L-Systems

Aristid Lindenmayer (a biologist) proposed in 1968 an axiomatic foundation for biological development called L-systems [21]. More recently, L-systems have found several applications in computer graphics [28]; two principal areas include generation of fractals and realistic modeling of plants. Central to L-systems, is the notion of rewriting, where the idea is to define complex objects by successively replacing parts of a simple object using a set of rewriting rules or productions. The rewriting can be carried out recursively. The most extensively studied and best understood rewriting systems operate on character strings. The essential difference between the most known Chomsky grammars and L-systems lies in the method of applying productions. In Chomsky grammars productions are applied sequentially, whereas in L-systems they are applied in parallel, replacing simultaneously all letters in a given word. This difference reflects the biological motivation of L-systems. Productions are intended to capture cell divisions in multicellular organisms, where many divisions may occur at the same time.

3.1 D0L-Systems

The simplest class of L-systems is the *D0L-systems* (deterministic and context free). To provide an intuitive understanding of the main idea, let us consider the example given by Prusinkiewicz and Lindenmayer [28] (See Figure 2.).

"Lets us consider strings built of two letters a and b (they may occur many times in a string). For each letter we specify a rewriting rule. The rule $a \rightarrow ab$ means that the letter a is to be replaced by the string ab, and the rule $b \rightarrow a$ means that the letter b is to be replaced by a. The rewriting process starts from a distinguished string called the axiom. Let us assume that it consist of a single letter b. In the first derivation step (the first step of rewriting) the axiom b is replaced by a using production $b \rightarrow a$. In the second step a is replaced by ab using the production $a \rightarrow ab$. The word ab consist of two letters, both of which are simultaneously replaced in the next derivation step. Thus, a is replaced by ab , b is replaced by a, and the string aba results. In a similar way (by the simultaneous replacement of all letters), the string aba yields abaab which in turn yields *abaababa*, then *abaababaabaab*, and so on."

Fig. 2. A D0L-system derivation example

4   Method

Our proposed approach uses an evolutionary algorithm that, given a target structure in internal relative coordinates *(input)*, will evolve an L-system L *(output)* that, once evaluated, would produce a string that matches the original target. For instance, the end-product of the EA run for the structure in Figure 1 would be an L   whose termination word is *RFRRLLRLRRFRLLRRFR*.

A generational EA with linear ranking selection and elitism was used to evolve sets of rewriting rules or L-systems that capture a target structure. As the variation operators, a recombination and three mutation operators were implemented. Two stopping criteria were considered: (i) if an individual arises the maximum fitness, that is, its L-system grammar exactly represents the target folding; and (ii) a predefined maximum number of generations is reached. The genotype encoding, initial population, genetic operators, and fitness evaluation are described below. Furthermore, the specific values for the various algorithm's parameters used in the experiments, are listed in Section 5.

4.1 Genotype Encoding and Initial Population

The L-system's alphabet will depend on the lattice and coordinate system used. For the experiments reported here, we selected the square 2D lattice with relative coordinates. Thus, the terminal characters are the symbols $\{F, L, R\}$.

Genotypes are encoded using D0L-systems with the following characteristics:

$$\text{Alphabet: } \Sigma = \Sigma_t \cup \Sigma_{nt}$$

where $\Sigma_t = \{F,L,R\}$ terminal characters and $\Sigma_{nt} = \{0,1,2,...,m\text{-}1\}$ non-terminal characters representing rewriting rules

$$\text{Axiom: } \alpha = S \qquad S \in \Sigma^+$$

Rewriting rules: $W_{0,1,2,...,m\text{-}1}$: w, where $w \in \Sigma^+$

A string representing the axiom, the number of rewriting rules and the strings representing each rule, determine the genotype of an individual. The maximum lengths of the axiom and rules, as well as the number of rules are parameters that will depend on the length of the original folding. As the maximum values are held as parameters, the specific values for each individual within a population may differ.

Let *max_r*, *max_la*, and *max_lr* be the maximum number of rules, and maximum string lengths for the axiom and production rules respectively; an individual of the initial population is generated as follows: the number of rules is randomly selected in the range 1 to *max_r*, this define the non-terminal characters allowed for the individual. The axiom is a randomly generated string of symbols of maximum length *max_la* where each symbol is selected with uniform distribution from the alphabet Σ. Thereafter, each rule is generated in a similar way as the axiom, with a maximum length of *max_lr*.

4.2 Genetic Operators

Recombination takes two individuals, *p1* and *p2* as parents and creates one offspring, *o*, by copying the axiom of *p1* and selecting rules from either *p1* or *p2* with a probability of 0.5; this recombination operator resembles *uniform* crossover, where the interchanged genes are complete rules. To maintain consistency, if a selected rule to conform *o* makes reference to a symbol (rule) not defined in *o*, then a repair operator changes that symbol for a suitable symbol (either terminal or non-terminal). Fig. 3 shows an example of how this operator is applied.

A mate selection strategy (*dissasortative* mating) was also implemented as a mechanism for increasing the population genetic diversity. Dissasortative mating was implemented as follows: when selecting two individuals for a crossover, the first par

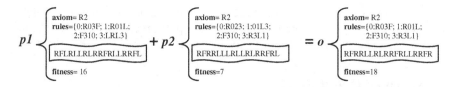

Fig. 3. Genotype, phenotype and fitness from parents and offspring in a recombination where *o* inherits the rules 0,1 from *p1* and 2,3 from *p2*

ent was selected as usual. To chose the second parent, a set of *s* (*scan* size) individuals were selected using the GA fitness-based selection method. Thereafter, the similarity between each of these *s* phenotypes and the first parent was computed, the phenotype with less similarity was chosen. For the experiments reported here, Hamming distance was used as the similarity measure, and the scan size *s* was set to 5.

Three mutation operators were implemented that perform: (i) addition, (ii) deletion, or (iii) modification of a single symbol that conforms either the axiom or the rewriting rules of each individual. When a mutation is to be performed, 60 % of times it will be a modification, 30 % an addition, and 10 % deletion.

4.3   Derivation Process, Post-processing and Fitness Calculation

For computing an individual's fitness, its L-system is derived. That is, starting from the axiom, the rewriting rules are applied in a parallel and iterated way, until either the number of terminal characters becomes equal to or greater than the string length of the target folding; or no more non-terminal characters are present in the string. Thereafter, a post-processing phase prunes the non-terminal symbols in the string to produce the phenotype. The fitness value will be the number of matches between the produced phenotype and the target folding, that is a generalized Hamming distance. So, the minimum fitness is 0 and the maximum is the length of the desired folding.

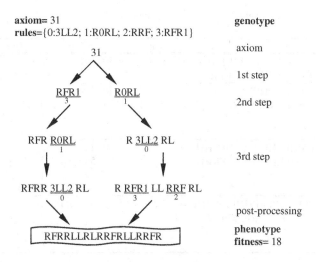

Fig. 4. Example of a derivation process

Figure 4 illustrates the derivation process for an individual (Solution 1 of Table 3) Three derivation steps, and the final result after a post-processing stage, are shown. In this case, the phenotype matches exactly the target *RFRRLLRLRRFRLLRRFR*.

5   Experiments and Results

We selected four protein instances from the HP benchmark available at http://www.cs.nott.ac.uk/~nxk/hppdb.html. Thereafter, their foldings embedded in the 2D square lattice with relative coordinates, were found using MAFRA (Memetic Algorithm FRAmework) [17]. Each of the obtained foldings was set as the target for our evolutionary approach, using the parameters listed in Table 1.

Table 1. Parameter values used for the experiments

Parameter	Value
Max. Number of Generations	2000
Population Size	50
Mating Strategy	Disassortative 5
Mutation rate (per symbol) Axiom	0.05
Mutation rate (per symbol) Rules	0.05
Recombination rate	1.00
Max. Number of Rules	4-5
Max. Length for Axiom	3
Max. Length for Rules	5

Table 2 summarises the results obtained for the selected four instances. The number of successes (runs that produced the target folding exactly) out of 50 runs, and a selected solution (L-system) are shown for each instance.

Table 2. Results for 4 instances (50 runs each)

Instance	Length	Successes	One Solution
HPHPPHHPHPPHHPPHPH→ RFRRLLRLRRFRLLRRFR	18	5/50 (4 rules)	See Table 3
HHHPPHPHPHPPHPHPHPPH → RRFRFRLFRRFLRLRFRR	18	3/50 (4 rules)	axiom = R2 4 rules = {0:RLR; 1:3F32L; 2:1FR33; 3:R102}
HHPPHPPHPPHPPHPPHPPHH → RLLFLFFRRFLLFRRRLRFFRRF	22	0/50 (4 rules) 1/50 (5 rules)	axiom = 1R 5 rules = { 0:4LF3; 1:RL243; 2:00F3; 3:RRFL; 4:0R14F}
PPHPPHHPPPPHHPPPPHHPPPPHH → FFRRFFFLLFFFFRRFFFFLLFF	23	1/50 (5 rules)	axiom= 32 4 rules = {0:20R2; 1:132F; 2:FF012; 3:0FLL}

Table 3. Some results obtained for the folding *RFRRLLRLRRFRLLRRFR*

Solution	Axiom	Rewriting rules			
1	31	0:3LL2	1:RORL	2:RRF	3:RFR1
2	31	0:3L23	1:ROL1	2:1LR	3:RFR1
3	31	0:3LLR	1:RO2L	2:23	3:RFR1
4	021	0:1R2LR	1:R1F1R	2:1LLR1	
5	11	0:2210L	1:RF30R	2:LR2	3:RRL
6 (bs)	01F	0:RFR1	1:2L2	2:ROL	
7	RF3	0:3RFR	1:312L (nu)	2:RRLLR	3:20L0R
8	RF3	0:R3L0	1:0L2R1	2:231RF	3:0R20L
9	RF0	0:R1LL0	1:0R2FR	2:LRR	
10	RF2	0:12RR0	1:RLL3R	2:R1F0	3:RL12R
11	12	0:RL10	1:RF2R	2:30L3L	3:12R1
12	30	0:RFR10	1:LL3R	2:3F13 (nu)	3:0R1LR
13	30	0:R32	1:01L2	2:030R	3:RFR1L

(bs: best solution, since it has fewer and shorter rules)
(nu: not used)

Table 3 shows results for the first target folding (length 18). Several L-systems (of 3 and 4 rules) that successfully capture the folded structure were found by the evolutionary algorithm. Some solutions (7 and 12) evolved rules that were not used in the derivation process. We distinguished solution 6 as the best obtained in this set, since it has fewer and shorter rules. Notice that some substrings that appear several times in the folded chain (e.g. *RFR*) also are present as part of the evolved rules. This supports the idea that the L-system captures the natural occurring substructures in the protein.

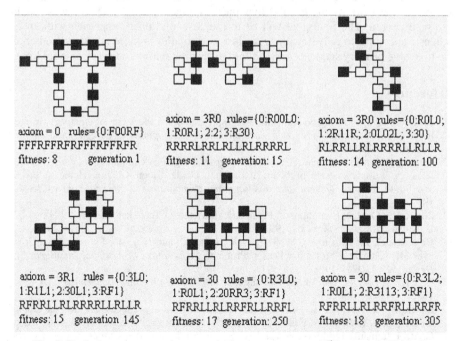

axiom = 0   rules={0:FOORF}
FFFRFFRFRFRFFFRFFRFR
fitness: 8   generation 1

axiom = 3R0   rules={0:ROOLO; 1:RORL; 2:2; 3:R30}
RRRRLRRLRLLRLRRRRL
fitness: 11   generation: 15

axiom = 3R0 rules={0:ROLO; 1:2R11R; 2:0L02L; 3:30}
RLRRLLRLRRRRLLRLLR
fitness: 14 generation: 100

axiom = 3R1   rules ={0:3L0; 1:R1L1; 2:30L1; 3:RF1}
RFRRLLRLRRRRLLRLLR
fitness: 15 generation 145

axiom = 30   rules = {0:R3L0; 1:ROL1; 2:20RR3; 3:RF1}
RFRRLLRLRRFRLLRRFL
fitness: 17 generation: 250

axiom = 30 rules={0:R3L2; 1:ROL1; 2:R3113; 3:RF1}
RFRRLLRLRRFRLLRRFR
fitness: 18 generation: 305

Fig. 5. Evolutionary progression towards the target structure (1st instance in Table 2)

Figure 5 shows the progression towards the target structure (1st instance in Table 2) as generations go by. The axiom, rules, fitness value, internal coordinates word, and graphical representation are displayed.

We would like to note that during the EA run, the production of illegal (not self-avoiding) structures was allowed (see for example the structure in generation 1 and 100 in Fig.5). However a successful L-system is only accepted when it is fully self-avoiding (like in generation 305). Also note that a given target structure may have various internal coordinates' representations (modulus rigid rotations), and that various distinct L-systems could produce the same internal coordinates word.

It is worth mentioning that the level of difficulty for evolving an adequate L-system widely varies with the instance selected. Additional to the folding's length; some instances seem more difficult than others. Our intuition is that the level of modularity and repetition within the protein folding varies across the space of possible structures.

6 Discussion

An evolutionary algorithm discovered L-systems that capture a target folding under the HP model in 2D lattices. These promising results set the foundations of a novel generative encoding for evolutionary approaches to both the protein structure prediction problem and inverse protein folding problem. We suggest that a generative encoding (i.e. a developmental approach for producing structures using a set of grammatical rewriting rules – L-system) may have better scaling properties than the direct internal coordinates encoding [1,2,10]. As noted in the previous section there are several symmetries that could be explicitly handled as to enhance the evolutionary search. Further work should test this hypothesis. Longer chains and 3D lattices should also be explored. The final goal will be to use an evolutionary approach with an L-system's encoding to solve challenging instances of the protein structure prediction and to evolve primary sequences which fold to specific native states (inverse folding).

References

1. Bentley, P. J.: Exploring component-based representations: the secret of creativity by evolution? *Fourth International Conference on Adaptive Computing in Design and Manufacture (ACDM 2000)*, 2000.
2. Bentley, P. J. and S. Kumar.: Three ways to grow designs: A comparison of embryogenies of an evolutionary design problem. In Banzhaf, Daida, Eiben, Garzon, Honavar, Jakiel, and Smith, editors, *Genetic and Evolutionary Computation Conference*, pages 35–43, 1999.
3. Curry, R.: On the Evolution of Parametric L-systems. Technical Report 1999-644-07. University of Calgary, Canadá, 1999.
4. De la Canal, E., Krasnogor, N., Marcos, D., Pelta, D. and Risi, W.: Encoding and Cross over Mismatch in a Molecular Design Problem. *Proceedings of Artificial Intelligence in Design '98 (AID98)*. 1998.
5. Dill, K.: Theory for the folding and stability of globular proteins. *Biochemistry*, 24:1501, 1985.
6. Ebner, M., Grigore, A., Heffer, A., y Albert, J.: Coevolution produces an arms race among virtual plants. In James A. Foster, Evelyne Lutton, Julian Miller, Conor Ryan, and Andrea G. B. Tettamanzi (editors): *Proceedings of the Fifth European Conference on Genetic Programming (EuroGP 2002)*, Kinsale, Ireland, pp. 316-325, Springer-Verlag, 2002.

7. Fraenkel, A.: Complexity of protein folding. *Bull. Math Biol*, 55:1199-1210, 1993.
8. Helling, R., Li, H., Miller, J., Mélin, R., Wingreen, N., Zeng, C., and Tang, C.: The Designability of Protein Structures. *J. Mol. Graph. Model.* 19, 157. 2001.
9. Hornby, G. and Pollack, J.: Evolving L-Systems to Generate Virtual Creatures. *Computers and Graphics.* 25:6, pp 1041-1048, 2001.
10. Hornby, G. and Pollack, J.: The advantages of Generative Grammatical Encodings for Physical Design. *Congress on Evolutionary Computation 2001 (CEC01)*, 2001.
11. Hornby, G., Lipson, H., and Pollack, J.: Evolution of Generative Design Systems for Modular Physical Robots. *IEEE International Conference on Robotics and Automation (ICRA)*, 2001.
12. Khimasia, M. and Coveney, P.: Protein structure prediction as a hard optimization problem: The genetic algorithm approach. In *Molecular Simulation*, volume 19, pages 205-226, 1997.
13. Kókai, G. Tóth, Z. and Ványi, R.: Evolving Artificial Trees described by Parametric L-Systems. In *Proc. IEEE Canadian Conference on Electrical & Computer Engineering*, Shaw Conference Centre, Canada, pages 1722-1728, 1999.
14. Kókai, G., Tóth, Z. and Ványi, R.: Modelling Blood Vesels of the Eye with Parametric L-Systems using Evolutionary Algorithms. In the *Proc Joint European Conference on Artificial Intelligence in Medicine and Medical Decision Making*, Denmark published by Springer-Verlag LNCS series 1620 pages 433-443, 1999.
15. Krasnogor, N. Pelta, D., Martinez, P. and De la Canal, E.: Genetic Algorithms for the Protein Folding Problem: a Critical View. *Engineering of Intelligent Systems (E.I.S. 98)*, Proceedings of the conference, 1997.
16. Krasnogor, N., Marcos, D., Pelta, D. and Risi, W.: Protein Structure Prediction as a Complex Adaptive System. *Proceedings of Frontiers in Evolutionary Algorithms (FEA98)*, 1998.
17. Krasnogor, N. and Smith, J.: MAFRA: A Java Memetic Algorithms Framework. In A. Wu, editor, Workshop Program, Proceedings of the 2000 Genetic and Evolutionary Computation Conference. Morgan Kaufmann, 2000.
18. Krasnogor, N., Blackburnem, B., Hirst, J. and Burke, E.: Multimeme Algorithms for Protein Structure Prediction. *Proceedings of Parallel Problem Solving From Nature. Lecture Notes in Computer Science*, 2002.
19. Krasnogor, N.: Studies on the Theory and Design Space of Memetic Algorithms. Ph.D. Dissertation, University of the West of England, 2002. Available [On-line] on http://www.cs.nott.ac.uk/~nxk/papers.html
20. Liang, F. and Wong, W.: Evolutionary Monte Carlo for protein folding simulations. *Journal of Chemical Physics*, 115(7):3374-3380, 2001.
21. Lindenmayer, A.: Mathematical models for cellular interactions in development, parts I-II. *Journal of Theoretical Biology* 18: 280-315, 1968.
22. Mock, K.: Wildwood: The Evolution of L-Systems Plants for Virtual Environments. In Proc ICEC 98. *IEEE-Press*, Anchorage, Alaska, 1998.
23. Noser, H., Stucki, P., Walser, H.: Integration of Optimization by Genetic Algorithms into an L-System Animation System. *Proceedings of Computer Animation* 2001, Seoul, Korea, November 7-8, 2001, pp. 106-112, 2001.
24. Ochoa, G.: On genetic algorithms and Lindenmayer Systems. In A. Eiben, T. Baeck, M. Schoenauer, y H.P. Schwefel, editors. *Parallel Problem Solving from Nature V*, pages 335-344. Springer-Verlag, 1998.
25. O'Neil M and Ryan C.: Grammatical Evolution: Evolutionary Automatic Programming in an Arbitrary Language. Series : Genetic Programming , Vol. 4. Springer, 2003.
26. Ortega A., Dalhoun A. and Alfonseca M.: Grammatical Evolution to Design Fractal Curves with a Given Dimension. IBM Journal Res & Dev, Vol7, Nro 47, 2003.

27. Patton, A., Punch, W. and Goodman, E.: A Standard GA approach to native protein conformation prediction. In *Proceedings of the Sixth International Conference on Genetic Algorithms*, pages 574-581. Morgan Kauffman, 1995.
28. Prusinkiewicz, P. and Lindenmayer, A.: *The Algorithmic Beauty of Plants*. Springer-Verlag, 1990.
29. Rudolph, S. and Alber, R.: An Evolutionary approach to the inverse problem in Rule-based design representations. *Proceedings of the 7th International Conference on Artificial Intelligence in Design (AID'02)*, Cambridge University, UK, 2002, Kluwer Academic Publishers.
30. Ryan C., Collins J.J. and O'Neil M.: Grammatical Evolution: Evolving Programs for an Arbitrary Language. Proceedings of the First European Workshop on Genetic Programming. 1998.
31. Ryan C., O'Neil M. and Collins J.J.: "Grammatical Evolution: Solving Trigonometric Identities". Proceedings of Mendel 1998: 4th International Mendel Conference on Genetic Algorithms, Optimisation Problems, Fuzzy Logic, Neural Networks, Rough Sets. 1998
32. Sadana, A. and Vo-Dinh, T.: Biomedical implications of protein folding and misfolding. *Biotechnol. Appl. Biochem.* 33, (7–16). Great Britain, 2001.
33. Unger, I. and Moult, J.: Finding the lowest free energy conformation of a protein is an NP-hard problem: Proof and implications. *Bull Math. Biol.*, 55:1183-1198, 1993.
34. Unger, I. and Moult, J.: Genetic Algorithms for protein folding simulations. *Journal of Molecular Biology*, 231 (1);75-81, 1993.

Evolving Rules for Document Classification

Laurence Hirsch[1], Masoud Saeedi[1], and Robin Hirsch[2]

[1] School of Management, Royal Holloway University of London, Surrey, TW20 OEX, UK
[2] University College London, Gower Street, London, WC1E 6BT, UK

Abstract. We describe a novel method for using Genetic Programming to create compact classification rules based on combinations of N-Grams (character strings). Genetic programs acquire fitness by producing rules that are effective classifiers in terms of precision and recall when evaluated against a set of training documents. We describe a set of functions and terminals and provide results from a classification task using the Reuters 21578 dataset. We also suggest that because the induced rules are meaningful to a human analyst they may have a number of other uses beyond classification and provide a basis for text mining applications.

1 Introduction

Automatic text classification is the activity of assigning pre-defined category labels to natural language texts based on information found in a training set of labelled documents. In recent years it has been recognised as an increasingly important tool for handling the exponential growth in available online texts and we have seen the development of many techniques aimed at the extraction of features from a set of training documents, which may then be used for categorisation purposes.

In the 1980's a common approach to text classification involved humans in the construction of a classifier, which could be used to define a particular text category. Such an expert system would typically consist of a set of manually defined logical rules, one per category, of type

if {DNF formula} **then** {category}

A DNF ("disjunctive normal form") formula is a disjunction of conjunctive clauses; the document is classified under a category if it satisfies the formula i.e. if it satisfies at least one of the clauses. An often quoted example of this approach is the CONSTRUE system [1], built by Carnegie Group for the Reuters news agency. A sample rule of the type used in CONSTRUE to classify documents in the 'wheat' category of the Reuters dataset is illustrated below.

```
if ((wheat & farm)      or
(wheat & commodity)     or
(bushels & export)      or
(wheat & tonnes) or
(wheat & winter & ¬ soft))
then
WHEAT  else  ¬ WHEAT
```

M. Keijzer et al. (Eds.): EuroGP 2005, LNCS 3447, pp. 85–95, 2005.

Such a method, sometimes referred to as 'knowledge engineering', provides accurate rules and has the additional benefit of being human understandable. That is, the definition of the category is meaningful to a human, thus producing additional uses of the rule including verification of the category. However the disadvantage is that the construction of such rules requires significant human input and the human needs some knowledge concerning the details of rule construction as well as domain knowledge [2].

Since the 1990's the machine learning approach to text categorisation has become the dominant one. In this case the system requires a set of pre-classified training documents and automatically produces a classifier from the documents. The domain expert is needed only to classify a set of existing documents. Such classifiers, usually built on the frequency of particular words in a document (sometimes called 'bag of words'), are based on two empirical observations regarding text:

1. the more times a word occurs in a document, the more relevant it is to the topic of the document.
2. the more times the word occurs throughout the documents in the collection the more poorly it discriminates between documents.

A well known approach for computing word weights is the term frequency inverse document frequency (tf-idf) weighting [3] which assigns the weight to a word in a document in proportion to the number of occurrences of the word in the document and in inverse proportion to the number of documents in the collection for which the word occurs at least once. A classifier can be constructed by mapping a document to a high dimensional feature vector, where each entry of the vector represents the presence or absence of a feature [4]. In this approach, text classification can be viewed as a special case of the more general problem of identifying a category in a space of high dimensions so as to define a given set of points in that space. Such sparse vectors can then be used in conjunction with many learning algorithms for computing the closeness of two documents and quite sophisticated geometric systems have been devised [5].

Although this method has produced accurate classifiers there are a number of drawbacks from the machine learning approach as compared to a rule based one.

1. All the word order information is lost; only the frequency of the terms in the document is stored.
2. The approach cannot normally identify word combinations, phrases or multi-word units e.g. 'information processing' [6].
3. If word stemming is used inflection information is also lost.
4. The classifier (the vector of weights) is not human understandable.

In this paper we describe a method to evolve compact human understandable rules using only a set of training documents. The system uses genetic programming (GP)[7] to produce a synthesis of machine learning and knowledge engineering with the intention of incorporating advantageous attributes from both. The rules produced by the GPs are based on N-Grams (sequences of N letters) and are able to use a wide variety of features including word combinations and negative information for discrimination purposes. In the next section, we review previous classification work with N-Grams and with phrases. We then provide information concerning the implementation of our application and the initial results we have obtained on a text classification task. Although GP has been used in a textual environment [8,9] it has not previously been used to evolve compressed classifiers based on evolving N-Gram patterns.

1.1 N-Grams

A character N-Gram is an N-character slice of a longer string. For example the word INFORM produces the 5-grams _INFO, INFOR, NFORM, FORM_ where the underscore represents a blank. The key benefit of N-Gram-based matching derives from its very nature: since every string is decomposed into small parts any errors that are present tend to affect only a limited number of those parts leaving the remainder intact. The N-Grams for related forms of a word (e.g., 'information', 'informative', 'informing', etc.) automatically have a lot in common. If we count N-Grams that are common to two strings, we get a measure of their similarity that is resistant to a wide variety of grammatical and typographical errors [10,11,12]. A useful property of N-Grams is that the lexicon obtained from the analysis of a text in terms of N-Grams of characters cannot grow larger than the size of the alphabet to the power of N. Furthermore, because most of the possible sequences of N characters rarely or never occur in practice for N>2, a table of the N-Grams occurring in a given text tends to be sparse, with the majority of possible N-Grams having a frequency of zero even for very large amounts of texts. Tauritz [13] and later Langdon [14] used this property to build an: adaptive information filtering system based on weighted trigram (N=3) analysis in which genetic algorithms were used to determine weight vectors. An interesting modification of N-Grams is to generalise N-Grams to substrings which need not be contiguous. Lodhi et al. [15] define a learning algorithm that uses non-contiguous substrings of N characters, but with a penalty for any gaps occurring between the N characters.

1.2 Phrases

The notion of N-Grams of words i.e. sequences or occurrences of N contiguous and non-contiguous words (with N typically equals to 2, 3, 4 or 5) has produced good results both in language identification, speech analysis and in several areas of knowledge extraction from text [16,17,18].   Pickens and Croft [6] make the distinction between 'adjacent phrases' where the phrase words must be adjacent and Boolean phrases where the phrase words are present anywhere in the document. They found that adjacent phrases tended to be better than Boolean phrases in terms or retrieval relevance but not in all cases. Restricting a search to only adjacent phrases means that some retrieval information is lost. The implementation described below is able to make use of both adjacent and Boolean phrases if they are found to aid discrimination between documents.

2 Our Genetic Programming Approach

When building text classifiers there are usually a variety of options regarding pre-processing of documents and particular parameters values. Examples include whether to remove stop words, to stem words to a common form, to use words or N-Grams as terms and whether to search for single terms, phrases or particular sequences of terms. Where N-Grams or phrases are used the length of the phrase or N-Gram must also be determined. Although many of these options have been researched [19] it is often the case that effects on the performance of the classifier will depend on the particular classifier and the particular text environment [20]. We have developed a GP system where many of these decisions are either made redundant or are taken by the individual GPs.

We summarise the key features below:

- The basic unit (or phrase unit) we use is an N-Gram (sequence of N characters).
- N-Gram based rules are produced by GPs and evaluate to true or false for a particular document.
- A classification rule must be evolved for each category c. Fitness is then accrued for GPs producing classification rules which are true for training documents in c but are not true for documents outside c. Thus the documents in the training set represent the fitness cases.

2.1 Data Set

The task involved categorising documents selected from the Reuters-21578 test collection, which has been a standard benchmark for the text categorisation tasks throughout the last ten years [20]. In our experiments we use the "ModApt´e split", a partition of the collection into a training set and a test set that has been widely adopted by text categorisation experimenters. The top 10 categories are also widely used and these are the categories we adopt here.

2.2 Pre-processing

Before we start the evolution of classification rules a number of pre-processing steps are made.

1. All the text in the document collection is placed in lower case.
2. Numbers are replaced by a special character and non-alphanumeric characters are replaced by a second special character.
3. All the documents in the collection are searched for N-Grams which are then stored in sets for size of N=2 to N=max_size (where max_size can be the longest word in the collection). The size of these sets is reduced by requiring that an N-Gram occur at least 4 times before being included in a set.

The use of N-Grams as features makes word stemming unnecessary and the natural screening process provided by the fitness test means that a stop list is not required. Note that only step 3 is actually essential for the GP system to run. Including upper case letters and numbers would significantly increase the search space of the GP system but could provide useful features for discriminating between documents in particular domains.

2.3 Fitness

GPs are set the task of assembling single letters into N-Gram strings and then combining N-Grams with Boolean functions to form a rule. The rule is then evaluated against the documents in the training set. Each rule can be tested against any document and will return a Boolean value indicating whether the rule is true for that document. An example of a rule produced by a GP evolving a classifier for the crude category of the Reuters 21578 is

(AND (EXISTS crude) (EXISTS (OR nerg barr)))

A classification rule must be evolved for each category c. Each rule is actually a binary classifier; that is it will classify documents as either in the category or outside the category. When evolving a rule for a particular category c the fitness depends on the number of documents in the category where the rule is true and the number of documents outside the category where the rule is true.

In information retrieval and text categorisation the F1 measure is commonly used for determining classification effectiveness and has the advantage of giving equal weight to precision and recall [21]. F1 is given by

$$F1(\pi, \rho) = \frac{2\pi\rho}{\pi + \rho} \tag{2}$$

where:

Recall (π)= the number of relevant documents returned/the total number of relevant documents in the collection

Precision (ρ)= the number of relevant documents returned/the number of documents returned.

F1 also gives a natural fitness measure for an evolving classifier. The fitness of an individual GP is therefore assigned in the following way:

1. evaluate the rule produced by the GP against all documents in the training set.
2. calculate precision, recall and F1 by counting the documents where the rule is true in the category and outside the category for which the classifier is being evolved.
3. compute standardised fitness as $1 - F1$ so that 0 is given to a perfect classifier for that category.

2.4 GP Types

We use a strongly typed tree based GP [22] system with types shown in Table 1.

Table 1. GP Types

GP Type	Description
String	A sequence of one or more characters.
Boolean	True/False: the return type of all GPs

2.5 GP Terminals

In our system we use the following character literals stored as string values.

> 26 lower case alphabetic characters (a-z).
> "~" meaning the space character
> "#" meaning any number.
> "^" meaning any non-alphanumeric character.

Note that for particular domains it may be useful to include numbers (still stored as strings), upper case characters and other special characters although this will increase the search space of the GP system.

2.6 GP Functions

The GPs are provided with protected string handling functions for combining characters into N-Gram strings and concatenating N-Grams into a longer N-Gram. Most combinations of letters above an N-Gram size of 2 are unlikely to occur in any text, with the majority of possible N-Grams having a frequency of zero even for very large amounts of texts. For example, 40 MB of text from the Wall Street Journal were found to contain only $2.7*10^5$ different 5-grams out of a possible $7.5*10^{18}$ based on an alphabet of 27 characters [23]. We guide the GPs through the vast search space of possible N-Gram patterns by the provision of protected 'EXPAND' function. The function initially forms a new N-Gram by appending one N-Gram to another. The EXPAND function checks if the new N-Gram is in the set of N-Grams of size N originally extracted from all the text in the all the training documents. If it is found the new N-Gram is returned. If it is not found, i.e. the N-Gram did not occur in the documents of the training set, the next N-Gram in the set (in alphabetical order) is returned.

We found that using an unprotected concatenation function it was quite rare for N-Grams of size greater than 2 to be evolved. However using the EXPAND function long N-Grams and words are easily and commonly evolved by combining shorter strings. For example the string 'wheat' could be evolved in the following way

```
(EXPAND w (EXPAND (EXPAND h e) (EXPAND a b)))
```

The function initially creates the string 'wheab'. This string is not found in the set of N-Grams of size 5 originally extracted from the collection. The next N-Gram in the set of 5-Grams is therefore returned ('wheat').

Table 2 shows a basic set of GP functions for evolving classification rules. Although the functions ANDSTR, ORSTR, and NOTSTR are not essential as they are definable by the other operators, we include them as a way reducing tree sizes.

Table 2. GP Functions

Function Name	No of Args	Type of Args	Return Type	Description
EXPAND	2	String	String	Concatenate 2 N-Grams and return the nearest N-Gram of the same length extracted from the training data. If found in the set of N-Grams extracted from the training data return that N-Gram else return the next N-Gram in the set.
EXISTS	1	String	Boolean	IF the N-Gram is found in a document return TRUE ELSE return FALSE
AND	2	Boolean	Boolean	Return arg1 AND arg2
OR	2	Boolean	Boolean	Return arg1 OR arg2
NOT	1	Boolean	Boolean	Return NOT arg1
ANDSTR	2	String	Boolean	IF arg1 AND arg2 are found in the document return TRUE ELSE return FALSE
ORSTR	2	String	Boolean	IF arg1 OR arg2 are found in the document return TRUE ELSE return FALSE
NOTSTR	1	String	Boolean	IF arg1 is NOT found in the document return TRUE ELSE return FALSE.

2.7 GP Parameters

The GP parameters used in our experiments are summarised in Table 3.

Table 3. GP Parameters

Parameter	Value
Population	800
Generations	40
Typing	Strongly typed
Creation Method	Ramped half and half
GP format	Tree Based
Selection type	Tournament
Tournament size	7
Mutation probability	0.1
Reproduction probability	0.1
Crossover probability	0.8
Elitism	No
ADF	No
Maximum tree depth at creation	9
Maximum tree depth	17
Maximum tree depth for mutation	4

3 Experiments and Results

The objective of our experiments were two fold:
1.   To evolve effective classifiers against the text dataset.
2.   To automatically produce compact human understandable rules with minimal features.

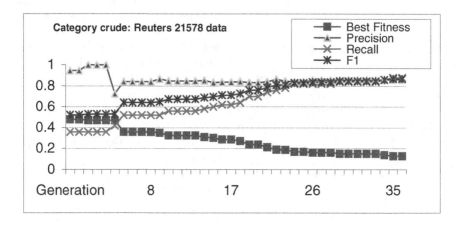

Fig. 1. Evolution of a rule for the Reuters 21578 Crude category

Fig. 1 shows a fairly typical pattern of evolution and in this case we see the emergence of a useful rule after approximately 20 generations. Precision is very high during the early evolution but is reduced as recall improves. In other cases we see recall starting very high and reducing as precision improves. In general we will see an improvement in F1 as measured against the training set and a corresponding but lower F1 as measured against the test set.

A classification rule was evolved for each category by using 4 GP runs and selecting the best rule to emerge from the 4 runs. The rule produced by the best individual for each category is shown in Table 4 together with the F1 measure (against the test set). Functions are shown in upper case and N-Grams are shown in lower case. The blank character is indicated by '~'.

The global macro-average F1 is 0.717 which compares favourably with other classifiers such as [18] although we should note that this is not a strictly controlled comparison. Indeed our intention at this point is not to produce the best classifier in terms of accuracy but to produce a good classifier which is based on a small number of features in a human understandable form. Comprehensibility may be improved by using various forms of parsimony pressure on the GP evolution and by favouring longer N-Grams or words.

Table 4. Rules evolved for Reuters top 10 categories

Name	F1	The Rule
Crude	0.826	(OR (OR (OR (ORSTR arrels~ rude~) (EXISTS opec~) (EXISTS energy) (EXISTS oleum))))
Corn	0.835	(ORSTR aize~ corn~)
Earn	0.857	(OR (ORSTR shr~ qt) (EXISTS ividend))
Grain	0.550	(OR (ORSTR ulture~ crop~) (EXISTS nnes~))
Interest	0.569	(OR (OR (AND (ORSTR engla deposit) (OR (NOTSTR vity) (EXISTS ny) (OR (AND (ORSTR lending epurcha) (ORSTR ~fut cut) (AND (OR (ANDSTR g-t ~l) (ORSTR ederal~ ~money~) (EXISTS further) (OR (AND (ORSTR epurc sbank) (NOT (EXISTS ny) (AND (OR (ANDSTR g-t bl) (ORSTR ngland~ ~money~) (NOT (EXISTS ny))))
money-fx	0.612	(ORSTR ~mone dollar~)
Ship	0.745	(OR (OR (ORSTR trike hips~) (ORSTR vesse river) EXISTS ipping~)))
Trade	0.761	(AND (ORSTR kore rade~) (OR (OR (AND (ORSTR ~yeu rade~) (ORSTR oods ficit) (ORSTR ~yeu domes) (ORSTR ~bil rplus)))
Wheat	0.663	(AND (NOTSTR prio) (AND (NOTSTR opme) EXISTS wheat))
Acq	0.755	(ORSTR cqui hares)

4 Discussion

Previous text classification systems have used various sets of features including words, word combinations and N-Grams. The system described here is capable of including any or all of these where they are found to be useful for classification purposes. In addition the system can easily make use of negative information via the inclusion of Boolean NOT functions in the rule. The rule produced can be reformulated and fed directly into a database or Internet search engine to retrieve similar texts. The rule is produced automatically but is somewhat similar to rules produced by knowledge engineering systems using human experts. For example the following rule was evolved for the Reuters Trade category happened to be in DNF form although it was not the most effective classifier (F1 0.692).

> (OR (OR (OR (ANDSTR llion export) (OR (ANDSTR llion surpl)
> (ANDSTR ~trad mport) (ANDSTR ~trad vis) (ANDSTR ~trad yeutt)))

The rule created may also be used for purposes beyond classification such as text mining. For example, the regular occurrence of synonyms (different words with the same meaning) and homonyms (words with the same spelling but with distinct meanings) are key problems in the analysis of text data: in the language of relational databases this is a classic many-to-many relationship. There is evidence that the rules evolved in our current system are using synonyms to improve the effectiveness of a rule, e.g.:

```
(ORSTR aize~ corn~)
```

Furthermore we suggest that homonyms are best discriminated by the use of contextual evidence, i.e. by an analysis of nearby strings in the text. Much of this contextual evidence can be detected simply by the use of the Boolean operators AND, OR and NOT, though it may be that additional operators that impose constraints on the relative positions of two N-Grams in the text will allow an improved discrimination

5 Conclusions and Future Work

We have produced a system capable of discovering rules based on a rich and varied set of features, which are useful to the task of discriminating between text documents. We suggest that there may a number of areas within automatic text analysis where the basic technology described here may be of use.

We are investigating the usefulness of new GP functions:

- Special functions for identifying word order. For example FOLLOWS X Y [9] indicates that the word matched by N-Grams Y must follow the word matched by N-Gram X in the text of a document.
- Kleene's star (*) could be included as a marker for an arbitrary sequence of characters, e.g. a*t matches any of "at", "ant" or "agony aunt" within an N-Gram. We will also investigate the use of full regular expressions for the rules evolved by the GPs.
- Functions for identifying words that are ADJACENT in the text or NEAR one another.
- New functions together with numeric terminals for identifying frequency information may be introduced [8]. Functions such as '>' return a Boolean value based on the frequency of a particular N-Gram in comparison to an integer terminal. This frequency could be a simple count of the occurrence of an N-Gram in a document or a more sophisticated measure such as the term frequency inverse document frequency (tf-idf) described above.

We believe that the system described here may be of particularly value when used in conjunction with other classification systems in a classification committee [20] because the method of producing the classifier is quite different to other automatic classifiers based on vectors of weights.

References

1. Hayes, P. J., Andersen, P.M., Nirenburg, I.B., Schmandt, L.M.: Tcs: a shell for content-based text categorization. In Proceedings of CAIA-90, 6th IEEE Conference on Artificial Intelligence Applications, Santa Barbara,CA, (1990), 320–326.
2. Apt'e, C., Damerau, F.J., Weiss, S.M.: Automated learning of decision rules for text categorization. ACM Trans. on Inform. Syst. 12, 3, 233–251.ATTARDI (1994)

3. Salton, G., McGill M.J.: An Introduction to Modern Information Retrieval, McGraw-Hill. (1983)
4. Joachims, T.: Text categorization with support vector machines: learning with many relevant features. In Proceedings of the 10th European Conference on Machine Learning (ECML 1998), pp 137-142.
5. Bennet K., Shawe-Taylor, J., Wu. D.: Enlarging the margins in perceptron decision trees. Machine Learning 41 (2000), pp 295-313
6. Pickens, J., Croft, W.B.: An Exploratory Analysis of Phrases in Text Retrieval. In Proceedings of RIAO Conference, Paris, France (2000).
7. Koza, J.R.: Genetic Programming: On the Programming of Computers by Means of Natural Selection. The MIT Press, Cambridge MA (1992).
8. Clack, C., Farrington, J., Lidwell, P., Yu, T.: Autonomous Document Classification for Business, in Proceedings of The ACM Agents Conference (1997).
9. Bergström, A., Jaksetic, P. Nordin, P.: Enhancing Information Retrieval by Automatic Acquisition of Textual Relations Using Genetic Programming. In Proceedings of the 2000 International Conference on Intelligent User Interfaces, ACM Press. (2000) pp. 29-32,
10. Cavnar, W., Trenkle, J.: N-Gram-Based Text Categorization In Proceedings of SDAIR-94, 3rd Annual Symposium on Document Analysis and Information Retrieval (1994).
11. Damashek, M.: Gauging similarity with n-grams: Language-independent categorization of text, Science, 267 (1995) pp. 843 . 848.
12. Biskri I., Delisle, S. Text Classification and Multilinguism: Getting at Words via N-grams of Characters. In Proceedings of the 6th World Multiconference on Systemics, Cybernetics and Informatics (SCI-2002), Orlando (Florida, USA), Volume V, (2002) 110-115.
13. Tauritz D.R., Kok, J.N., Sprinkhuizen-Kuyper I.G.: Adaptive information filtering using evolutionary computation, Information Sciences, vol.122/2-4, (2000) pp.121-140.
14. Langdon, W.B., Natural Language Text Classification and Filtering with Trigrams and Evolutionary Classifiers, Late Breaking Papers at the 2000 Genetic and Evolutionary Computation Conference, Las Vegas, Nevada, USA, editor Darrell Whitley, (2000) pages 210—217.
15. Lodhi H., Shawe-Taylor, J., Cristianini, N., Watkins, C.: Text classification using string kernels. In Leen, T.K., Dietterich, T.G., Tresp, V. editors, Advances in Neural Information Processing Systems 13, pages 563--569. MIT Press (2001).
16. Feldman R., Fresko M., Kinar Y., Lindell, O., Liphstat, M., Rajman, Y., Schler, O., Zamir, O.: Text mining at the term level. In Proceedings of the Second European Symposium on Principles of Data Mining and Knowledge Discovery, pages 65--73, Nantes, France (1998).
17. Ahonen-Myka, H.: Finding All Maximal Frequent Sequences in Text. In Proceedings of the 16th International Conference in Machine Learning ICML Bled, Slovenia, (1999).
18. Tan, C.M., Wang, Y. F., Lee, C.D.: The use of bigrams to enhance text categorization In Information Processing and Management: an International Journal, Vol 38, Number 4 (2002) Pages 529-546
19. Berleant, D., Gu, Z.: Hash table sizes for storing n-grams for text processing, Technical Report 10-00a, Software Research Lab, 3215 Coover Hall, Dept. of Electrical and Computer Engineering, Iowa State University. (2000).
20. Sebastiani, F.: Machine learning in automated text categorization, ACM Computing Surveys, 34(1), (2000), pp. 1-47.
21. Van Rijsbergen, C.J.:Information Retrieval, 2nd edition, Department of Computer Science, University of Glasgow (1979).
22. Montana, D.: Strongly Typed Genetic Programming. In Evolutionary Computation. 3:2, 199--230. The MIT Press, Cambridge MA. (1995).
23. Ebert, D., Shaw, D., Zwa, A. Miller, E. Roberts, D., Interactive Volumetric Information Visualization for Document Corpus Management, Proceedings of Graphics Interface .97, Kelowna, B.C., May 1997, 121-128

Genetic Programming in Wireless Sensor Networks

Derek M. Johnson, Ankur M. Teredesai, and Robert T. Saltarelli

Computer Science Department, Rochester Institute of Technology,
Rochester NY 14623, USA
{dmj3538, amt, rts9020}@cs.rit.edu

Abstract. Wireless sensor networks (WSNs) are medium scale mani-
festations of a paintable or amorphous computing paradigm. WSNs are
becoming increasingly important as they attain greater deployment. New
techniques for evolutionary computing (EC) are needed to address these
new computing models. This paper describes a novel effort to develop
a variation of traditional parallel evolutionary computing models to enable their use in the wireless sensor network. The ability to compute
evolutionary algorithms within the WSN has innumerable advantages
including intelligent-sensing, resource-optimized communication strategies, intelligent-routing protocol design, novelty detection, etc. In this
paper we develop a parallel evolutionary algorithm suitable for use in
a WSN. We then describe the adaptations required to develop practicable implementations to effectively operate in resource constrained
environments such as WSNs. Several adaptations including a novel representation scheme, an approximate fitness computation method and a
sufficient statistics based data reduction technique. These adaptations
lead to the development of a GP implementation that is usable on the
low-power, small footprint architectures typical to wireless sensor motes.
We demonstrate the utility of our formulations and validate the proposed
ideas using the algorithm to compute symbolic regression problems.

1   Introduction

Amorphous or paintable computers are very large arrays of low powered computers. Computers with a few hundred kilobytes of RAM and short range wireless
communications are deployed with a density of tens to hundreds of elements
per square centimeter. These computers are unreliable and have no global addressing scheme. This new genre of computing poses many new and interesting
problems to the programmer and algorithm designer. How do you take advantage of a massively distributed computer whose individual elements are very
resource constrained? How do you write distributed algorithms without a global
addressing scheme or predictable topology? William Butera, V. Michael Bove,
and James McBride of the MIT Media Lab proposed a series of algorithms
for performing media processing and storage on paintable computers [1]. Their
paintable computer architecture is the basis for the one used in this research.

M. Keijzer et al. (Eds.): EuroGP 2005, LNCS 3447, pp. 96–107, 2005.

A paintable computing element runs process fragments (pfrags), containing code and data used as part of a global program. These fragments have read-write access to the home-page of the processor they are running on. A home-page contains key value pairs and is somewhat analogous to a tuple space. Process fragments have read-only access to the home-pages of proximal processing elements. Process fragments can migrate to neighboring processing elements.

Currently, no known implementations of paintable computing exist except validated simulations. On the other hand there is the wireless sensor network technology that has been gaining tremendous importance in recent years. Several problems that theoretically present themselves in paintable computing often can be manifested as challenges in wireless sensor network environments due to their similarity in terms of being highly resource constrained. In this paper we develop genetic programming solutions that effectively work in the wireless sensor network environment and demonstrate the utility of continuing this direction of research to realize the goals towards paintable computing.

There are innumerable technological hurdles that must be overcome for ad-hoc sensor networks to become practical. A single unit of WSNs is often termed as a 'mote'; these individual motes are incredibly resource constrained. They are characterized by a limited processing speed, storage capacity, and communication bandwidth. Moreover, their lifetime is determined by their ability to conserve power. Everything we take for granted in personal computing at the PC or desktop level comes at a huge premium in WSNs. All things considered, such constraints demand new hardware designs, network architectures, software applications, and therefore new learning algorithms that maximize the motes capabilities while keeping them inexpensive to deploy and maintain.

Developing GP solutions to work in WSNs is the primary focus of this work. Specifically, we describe the following contributions:

– A novel framework for performing genetic programming on a wireless sensor mote.
– A continuous algorithm to effectively evolve an in-network GP solution.

This paper is organized as follows: In the next section we outline the necessity for developing effective evolutionary computing solutions for wireless sensor networks. In section 3 we outline the changes and adaptations required to develop small-footprint GP. In section 3.1 we outline the details for developing a continuous algorithm that asynchronously computes a symbolic regression solution along with the distributed architecture it follows to do this computation. We then demonstrate the utility of our proposed algorithms by conducting experiments on a variety of problem sets and present the results. A brief discussion of these results concludes the paper.

2   Background and Related Work

There is significant interest within the GP community to derive effective formulations that help solve real world problems. The domain of wireless sensor networks and amorphous computing is one such real world domain that is gaining increasing attention. GP solutions have been previously proposed for several problems that manifest themselves in sensory computing systems. Seok et al. describe a technique to perform calibration of sensors using Genetic Programming on evolvable hardware [7]. Ziegler and Banzhaf proposed to use evolutionary techniques to develop a sensory nose for a robot [9]. The Mate system proposed by Levis is a tiny virtual machine for sensor networks [2]. The contribution in this paper is geared more toward adapting the GP system to work in a resource constrained environment and make decisions on the sensory data. For example, consider the problem of determining correlation between light and temperature signals in a sensor network. It is known within the signal processing community that these two parameters display similar variations under most environmental conditions. By adapting GP to work on an intelligent sensor node (termed mote) one can compute the exact correlation function that best describes the relationship between sensory attributes based on input data. Another example of a problem the proposed architecture can help address is the problem of optimized routing to save communication costs in WSNs. In this case, the proposed system can be instructed to compute an optimal routing path computed locally using the signal strength as an input parameter. Distributed systems that compute an optimization problem have been studied extensively and GP solutions have been proposed to solve problems in those domains [8]. Along similar lines, effective decision making using multi-agent teams was proposed by Luke et al. [3]. From a computation environment perspective, Nordin et al. explore ways to evolve machine code for embedded systems [6]. Our work extends these efforts and focuses on developing a GP system that effectively works in resource constrained environments.

Due to the Micro Electro-Mechanical Systems (MEMS) revolution microsensors are now following manufacturing curves that are at least related to Moore's Law. Such current trends in paintable or spray computing are summarized by Mamei [4]. They also highlight the need for intelligent in-network processing architectures such as the one we propose in this paper. Nagpal et al. present a programming methodology for self-assembling complex structures from vast numbers of locally-interacting identically-programmed agents, using techniques inspired by developmental biology [5]. Our work is inspired by this effort to develop a GP paradigm that can later be extended to include self-assembly type optimization problems.

Basic and Parallel Evolutionary Algorithm

The Basic Evolutionary Algorithm (BEA) is the most common model for evolutionary computation, however, it is clearly inappropriate for a wireless sensor network. If each mote were running a BEA they would likely take too long to converge to be effective, nor would the algorithm exploit the parallel nature of a WSN.

Evolutionary algorithms tend to be highly parallelizable and many specific algorithms have been developed which take advantage of this. There are a number of common parallel evolutionary algorithms:

Controller/Client Model. uses a global population, operations are spread among the clients by the Controller as necessary to perform evaluations on the global population members. Has a high communication, but may be appropriate for some types of parallel architectures.

Island Model. has a separate evolutionary algorithm running on each available processing node. Exchanges between processing nodes depend on the island (generally based on the network) topology. This technique shares the best local solutions, which usually contain good partial solutions, with other nodes in the same parallel algorithm. The good partial solutions are either introduced to different gene pools or, if already present, reinforce the good partial solution by increasing their influence.

Cellular Model.[1] places each potential solution on a separate processing node. Each individual node can choose potential mates from neighboring nodes using some selection method. As in the island model, the neighbors are generally dependent on network topology which can vary greatly depending on the particular implementation. Due to its simplicity a matrix cellular model is particularly effective on massively parallel computers.

3   Evolutionary Algorithms on a Mote

WSNs are an excellent target for distributed evolutionary computing. WSNs require learning algorithms that are capable of learning independent of the operation of other motes, but are also capable of using information available globally within the network to better optimize for local conditions. A distributed evolutionary algorithm can achieve both of these goals. Each mote can independently evolve, yet recombine genetic information from the surrounding motes to improve its suitability to the local environment.

In addition evolutionary algorithms, or learning algorithms in general, must be designed to address the resource constraints present in a WSN while also taking advantage of its unique properties. In our algorithm specifically, the ability to wirelessly broadcast information is used, a feature of WSNs not generally exploited in traditional distributed computing systems.

The parallel evolutionary algorithm described in this paper is based on traditional parallel algorithms like the Island Model described in section 2, but modifications are made to allow the algorithm to operate in a resource constrained WSN. Motes in a WSN communicate wirelessly, in effect all communication is necessarily broadcast, although not all receivers may choose to observe the message. Our algorithm uses broadcast transmission exclusively instead of relying on point-to-point communication. This reduces the total bandwidth requirements, as well as conserves the limited battery power of the wireless modes, as wireless transmitting is generally the most power hungry mode of these devices.

---

[1] The cellular model is so named because it has been shown to simulate a certain class of cellular automata.

3.1   Broadcast-Distributed Parallel Genetic Programming

We propose the Broadcast-Distributed Parallel genetic programming model (BDP) to address WSNs and paintable computing architectures. The primary phases of the algorithm are fitness evaluation, genetic reproduction with local and re- mote genetic information and broadcast of genetic information. Because the controller/client model is inappropriate for the ad-hoc nature of WSNs, and be- cause the benefits of the cellular model are not applicable in our domain, we chose to base the BDP algorithm on the island model.

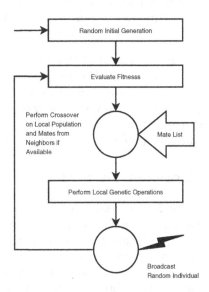

Fig. 1. Broadcast-Distributed Parallel Algorithm

In BDP each mote carries its own population, and distributes genetic infor- mation in an asynchronous fashion. There is no need to have any physical clock on the motes running the algorithm. Exchanges of genetic information are made asynchronously. Each generation involves local reproduction, reproduction with remote genetic information, calculating the fitness for the entire population, and broadcasting local genetic information. Conceptually, if the motes in a WSN running the BDP algorithm are too far apart to be able to communicate, each will operate as if they were running the BEA because they are unable to inject any external genetic information. See Figure 1 for an illustration of the BDP generation.

After each generation of the BEA on a mote M_b a random member of the population $\{M_b^{p_i} \mid i < |M_b^p|\}$ is selected and broadcast to remote motes[2]. The

---

[2] M_b is a broadcasting mote, M_r is a receiving mote, M^p is the population on a mote, M^m is the mate list on a mote.

entire individual is sent. However, the size of any given individual is particularly small, on the order of a few bytes.

Each mote M_r within wireless range of a broadcasting mote (henceforth referred to as *neighbors*) receives the broadcast $M_b^{p_i}$ and appends it to a list of incoming genetic information (the *mate* list), M_r^m. When enough genetic material is received by a mote, that is when $|M_r^m| > a$, where a is some arbitrary constraint based on the available memory on the mote, a selection is performed on M_r^m and the internal population M_r^p and a crossover operation is performed.

$$crossover(M_r^{p_j}, M_r^{m_k}) \qquad \text{for some } j < |M_r^p|, k < |M_r^m|$$

No data is exchanged in the reverse direction from M_r to M_b as a result of this crossover operation. However, at the end of the next internal generation, mote M_r will broadcast a random mote as described above. Because M_r was a neighbor of M_b so that it received $M_b^{p_i}$ during a broadcast transmission, it is likely that M_b will remain a neighbor when their roles are reversed. Thusly, crossover will generally be equilibrious as exchanges will likely occur in both directions, although not necessarily in the same discreet generation.

It is worth noting that some motes will have an advantage if they have more neighbors. They will not only exchange more of their own genetic information, but they also will receive more genetic information, therefore they may exhibit quicker fitness improvements when compared with a lone mote that lacks many neighbors.

The number of neighbors may vary within the set of motes with time. This modifies the chance a particular member is selected given its current position and the current time. This makes an examination of the survivability of a particular member in time difficult and has not yet been addressed but the overall result of selection is not affected. That is, the most fit member of a motes population is selected and its partial solutions survive. This maintains the selective pressure of the algorithm that causes the overall fitness of the population to improve. The individuals involved in crossover depend entirely and only on the results of selection as selection is fundamentally the same in a BDP and an BEA, crossover is also comparable.

3.2   Resource Constraints

The motes in a wireless sensor network are typically very low power compared to traditional PCs. They generally have far less total storage, perhaps no secondary storage (i.e. disk storage), and may rely on the operating system, software and data being able to fit in a small amount of solid-state primary storage. Many enhancements can be made to reduce the space requirements of both the software binary, as well as the data representation and run-time memory requirements.

There are obvious memory usage improvements achievable by using a steady state evolutionary algorithm with an in-place replacement strategy instead of a generational algorithm that replicates the entire population with each generation before replacing the old population with the new; roughly half the amount of memory is needed when using a steady state algorithm.

The encoding strategy for individuals can significantly alter the memory requirements for a single mote in a BDP network. Particularly for evolutionary algorithms such as genetic programming, where the size of an individual is not fixed, it is necessary to set reasonable upper bounds on the allowable size of an individual. This is achieved by adding a limit to the allowable tree depth of a candidate solution.

It is also necessary to weigh the differences in using an interpreted language versus a compiled language to develop an implementation. An interpreted language (e.g. Perl, Lisp or Java) requires an entire virtual machine to be running on a mote; this would lead to a very sizeable increase in the memory footprint as well as executable code size. However a compiled language has the advantage of being targeted to a specific platform, and therefore omits any penalties (in performance or memory usage) introduced by having a virtual machine present.

The obvious approach in a language offering dynamic memory allocation such as C is to store individuals as trees of dynamically allocated nodes in the program heap. Under this scheme each node is composed of a datum and two pointers to its left and right children. On the target system in question (4 byte memory address) this solution weighs in at nine bytes per node.

Maintaining the program trees in blocks of statically allocated memory is attractive because the code for dynamically allocating memory (`malloc` and `free` in C) can be omitted from the final binary, provided dynamically allocated memory is not used elsewhere. Dynamic functionality is not available at all in the standard libraries of the smallest conceivable target platforms; it would require a significant increase in source code size.

The most compact memory usage for program tree storage is to using a constant size for each operator or terminal in the program tree. The offset for the right sub-tree is related to the size of the left sub-tree and can be calculated by recursing down the left sub-tree. Since most operations already involve an in-order traversal of the program tree this representational scheme requires little additional code. This method is essentially a form of prefix notation. Because operators and terminals are well defined and each operator requires exactly two operands it is possible to evaluate the tree without any additional structure other than order. With a more complicated operator set, it may require a small amount of additional effort to achieve this effect.

The downside of this strategy is that any operation such as mutation or crossover which adjusts the size of the tree at anything other that the right most leaf node will require a resizing of the entire data structure. This is a reasonable tradeoff in severely memory constrained motes.

By using a prefix notation and reducing the size of symbols in the program tree, we are able to reduce the run-time memory requirements to just 5.6% of the memory consumed by most traditional GP representations using dynamic memory allocation. In terms of computation this method is no more expensive than any other representation for operations which otherwise require a traversal of the tree.

3.3   Memory Usage Requirements

By using the techniques described above the memory requirements of the BDP algorithm was significantly reduced. Equation 1 shows the total memory requirements for each node[3].

$$mem = b + v + p + i \tag{1}$$

where:

$$b \simeq 6\text{kB} \qquad \qquad \text{program binary}$$

$$v = |vars| \times \text{sizeof(float)} \qquad \qquad \text{input variables}$$

$$p = |pop| \times (2^{depth_{max}} - 1)$$
$$\times \text{sizeof(tree node)} + (2 \times \text{sizeof(uint)}) \quad \text{population}$$

$$i = |mates| \times (2^{depth_{max}} - 1)$$
$$\times \text{sizeof(tree node)} + (2 \times \text{sizeof(uint)}) \quad \text{mate list}$$

For typical parameter sizes the memory requirements of the algorithm are tenable even on very low-powered WSN devices. The memory requirements are even more impressive when compared with parallel GP implementations that make little or no attempt to restrict memory usage. See Table 1.

Table 1. Typical Memory Requirements

population on mote	typical parallel GP memory usage[4]	BDP memory usage	percentage of typical
50	187kB	14.4kB	7.7%
100	296kB	20.4kB	6.8%
200	514kB	32.5kB	6.3%
500	1168kB	68.9kB	5.9%

3.4   Improving Training Efficiency

The major impediment to compute an evolutionary algorithm over large datasets is the the amount of data required to be present in-memory while training. We attempt to alleviate this problem by reducing the amount of training data necessary to achieve convergence. This involves reducing the training set to a minimal set of variant data. The training set must be diverse enough to encompass the entire search space, but also sparse enough to avoid over training on any particular area of the search space.

In an attempt to minimize memory usage due to training data storage while maintaining speed and scalability we examined a clustering approach to

---

[3] For the TinyOS architecture: sizeof(float) = 32 bits, sizeof(tree node) = 4 bits, sizeof(uint) = 16 bits.

[4] For a standard parallel GP implementation with program binary ~80kB and using 9B per tree node.

training data sampling. Test data was generated by choosing cluster centroids and creating a random set of points within a certain distance of those centroids. The number of points per cluster and the number of clusters are variable.

Figure 2 shows a visualization of the full data from one test run, consisting of 500 training data points over four variables. In creating the three dimensional image the fourth value and correct answer are ignored, however the images serve to represent both the clustered nature of the training data and the significant reduction in the number of points thanks to the clustering approach.

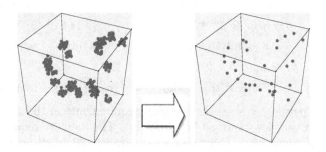

Fig. 2. Result of Clustering to Reduce Training Data

4   Performance Analysis

Both the BEA and BDP algorithms ran a symbolic regression task using a variety of equations, ranging from simple 3 variable to large 10 variable problems. The convergence properties of both algorithms were experimentally measured. The number of motes in the network was varied as well as the population on each mote.

It is clear that the broadcast distribution does not have a negative effect on the ability of the algorithm to converge on a solution. The evolutionary pressure is still sufficient, and the distributed algorithm significantly outperforms a single BEA running with the same population size indicating that the effect of broadcast distribution is positive.

Because generations is no longer a valid term when referring to a BDP we instead refer to the count of genetic operations (mutation and crossover) either on a single mote or all of the motes in the network.

Figure 3 shows that more genetic operations are performed before converging on the solution when the number of motes in the network is increased. This is intuitive; when there are more motes in the network, each is independently computing, resulting in a greater number of genetic operations as a whole, before a solution is found. This does not mean that the time required for the network to converge increases, because as more motes are added, more of the genetic operations are being done in parallel. Each data point in the figure is based

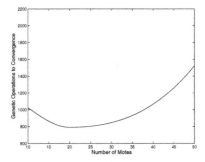

Fig. 3. Varying the Number of Motes in Network

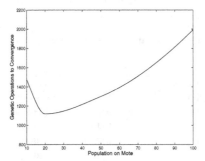

Fig. 4. Varying the Population on Each Mote

on an average of several runs on the same problem with the same number of motes.

The same effect is also observed with the population size is increased on each mote. Figure 4 shows this behavior. The number of motes is kept constant in this experiment. Again, each data point is an average of several runs all operating on the same problem.

Figure 5 show the convergence properties of the BEA and BDP. The graphs are an aggregate of the run results of several different problems. The BDP improves fitness more gradually, but reaches the solution in approximately 25,000 genetic operations, versus 85,000 using the BEA. The total population sizes used in each algorithm was the same, in the BDP the total population was distributed across multiple motes in a fixed environment.

It is also worth noting that we observed the BDP algorithm with total population size p where each mote has a population of $|M|/p$ to be less prone to stagnation than a population of size p running the BEA algorithm. This is due to the propensity of good solutions to distribute slowly throughout the network, this mitigates factors that can occasionally over-emphasize highly fit solutions in the BEA. The effect of this is shown in Table 2, the problems that could not be solved by the BEA in a reasonable amount of time were due to over-emphasis

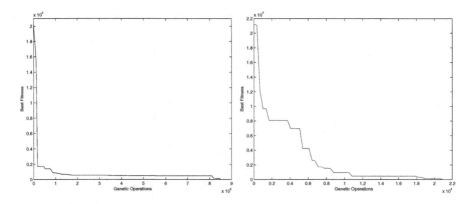

Fig. 5. Convergence of the BEA (left) and BDP (right) Algorithms

Table 2. Aggregate Genetic Operations Until Convergence Over Various Symbolic Regression Problems

Problem	BEA	BDP
P1	6692	8350
P2	7207	9208
P3	84942	18828
P4	130759	23649
P5	315275	43997
P6	DNF[5]	45756
P7	DNF[5]	99893

of highly fit but sub-optimal solutions. The problems increase in difficulty from top to bottom, they range from equations of 3 variables to 8.

5   Conclusions and Future Work

In this work we argue that broadcast-distributed parallel genetic programming shows promise as a model for evolutionary computing on wireless sensor networks and by extension future amorphous or paintable computing architectures. We show empirically via simulations that it is possible to use the broadcast nature of communication between motes to improve the ability of single motes to find a near ideal solution in a reduced number of operations when compared to non-parallel algorithms. We also demonstrate possible ways to build GP implementations that are practicable on resource constrained WSN motes.

Future work will attempt to determine how well BDP will allow motes to evolve solutions that are ideal for their local conditions, and whether such motes will benefit from the receipt of genetic information for neighboring motes that

---

[5] Did not finish computation in a reasonable amount of time.

likely share similar conditions. These experiments will also focus on obtaining data from actual motes in real environments.

References

1. William Butera, V. Michael Bove Jr., and James McBride. Extremely distributed media processing. In *Proceedings of SPIE Media Processors*, 2002.
2. P. Levis and D. Culler. Mate: A tiny virtual machine for sensor networks. In *International Conference on Architectural Support for Programming Languages and Operating Systems, San Jose, CA, USA*, Oct. 2002. To appear.
3. Sean Luke and Lee Spector. Evolving teamwork and coordination with genetic programming. In John R. Koza, David E. Goldberg, David B. Fogel, and Rick L. Riolo, editors, *Genetic Programming 1996: Proceedings of the First Annual Conference*, pages 150–156, Stanford University, CA, USA, 28–31 July 1996. MIT Press.
4. Marco Mamei and Franco Zambonelli. Spray computers: Frontiers of self-organization for pervasive computing. Web: http://polaris.ing.unimo.it/Zambonelli/spray.html.
5. R. Nagpal, A. Kondacs, and C. Chang. Programming methodology for biologically-inspired self-assembling systems. In *AAAI Spring Symposium on Computational Synthesis, March 2003.*, 2003.
6. Peter Nordin, Wolfgang Banzhaf, and Francone Francone. Efficient evolution of machine code for CISC architectures using instruction blocks and homologous crossover. In Lee Spector, William B. Langdon, Una-May O'Reilly, and Peter J. Angeline, editors, *Advances in Genetic Programming 3*, chapter 12, pages 275–299. MIT Press, Cambridge, MA, USA, June 1999.
7. Ho-Sik Seok and Byoung-Tak Zhang. Evolutionary calibration of sensors using genetic programming on evolvable hardware. In *Proceedings of the 2001 Congress on Evolutionary Computation CEC2001*, pages 630–634, COEX, World Trade Center, 159 Samseong-dong, Gangnam-gu, Seoul, Korea, 27-30 May 2001. IEEE Press.
8. Ivan Tanev and Katsunori Shimohara. On role of implicit interaction and explicit communications in emergence of social behavior in continuous predators-prey pursuit problem. In *Genetic and Evolutionary Computation – GECCO-2003*, volume 2724 of *LNCS*, pages 74–85, Berlin, 12-16 July 2003. Springer-Verlag.
9. Jens Ziegler and Wolfgang Banzhaf. Evolving a "nose" for a robot. In *Evolution of Sensors in Nature, Hardware, and*, pages 226–230, Las Vegas, Nevada, USA, 8 July 2000.

Genetic Transposition in Tree-Adjoining Grammar Guided Genetic Programming: The Duplication Operator

Nguyen Xuan Hoai[1], Robert Ian Bob McKay[2], Daryl Essam[2], and Hoang Tuan Hao

School of IT&EE, Australian Defence Force Academy,
University of New South Wales, ACT 2600 Australia
x.nguyen@adfa.edu.au, {rim, dary}l@cs.adfa.edu.au

Abstract. We empirically investigate the use of dual duplication/truncation operators both as mutation operators and as generic local search operators, in combination with genetic search in a tree adjoining grammar guided genetic programming system (TAG3P). The results show that, on the problems tried, duplication/truncation works well as a mutation operator but not reliably when the complexity of the problem was scaled up. When using these dual operators as a generic local search operator, however, it helped TAG3P not only to solve the problems reliably but also cope well with scalability in problem complexity. Moreover, it managed to solve problems with very small population sizes.

1 Introduction

Tree adjoining grammar guided genetic programming (TAG3P) [20] is a genetic programming system that uses tree-adjoining grammars (TAGs) as the formalisms to define its language bias. It was argued in [20] that one of the advantages of using TAG-based representation is the 'feasibility' (described in section 3) in TAG derivation trees, which allows us to design many types of general-purpose search operators on syntactically-constrained domains [20]. In recent works [21-23], we have shown the usefulness of some of these operators. In particular, in [23], relocation, which arises naturally as an operator for doing genetic transposition in a TAG-based representation, was investigated.

In this paper, in the context of tree adjoining grammar guided genetic programming (TAG3P), we empirically investigate the use of duplication, another kind of genetic transposition, both as a mutation operator and as a generic local search operator for genome evolution. For that purpose, we compare results to TAG3P using sub-tree crossover and sub-tree mutation operators as in [20], as well as with standard GP.

The paper is organized as follows. In section 2, we review the concepts of genetic transposition in biological evolution and its possible roles in the field of genetic programming. Section 3 briefly reintroduces the definitions of tree adjoining grammars and TAG3P as well as the relocation operator. The experiments to investigate the roles of relocation in TAG3P are presented and discussed in section 4. Finally, section 5 concludes the paper and highlights some future work.

M. Keijzer et al. (Eds.): EuroGP 2005, LNCS 3447, 2005, pp. 108–119, 2005.
© Springer-Verlag Berlin Heidelberg 2005

2   Genetic Transposition and Genetic Programming (GP)

Genetic transposition in genome evolution is a phenomenon whereby a region of DNA copies itself to another place on the genome. These mobile genetic sequences are called transposable elements [3, 28], transposons, or more informally jumping genes [3]. It is conjectured that genetic transposition plays an important role in forming scattered clusters of related genes in the genome of organisms.   Some researchers ([25]) even went futher, in arguing that genetic transposition (replicative transposition or duplication) should be considered as one of the main workhorses of genome evolution.

There are two types of genetic tranposition, namely, replicative transposition and conservative transposition [28]. In the former, a transposon makes a repeated copy of itself to elsewhere using reverse transcription on an RNA, while in the latter a transposon moves to another place by copying itself [28]. In a recent work [23], we have investigated the metaphor of conservative genetic transposition, which we call genetic relocation. In this paper, the metaphor of replicative transposition, called duplication, is studied in the context of a genetic programming system - TAG3P.

In the field of evolutionary algorithms (EAs), Schwefel ([26]) was probably the first researcher to use gene duplication in solving some real-world problems in industry. In [11], the concept of gene duplication was also proposed in order to raise the power of EAs. Gene duplication appears useful because it can be used to multiply useful building blocks within one individual, and then later the copied building blocks can be subjected to change at the new places by the subsequent gene operations.

In the field of genetic programming [2, 14-16], gene duplication has been studied in several forms. In [18, 19], gene duplication was implemented by copying automatically defined function branches in multi-part programs. Haynes ([9,10]) also implemented a kind of gene duplication for evolving collective behaviours by exchanging codes between individuals in the population. However a general-purpose gene duplication operator acting directly on standard GP expression trees (or the executing branch of GP with ADFs) has not previously been implemented. Although in [8], a gene duplication operation was defined for GP expression trees and was shown to be useful, the implementation is ad-hoc and problem dependent. We believe that this difficulty comes from the fixed-arity property in GP expression tree.

For GEP, a version of GP, the linear representation facilitates the design of genetic duplication [4,5]. However, just as in the case of relocation, the duplication of any trunk of genes (subcode) in GEP can potentially affect the positions as well as the expressiveness (i.e. coding or non-coding) of many other genes, not just at the source and destination positions. Thus it creates a global random side effect on the phenotype. Nevertheless, it was shown to be useful for GEP in some cases [5].

In grammar guided genetic programming (GGGP) [6,7,27,29,30], where the structure of the programs is constrained by grammar rules, it is more difficult than in GP to implement genetic duplication on the genotypic level (usually derivation trees of the grammars) because of the rule-based nature of the formalism. GE [27] is an exception, because of the linear structure of the genotype, it is easy to implement genetic duplication. However, as with GEP, since the GE genotype-to-phenotype map does not posses the locality property (i.e. small change in genotype cause small change in phenotype), the duplication of sections of genes (subcodes) in GE can completely change the meaning and the expressiveness of the genes following the destination position. Depending on the context before the destination position, the meaning of the duplicated genes might also vastly change. Therefore, as with GEP, it creates a global random side effect on the phenotype.

3 TAG-Based Representation for GP

In this section, we first give the definitions of tree adjoining grammars (TAGs) and their derivation trees. Then, we describe TAG3P and the duplication and truncation operators on TAG-derivation trees, which possess the locality property.

3.1 Tree Adjoining Grammars

Joshi and his colleagues in [12] proposed tree-adjunct grammars, the original form of tree adjoining grammars (TAG). Adjunction was the only tree-rewriting operation. Later, the substitution operation was added but it does not change the power of the formalism and therefore will be ignored here.

TAGs are tree-rewriting systems, defined in [13] as a 5-tuple (T, V, I, A, S), where T is a finite set of terminal symbols; V is a finite set of non-terminal symbols (T $\cap$ V = $\emptyset$); S $\in$ V is a distinguished symbol called the start symbol; and E = I $\cup$ A is a set of elementary trees (initial and auxiliary respectively). In an elementary tree, interior nodes are labeled by non-terminal symbols, while nodes on the frontier are labeled either by terminal or non-terminal symbols. The frontier of an initial tree contains all terminal symbols, while the frontier of an auxiliary tree contains all terminal symbols but a distinguished node, the foot node, labeled by the same non-terminal as the root. Initial and auxiliary trees are denoted α and β respectively. A tree whose root is labeled by X is called an X-type tree.

The key operation used with tree-adjoining grammars is the adjunction of trees. Adjunction builds a new (derived) tree γ from an auxiliary tree β and a tree α (initial, auxiliary or derived). If tree α has an interior node labeled A, and β is an A-type tree, the adjunction of β into α to produce γ is as follows: Firstly, the sub-tree α_1 rooted at A is temporarily disconnected from α. Next, β is attached to α to replace the sub-tree. Finally, α_1 is attached back to the foot node of β. γ is the final derived tree achieved from this process.

The tree set of a TAG can be defined as follows [13]:

T_G = {all tree t: t is completed and t is derived from some initial S-trees through adjunctions}

Where a tree t is completed if all of the leaf nodes of t are labeled by terminal symbols. The language generated by the TAG G is defined as

L_G = {w $\in$ T*: w is the yield of some tree t $\in$ T_G}.

In TAG, there is a distinction between derivation and derived trees. A derivation tree in TAG [13] is a tree-structure, which encodes the history of derivation (substitutions and adjunctions) used to produce the derived tree. Each node is labelled by an elementary tree name: the root must be labelled by an α-tree name, and the other nodes with either an α or β tree. The links between a node and its offspring are marked by addresses for adjunctions.

Figure 1 illustrates the derivation and derived trees in TAGs

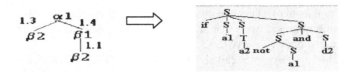

Fig. 1. Examples of a derivation tree and derived tree in TAGs

One special class of TAGs is lexicalized TAGs (LTAGs) [13], in which each elementary tree of an LTAG must have at least one terminal node. It has been proven that there is an algorithm, which, for any context-free grammar G, generates a corresponding LTAG G_{lex} that generates the same language and tree set as G [13]. In this case, the derivation trees in G are equivalent to the derived trees of G_{lex}.

3.2 Tree Adjoining Grammar Guided Genetic Programming (TAG3P)

In TAG3P [20], the derivation tree in LTAG (G_{lex}) was used as the genotype. The phenotype is the derived tree of G_{lex}, which is a derivation tree in the corresponding context-free grammar G. In [29], it was shown that when solving type-less problems like GP, there is a one-to-one map between derivation trees in G and expression trees in GP. The mapping schema in our TAG-based representation, therefore, can be summarized in figure 2 as follows where the second phase of the map is optional.

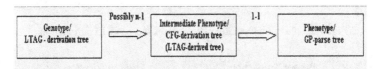

Fig. 2. Schema for Genotype-to-Phenotype map in TAG-based Representation

Other components of TAG3P are as follows [20]:

Parameters: minimum size of genomes (MIN_SIZE), maximum size of genomes (MAX_SIZE), size of population (POP_SIZE), maximum number of generations (MAX_GEN) and probabilities for genetic operators.

Initialization procedure: Each individual is generated by randomly growing a derivation tree in G_{lex} to a size randomly chosen between size bounds.

Fitness Evaluation: an individual derivation tree is first mapped to its derived tree (CFG derivation tree). The expression defined by the derived tree is then semantically evaluated as in grammar guided genetic programming (GGGP) [29], or translated further into the parse tree to, then, be evaluated as in GP [2].

Main Genetic operators: sub-tree crossover and sub-tree mutation [20].

3.3 Duplication and Truncation Operators

The derivation tree structure in LTAG has an important property: when growing it, one can stop at any time, and the derivation tree and the corresponding derived tree are still valid. In other words, the derivation tree in LTAG is a non-fixed-arity tree structure (Catalan tree). The maximal arity (number of children) of a node is the number of adjoining addresses that are present in the elementary tree of that node. If this arity is n, the node can have 0, 1,... or n children.

In [20], this property was called feasibility. Feasibility allows us to design and implement many other new search operators, including bio-inspired ones, in TAG3P which would not be possible in standard GP and other GGGP systems [20]. In particular, the duplication operator arises naturally from this TAG-based representation. To implement duplication in TAG-based representation, a random sub-code (sub-tree) is copied from the tree and is then connected at a random NULL node, provided that the adjunction is valid. It is noted that the copying of this sub-code does not affect the meaning or the expressiveness (coding or non-coding) of the other sub-codes in the tree. Therefore, the effect of the change in phenotype is local (at the source and destination positions only). It is noted that duplication changes the size of the tree. Figure 3 illustrates how relocation works.

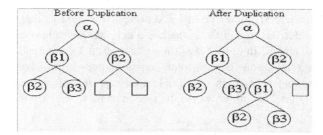

Fig. 3. Duplication Operator. The squares mean NULL adjunction

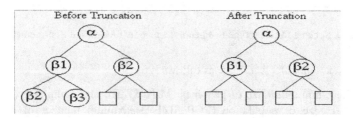

Fig. 4. Truncation Operator. The squares mean NULL adjunction

The dual operator to duplication is truncation, whereby a subtree (subcode) in the tree is chosen at random and is subsequently removed from the tree. Since a TAG-derivation tree is non-fixed arity, the removal of any of its subtrees does not affect the validity of the tree. In other words, the tree resulting from truncation is still a valid TAG-derivation tree, and its derived tree is also valid. Figure 4 shows how truncation works on a TAG-based representation.

4 Experiments and Results

Although gene duplication is bio-inspired and potentially useful for genetic programming in general, some preliminary experiments indicated that a direct application of duplication operators in TAG3P might not work. The reason is that the duplication operator increases the size of the individuals that it is applied to. Some preliminary runs suggested that the direct use of the duplication operator without any size control strategy usually resulted in very rapid bloat in code size. Consequently, in the experiments in this paper, we partnered duplication with the truncation operator as a dual operator.

As in the cases of the insertion/deletion and relocation operators in [22, 23], two possible roles for duplication and truncation are investigated in this paper, namely, as mutation operators, and as dual generic local search operators (i.e. they are used as a combined operator with equivalent probability of being chosen each time the "dual operator" is called). The term "generic local search operator" is used here to distinguish it from another class of local search operators in the literature [1], where problem dependent heuristics are usually involved in the design of the operators, e.g the well-known Lin-Kerninghan local search operator, using 2-opt and 3-opt heuristics, in the travelling salesman problem [19]. In our case, the design of duplication (and truncation) is solely dependent on the (TAG-based) representation, not on any particular application.

4.1 Test Problems

The test problems used in this section are simple symbolic regression problem. In our general symbolic regression problem [14, 24], the task is to learn a function of one independent variable X from 20 sample points in [-1..1]; the function and terminal set are $F=\{+,-,*,/,\sin,\cos,\exp,\text{rlog}\}$ and $T=\{X\}$. The target functions used in this paper are the family of 6 polynomial functions of increasing order of structural complexity: $F_1=X^4+X^3+X^2+X$, $F_2=X^5+X^4+X^3+X^2+X$, $F_3=X^6+X^5+X^4+X^3+X^2+X$, $F_4=X^7+X^6+X^5+X^4+X^3+X^2+X$, $F_5=X^8+X^7+X^6+X^5+X^4+X^3+X^2+X$, $F_6=X^9+X^8+X^7+X^6+X^5+X^4+X^3+X^2+X$.

We note that there is considerable self-similarity in the structures of the above polynomials, and that on the interval of interest ([-1..+1]), the higher-degree polynomials can be approximated well by those with lower degree. Moreover, the structural complexity of the target function increases from F_1 to F_6. In fact, the higher degree polynomial can be recursively generated by: $F_i=F_{i-1}*X+X$ (i=2,...,6). Therefore, the multiplication of building blocks by copying useful subcodes within one polynomial with low degree, might help it to accumulate more partial polynomial parts (such as X+X*) to become (or approximate well) a higher degree polynomial.

4.2 Experimental Setup

We designed a set of base runs for TAG3P and GP with typical population sizes. Table 1 summarises their experiment settings. The grammar G and G_{lex} are similar to those in [24].

4.3   Experiment 1

In the first experiment, duplication and truncation operators are used as the mutation operators (TAG3PM) and the results are compared with standard GP and with TAG3P using subtree mutation. To separate out the effect of using duplication and truncation as mutation operators from the pure effect of subtree crossover, one set of runs was allocated to TAG3P (TAGCROSS) using subtree crossover as the sole genetic operator.

For each set of system and problem instances, 100 runs was allocated, making a total of 2400 runs. Table 2 that follows shows the proportion of success for all systems on the six problem instances. Figure 5 shows the cumulative frequencies of GP, TAG3P, TAGCROSS, and TAG3PM on these problem instances.

Table 1. Experiment Setup for base runs

Objective	Find a function of one independent variable and one dependant variable that fits a given sample of 20 (x_i, y_i) data points, where the target functions are F_1-F_6.
Terminal Operands	X (the independent variable);
Terminal Operators	The binary operators are +,-,*,/. The unary operators are sin, cos, exp and rlog .
Fitness Cases	The sample of 20 points in the interval [-1..+1]
Raw fitness	The sum, taken over 20 fitness cases, of the errors.
Standardized Fitness	Same as raw fitness.
Hits	The number of fitness cases for which the error is less than 0.01.
Genetic Operators	Tournament selection of size 3, subtree crossover and subtree mutation for both GP and TAG3P.
Parameters	POP_SIZE=500, MAX_GEN=51, MAX_SIZE=40 (for TAG3P), MAX_DEPTH=15 (for GP), Crossover rate=0.9, mutation rate=0.1
Success predicate	An individual scores 20 hits.

Table 2. Proportion of success for all systems on the six problem instances

Problem	GP	TAG3P	TAGCROSS	TAG3PM
F_1	9%	93%	93%	93%
F_2	3%	82%	87%	90%
F_3	1%	43%	61%	64%
F_4	2%	48%	43%	53%
F_5	2%	12%	27%	29%
F_6	0%	22%	22%	20%

4.4 Experiment 2

In the second experiment, duplication and truncation were used as a dual generic local search operator, in combination with genetic search in TAG3P using subtree crossover and subtree mutation. The results are compared to TAG3P (using full population size - 500). To compensate for the fitness evaluations taken for the local search, the population sizes were set as 50 (LSTAG3P50) and 10 (LSTAG3P10). To balance this, the numbers of local search steps were 10 and 50 respectively. In other words, the maximal number of fitness evaluations is the same as for the TAG3P runs (using POPSIZE=500). Other parameter settings of TAG3P50 and TAG3P10 are similar to those for TAG3P. The local search strategy was stochastic hill-climbing, with Lamarckian inheritance (i.e. when local search finds a better individual in the neighbourhood of an individual, it will be replaced by the newly found individual). On each problem instance, each system was allocated 100 runs, making a total of 1200 runs in this experiment.

The following Table 3 shows the proportion of success on the six problems instances for all systems. Figure 6 depicts their cumulative frequencies.

Table 3. Proportion of success for all systems on the six problem instances

Problem	GP	TAG3P	LSTAG3P50	LSTAG3P10
F_1	9%	93%	93%	79%
F_2	3%	82%	91%	80%
F_3	1%	43%	85%	69%
F_4	2%	48%	74%	71%
F_5	2%	12%	59%	61%
F_6	0%	22%	67%	60%

From the results in Table 3 and Figure 6, it is obvious that, on the problem instances tried, LSTAG3P outperformed TAG3P significantly, especially for target function with high complexity in structure (i.e with high repeated and self-similar patterns). It indicates that duplication and truncation, when used as a dual generic local search operator in combination with genetic search, can lead to significant improvements in TAG3P performance. Moreover, the performance is still very reliable when the structural complexity of the target function is scaled up. It is noted that the superior performance of LSTAG3P50 and LSTAG3P10 is achieved by very small population sizes. Furthermore, the similar performances of LSTAG3P50 and LSTAG3P10 on F_4, F_5, and F_6 suggest that, under some circumstances, it is possible to use a smaller population, with longer local search, without affecting the performance of the system. It is particularly useful in reducing the sometimes huge population sizes of GP [16].

4.5 Experiment 3

The purpose of the third experiment is to investigate wether the superior performance of LSTAG3P came from the power of duplication/truncation or from favouring exploitation in the evolutionary search balance (by reducing the population sizes). To

Table 4. Proportion of success for all systems on the six problem instances

Problem	GP50	TAG3P50	LSTAG3P50	GP10	TAG3P10	LSTAG3P10
F_1	12%	70%	**93%**	4%	51%	**79%**
F_2	6%	66%	**91%**	2%	46%	**80%**
F_3	0%	48%	**85%**	4%	36%	**69%**
F_4	2%	47%	**74%**	4%	26%	**71%**
F_5	2%	32%	**59%**	2%	27%	**61%**
F_6	3%	29%	**67%**	2%	28%	**60%**

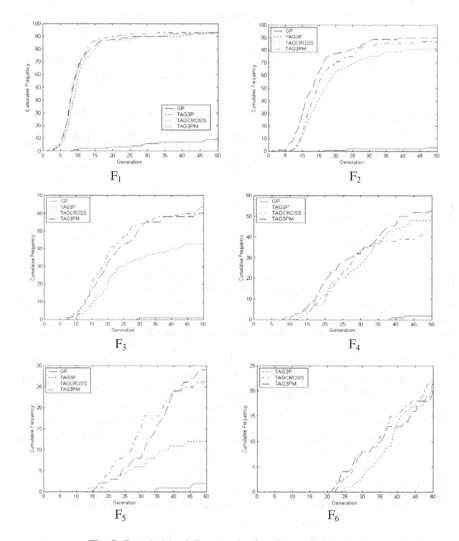

Fig. 5. Cumulative of Frequencies for all systems in experiment 1

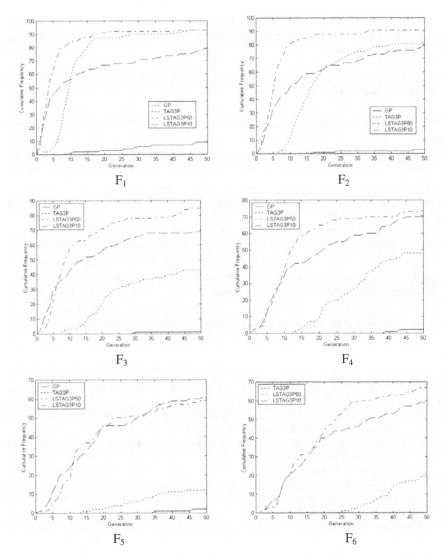

Fig. 6. Cumulative of Frequencies for all systems in experiment 2

accomplish the task, the same population sizes (50 and 10) were used for TAG3P (TAG3P50, TAG3P10) and GP (GP50, GP10) - using subtree crossover and subtree mutation), with the maximal numbers of generations (MAX_GEN) being correspondingly increased (511 and 2551).

Each system was allocated 100 runs for each problem instance, making a total of 2400 runs for this experiment. Table 4 shows the results of TAG3P50 and TAG3P10 compared with LSTAG3P50, LSTAG3P10.

Results in table 4 show that LSTAG3P outperformed TAG3P50 and TAG3P10 (as well as GP50 and GP10) by a very large margin. Thus the superior performance of LSTAG3P in experiment 2 was certainly a result of the use of the dual duplication and truncation local search operator in combination with TAG3P genetic search, and not from the effect of reducing the population size and therefore increasing the exploitation of the search.

5 Conclusion and Future Work

In this paper, we empirically investigated two possible roles for the duplication and truncation operators in TAG3P. The results show that, on a problem where duplication of subcodes is likely to cause fitness improvement (learning a family of polynomials of increasing degree), duplication and truncation are slightly better mutation operators than the more standard subtree mutation. However when used as mutation operators, they did not scale particularly well with increasing structural complexity of the target function.

By contrast, when duplication and truncation were used in a second role as a dual generic-local search operator, they not only improved the performance of TAG3P, but also enabled TAG3P to scale well with the problem complexity. Moreover, they enabled TAG3P to solve problems with very small population sizes.

The local search strategy used in this paper is still very naive. In future, we will investigate the use of duplication/truncation as generic local search operators using other adaptive search strategies, such as simulated annealing or tabu search.

References

1. Aarts E. and Lenstra J.K.: *Local Search in Combinatorial Optimization*, John Wiley and Sons (1997).
2. Banzhaf W., Nordin P., Keller R.E., and Francone F.D.: *Genetic Programming: An Introduction.* Morgan Kaufmann Pub (1998).
3. Drlica K.: *Understanding DNA and Gene Cloning: A Guide for the Curious*, John Wiley & Sons, USA (1984).
4. Ferreira C.: Gene Expression Programming: A New Adaptive Algorithm for Solving Problems, *Complex Systems* 3 (2), (2001) 87-129.
5. Ferreira C.: Mutation, Transposition, and Recombination: An Analysis of the Evolutionary Dynamics, in *Proceedings of the 4th Int. Workshop on Frontiers in Evolutionary Algorithms*, (2002), 614-617.
6. Geyer-Schulz A.: *Fuzzy Rule-Based Expert Systems and Genetic Machine Learning*, Physica-Verlag (1995).
7. Gruau F.: On Using Syntactic Constraints with Genetic Programming, In: *Advances in Genetic Programming II*, The MIT Press, (1996) 377-394.
8. Haynes T.: Duplication of Coding Segments in Genetic Programming, *Technical Report UTULSA-MCS-96-03*, The University of Tulsa (1996).
9. Haynes T.: Collective Adaptation: The Exchange of Coding Segments, *Evolutionary Computation*, 6 (4), (1998) 311-338.
10. Haynes T.: Collective Adaptation: The Sharing of Building Blocks, PhD Thesis, Department of Mathematical and Computer Sciences, University of Tulsa, 1998.
11. Holland J.: Adaptation in Natural and Artificial Intelligence: An Introductory Analysis with Application in Biology, Control, and Artificial Intelligence, Michigan University Press (1975).
12. Joshi A.K., Levy L.S., and Takahashi M.: Tree Adjunct Grammars, *Journal of Computer and System Sciences*, 10 (1), (1975) 136-163.

13. Joshi A.K. and Schabes Y.: Tree Adjoining Grammars, in *Handbook of Formal Languages*, Springer-Verlag, (1997) 69-123.
14. Koza J.: *Genetic Programming: On the programming of Computers by Means of Natural Selection*, MIT Press (1992).
15. Koza J.: *Genetic Programming II: Automatic Discoveries of Reusable Programs*, MIT Press, 1994.
16. Koza J., Andre D., Bennett III F.H. , and Kean M.: *Genetic Programming III: Darwinian Invention and Problem Solving*, Morgan Kaufmann (1999).
17. Koza J.: Gene Duplication to Enable Genetic Programming to Concurrently Evolve Both the Architecture and Work-Performing Steps of a Computer Program, in *Proceedings of the Fourteenth International Joint Conference on Artificial Intelligence*, (1995) 734-740.
18. Koza J. and Andre D.: Classifying Protein Segments as Transmembrane Domains Using Architecture-Altering Operations in Genetic Programming, in *Advances in Genetic Programming 2*, Chapter 8, MIT Press (1996).
19. Lin S. and Kerninghan B.W.: An Effective Heuristic Algorithm for the Traveling Salesman Problem, *Operation Research*, 21, (1973) 458-516.
20. Nguyen Xuan Hoai, McKay R.I., and Abbass, H.A.: Tree Adjoining Grammars, Language Bias, and Genetic Programming, in *Proceedings of the 6th European Conference on Genetic Programming (EuroGP 2003)*, LNCS 2610, Springer-Verlag, (2003), 335-344.
21. Nguyen Xuan Hoai and McKay R.I.: Softening the Structural Difficulty with TAG-based Representation and Insertion/Deletion Operators, in *Proceedings of Genetic and Evolutionary Computation Conference (GECCO 2004)*, LNCS 3103, Springer-Verlag, (2004), 605-616.
22. Nguyen Xuan Hoai and McKay R.I.: An Investigation on the Roles of Insertion and Deletion Operators in Tree Adjoining Grammar Guided Genetic Programming, in *Proceedings of Congress on Evolutionary Computation (CEC 2004)*, IEEE Press, (2004), 472-477.
23. Nguyen Xuan Hoai, McKay R.I., and Essam D.: Genetic Transposition in Tree-Adjoining Grammar Guided Gentic Programming: The Relocation Operator, *in Proceedings of the 5th International Conference on Simulated Evolution and Learning (SEAL 2004)*, IEEE Press (2004).
24. Nguyen Xuan Hoai, McKay R.I., and Essam D.: Solving Symbolic Regression Problem with Tree Adjoining Grammar Guided Genetic Programming, Australian Journal of Inteligent Information Processing Systems, 7(3), (2002), 114-121.
25. Ohno S.: *Evolution by Duplication*, Springer-Verlag, 1970.
26. Schwefel H.P.: Projekt MHD-Staustrahlrohr: Experimentelle Optimierung einer Zweiphasenduse Teil I Technischer Bericht 11.034/68, 35, AEG Forschungsinstitut, Berlin, 1968.
27. O'Neil M. and Ryan C.: Grammatical Evolution, *IEEE Trans on EC*, 4 (4), (2000) 349-357, 2000.
28. Ridley M.: *Evolution*, Second Edition, Blackwell Science, USA (1996).
29. Whigham P.A.: Grammatical Bias for Evolutionary Learning, Ph.D Thesis, UNSW, Australia, (1996).
30. Wong M.L. and Leung K.S.: Evolutionary Program Induction Directed by Logic Grammars, *Evolutionary Computation*, 5, (1997) 143-180.

GP-EndChess: Using Genetic Programming to Evolve Chess Endgame Players

Ami Hauptman and Moshe Sipper

Department of Computer Science, Ben-Gurion University, Israel
{amiha, sipper}@cs.bgu.ac.il
www.moshesipper.com

Abstract. We apply genetic programming to the evolution of strategies for playing chess endgames. Our evolved programs are able to draw or win against an expert human-based strategy, and draw against CRAFTY—a world-class chess program, which finished second in the 2004 Computer Chess Championship.

1 Introduction

Developing intelligent (or at least pseudo-intelligent) computer players of strategy games is a problem which AI research have been addressing since the field's onset. Because excelling at strategy games has often been considered to be a sign of intellectual excellence, many have felt that developing an intelligent game player would represent a big step towards developing a more generally intelligent machine [1].

The game of chess has always been viewed as an intellectual game par excellence, "a touchstone of the intellect," according to Goethe.[1] The game's complexity stems from two main sources. First, the size of the search space: after the opening phase, each player has to select the next move from approximately 50 possible moves on average. Since a single game typically consists of a few dozen moves, the search space is enormous. A second source of complexity stems from the amount of information contained in a single board. Since each player starts with 16 pieces of 6 different types, and as the board comprises 64 squares, evaluating a single board (a "position") entails elaborate computation, even without looking ahead.

Computer programs capable of playing the game of chess have been designed for more than 40 years, starting with the first working program that was reported in 1958 [2]. According to Russell and Norvig [3], from 1965 to 1994 there was an almost linear increase in the strength of computer chess programs—as measured in their performance in human-rated tournaments. This increase culminated in the defeat in 1997 of Gary Kasparov—the former World Chess Champion—by IBM's special-purpose chess engine, Deep Blue (see [4])

---

[1] Some basic chess terms are explained in the appendix.

M. Keijzer et al. (Eds.): EuroGP 2005, LNCS 3447, pp. 120–131, 2005.

Deep Blue, and its offspring Deeper Blue, rely mainly on brute-force methods to gain an advantage over the opponent, by traversing as deeply as possible the game tree [5]. Although these programs have achieved amazing performance levels, Noam Chomsky [6] has criticized this aspect of game-playing research as being "about as interesting as the fact that a bulldozer can lift more than some weight lifter."

The number of feasible games possible (i.e., the size of the game tree), given a board configuration, is astronomical, even if one limits oneself to endgames. While endgames typically contain but a few pieces, the problem of evaluation is still hard, as the pieces are usually free to move all over the board, resulting in complex game trees—both deep and with high branching factors. Thus, we cannot rely on brute-force methods alone. We need to develop better ways to approximate the outcome of games with "smart" evaluation functions. The automated learning of evaluation functions is a promising research area if we are to produce stronger artificial players [5].

We will use the *Genetic Programming* (GP) paradigm to evolve board-evaluation functions, the basic idea of GP being to breed computer programs to solve a particular problem [7]: Start with a population of random, (usually) low-fitness individuals. Every individual plays a few games with its peers, and is assigned a score according to its level of success (or failure), i.e., its fitness. The next generation is stochastically constructed, based on individuals' fitness values. This process repeats itself until the single best individual is returned as the solution, at the time of the evolutionary program's termination.

This paper is organized as follows: In the next section we describe previous work on on automated methods for developing chess endgame strategies. Section 3 describes our GP setup for the evolution of chess endgame players, followed by results in Section 4. Finally, we end with concluding remarks and future work in Section 5.

2   Previous Work

GP has recently been argued to deliver "high-return, human-competitive machine intelligence" [8]. Indeed, over the years, several strategies or agents that play games have been evolved using GP (or some other form of evolutionary algorithm).

Ferret and Martin [1] had a computer play the ancient Egyptian board game of Senet, by evolving board-evaluation functions using tournament-style fitness evaluation. Gross *et al.* [9] introduced a system that integrates GP and Evolutionary Strategies to learn to play chess. This system did not learn from scratch, but instead a "scaffolding" algorithm that could perform the task already was improved by means of evolutionary techniques.

Kendall and Whitwell [5] used evolutionary algorithms to tune evaluation-function parameters. The resulting individuals were successfully matched against commercial chess programs, but only when the lookahead for the commercial program was strictly limited.

Previous works only used simple board-evaluation functions as the building blocks for the evolutionary algorithm. For example, some typical functions used by Gross *et al.* [9] are: material values for the different pieces, penalty for bishops in initial positions, bonus for pawns in center of chessboard, penalty for doubled pawns and for backward pawns, castling bonus if this move was taken and penalty if it was not, and rook bonus for an open line or on the same line of a passed pawn. Kendall and Whitwell [5] used fewer board-evaluation functions, and focused on the weights of the remaining pieces.

3   Evolving Chess Endgame Strategies Using Genetic Programming

We evolve chess endgame strategies using Koza-style GP [7]. Each individual— a LISP-like tree expression—represents a strategy, the purpose of which is to evaluate a given board configuration and generate a real-valued score. The tree's internal nodes are called *functions*, and the leaves—*terminals*. We used simple Boolean functions (AND, OR, NOT), and IF functions; terminals were used to analyze certain features of the game position. We included a large number of terminals, varying from simple ones (such as the number of moves for the player's king), to more complex features (for example, the number of pieces attacking a given piece). A full description of functions and terminals used is given in Section 3.3.

In order to better control the structure of our programs we used *Strongly Typed Genetic Programming* (STGP) [10]. This method allows the user to assign a type to a tree edge. Each function is assigned both a return type and a type for each of its arguments; each terminal is assigned a return type. Assigning more than one type per edge is also possible. All trees must be constructed according to these conventions, and only compatible types are allowed to interact. Thus, a user-defined typing scheme is imposed, although in fact all data passed within the tree consists of real numbers. We used the ECJ GP System of Luke [11].

3.1   Board Evaluation

We wish to develop evaluation strategies that bear similarity to human board analysis. Thus, instead of looking deep into the game tree, we traverse less nodes, but consider each node more thoroughly. As such, our strategies use only limited lookahead.

The current player receives as input all possible board configurations reachable from the current position by making one legal move (this is quite easy to compute). After these boards are evaluated, the one that received the highest score is selected, and that move is made. Thus, an artificial player is had by combining an (evolved) board evaluator with a program that generates all possible next moves.

Although this approach has been successfully used in several game-strategy evolution scenarios (see [1]), it has not yet been applied to chess endgames.

3.2   Tree Topology

Our programs play chess endgames consisting of kings, queens, and rooks (in the future we shall also consider bishops and knights). Each game starts from a different (random) legal position, in which no piece is attacked, e.g., two kings, two rooks, and two queens in a KQRKQR endgame. Although at first each program was evolved to play a different type of endgame (KRKR, KRRKRR, KQKQ, KQRKQR, etc.), which implies using different game strategies, the same set of terminals and functions was used for all types. Moreover, this set was also used for our more complex runs, in which GP chess players were evolved to play several types of endgames. Our ultimate aim is the evolution of general-purpose strategies.

Still, as most chess players would agree, playing a winning position (e.g., with material advantage) is very different than playing a losing position, or an even one (see Appendix). For this reason, each individual contains three trees: an advantage tree, an even tree, and a disadvantage tree. These trees are used according to the current status of the board. The disadvantage tree is smaller, since achieving a stalemate and avoiding exchanges requires less complicated reasoning.

3.3   Tree Nodes

While evaluating a position, an expert chess player considers various aspects of the board. Some are simple, while others require a deep understanding of the game. Chase and Simon found that experts recalled meaningful chess formations better than novices [12]. This lead them to hypothesize that chess skill depends on a large knowledge base, indexed through thousands of familiar chess patterns.

We assumed that complex aspects of the game board are comprised of simpler units, which require less game knowledge, and are to be combined in some way. Our chess programs use terminals, which represent those relatively simple aspects, and functions, which incorporate no game knowledge, but supply methods of combining those aspects. As we used STGP, all functions and terminals were assigned one or more of two data types: *Float* and *Boolean*. We also included a third data type, named *Query*, which could be used as any of the former two.

The function set used included the If function, and simple Boolean functions. Although our tree returns a real number, we omitted arithmetic functions, for several reasons. First, a large part of contemporary research in the field of machine learning and game theory (in particular for perfect-information games) revolves around inducing logical rules for learning games (for example, see [13], [14] and [15]). Second, according to the players we consulted, while evaluating positions involves considering various aspects of the board, some more important than others, performing logical operations on these aspects seems natural, while mathematical operations does not. Third, we observed that numeric functions sometimes returned extremely large values, which interfered with subtle calculations. Therefore the scheme we used was a (carefully ordered) series of Boolean queries, each returning a fixed value (either an ERC or a numeric terminal, see below). See Table 1 for the complete list of functions.

Table 1. Function set of GP individual. B: Boolean, F: Float

F=If3(B_1, F_1, F_2)	If B_1 is non-zero, return F_1, else return F_2
B=Or2(B_1, B_2)	Return 1 if at least one of B_1, B_2 is non-zero, 0 otherwise
B=Or3(B_1, B_2, B_3)	Return 1 if at least one of B_1, B_2, B_3 is non-zero, 0 otherwise
B=And2(B_1, B_2)	Return 1 only if B_1 and B_2 are non-zero, 0 otherwise
B=And3(B_1, B_2, B_3)	Return 1 only if B_1, B_2, and B_3 are non-zero, 0 otherwise
B=Smaller(B_1, B_2)	Return 1 if B_1 is smaller than B_2, 0 otherwise
B=Not(B_1)	Return 0 if B_1 is non-zero, 1 otherwise

We developed most of our terminals by consulting several high-ranking chess players [2]. The terminal set examines various aspects of the chessboard, and may be divided into 3 groups:

1. *Float values,* created using the ERC (*Ephemeral Random Constants*) mechanism (see [7] for details). An ERC is chosen at random to be one of the following six values $\pm 1 \cdot \{\frac{1}{2}, \frac{1}{3}, \frac{1}{4}\} \cdot MAX$ (MAX was empirically set to 1000), and the inverses of these numbers. This guarantees that when a value is returned after some group of features has been identified, it will be distinct enough to engender the outcome.

2. *Simple terminals,* which analyze relatively simple aspects of the board, such as the number of possible moves for each king, and the number of attacked pieces for each player. These terminals were derived by breaking relatively complex aspects of the board into simpler notions. More complex terminals belong to the next group (see below). For example, a player should capture his opponent's piece if it is not sufficiently protected, meaning that the number of attacking pieces the player controls is greater than the number of pieces protecting the opponent's piece, and the material value of the defending pieces is equal to or greater than the player's. Adjudicating these considerations is not simple, and therefore a terminal that performs this entire computational feat by itself belongs to the next group of complex terminals.

The simple terminals comprising this second group are derived by refining the logical resolution of the previous paragraphs' reasoning: Is an opponent's piece attacked? How many of the player's pieces are attacking that piece? How many pieces are protecting a given opponent's piece? What is the material value of pieces attacking and defending a given opponent's piece? All these questions are embodied as terminals within the second group. The ability to easily embody such reasoning within the GP setup, as functions and terminals, is a major asset of GP.

Other terminals were also derived in a similar manner. See Table 2 for a complete list of simple terminals. Note that some of the terminals are inverted—we would like terminals to always return positive (or true) values, since these values represent a favorable position. This is why we used, for example, a terminal

---

[2] The highest-ranking player we consulted was Boris Gutkin, ELO 2400, International Master (see appendix), and fully qualified chess teacher.

Table 2. Simple terminals. Opp: opponent, My: player

B=NotMyKingInCheck()	Is the player's king not being checked?
B=IsOppKingInCheck()	Is the opponent's king being checked?
F=MyKingDistEdges()	The player's king's distance form the edges of the board
F=OppKingProximityToEdges()	The player's king's proximity to the edges of the board
F=NumMyPiecesNotAttacked()	The number of the player's pieces that are not attacked
F=NumOppPiecesAttacked()	The number of the opponent's attacked pieces
F=ValueMyPiecesAttacking()	The material value of the player's pieces which are attacking
F=ValueOppPiecesAttacking()	The material value of the opponent's pieces which are attacking
B=IsMyQueenNotAttacked()	Is the player's queen not attacked?
B=IsOppQueenAttacked()	Is the opponent's queen attacked?
B=IsMyFork()	Is the player creating a fork?
B=IsOppNotFork()	Is the opponent not creating a fork?
F=NumMovesMyKing()	The number of legal moves for the player's king
F=NumNotMovesOppKing()	The number of illegal moves for the opponent's king
F=MyKingProxRook()	Proximity of my king and rook(s)
F=OppKingDistRook()	Distance between opponent's king and rook(s)
B=MyPiecesSameLine()	Are two or more of the player's pieces protecting each other?
B=OppPiecesNotSameLine()	Are two or more of the opponent's pieces protecting each other?
B=IsOppKingProtectingPiece()	Is the opponent's king protecting one of his pieces?
B=IsMyKingProtectingPiece()	Is the player's king protecting one of his pieces?

evaluating the player's king's *distance* from the edges of the board (generally a favorable feature for endgames), while using a terminal evaluating the *proximity* of the opponent's king to the edges (again, a positive feature).

3. Complex terminals. These are terminals that check the same aspects of the board a human player would. Some prominent examples include: the terminal OppPieceCanBeCaptured considering the capture of a piece; checking if the current position is a draw, a mate, or a stalemate (especially important for non-even boards); checking if there is a mate in one or two moves (this is the most complex terminal); the material value of the position; comparing the material value of the position to the original board—this is important since it is easier to consider change than to evaluate the board in an absolute manner. See Table 3 for a full list of complex terminals.

Since some of these terminals are hard to compute, and most appear more than once in the individual's trees, we used a memoization scheme to save time [16]: After the first calculation of each terminal, the result is stored, so

Table 3. Complex terminals. Opp: opponent, My: player. Some of these terminals perform lookahead, while others compare with the original board

F=EvaluateMaterial()	The material value of the board
B=IsMaterialIncrease()	Did the player capture a piece?
B=IsMate()	Is this a mate position?
B=IsMateInOne()	Can the opponent mate the player after this move?
B=OppPieceCanBeCaptured()	Is it possible to capture one of the opponent's pieces without retaliation?
B=MyPieceCannotBeCaptured()	Is it not possible to capture one of the player's pieces without retaliation?
B=IsOppKingStuck()	Do all legal moves for the opponent's king advance it closer to the edges?
B=IsMyKingNotStuck()	Is there a legal move for the player's king that advances it away from the edges?
B=IsOppKingBehindPiece()	Is the opponent's king two or more squares behind one of his pieces?
B=IsMyKingNotBehindPiece()	Is the player's king not two or more squares behind one of my pieces?
B=IsOppPiecePinned()	Is one or more of the opponent's pieces pinned?
B=IsMyPieceNotPinned()	Are all the player's pieces not pinned?

that further calls to the same terminal (on the same board) do not repeat the calculation. Memoization greatly reduced the evolutionary run-time.

3.4   Fitness Evaluation

As we used a competitive evaluation scheme, the fitness of an individual was determined by its success against its peers. We used the random-2-ways method (see [17] for full details), in which each individual plays against a fixed number of randomly selected peers (typically 5). Each of these encounters entails a fixed number of games, each starting from a randomly generated position. Since random starting positions can sometimes be uneven (for example, allowing the starting player to attain a capture position), every starting position was played twice, each player playing both black and white. This way a better starting position could benefit both players and the tournament was less biased. In addition, in each encounter several games were played, to further reduce the element of chance.

The score for each game is derived from the outcome of the game. Players that manage to mate their opponents receive more points than those that achieve only a material advantage. Draws are rewarded by a score of low value and losses entail no points at all.

The final fitness for each player is the sum of all points earned in the entire tournament for that generation. We used the standard reproduction, crossover, and mutation operators, as in [7]. The major parameters were: population size – 80, generation count – between 150 and 250, reproduction probability – 0.35, crossover probability – 0.5, and mutation probability – 0.15 (including ERC).

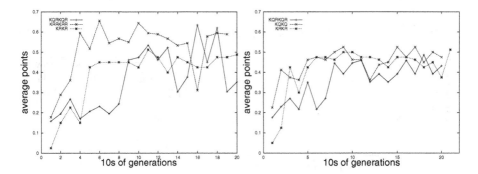

Fig. 1. Left: Results against a strategy defined by a chess Master. The three graphs show the average score over time of the best run of 50 runs carried out, for three types of endgames: KRKR, KRRKRR, KRQKRQ. A point represents the score of the best individual at that time, pitted in a 150-game tournament against the human-defined strategy. Right: Results against CRAFTY. The three graphs show the average score over time of the best run of 15 runs carried out, for three types of endgames: KRKR, KQKQ, KQRKQR. A point represents the score of the best individual at that time, pitted in a 50-game tournament against CRAFTY

4   Results

We conducted several experiments to test our evolving chess players. The scoring method was based on the one used in chess tournaments: victory—1 point, draw—$\frac{1}{2}$ point, loss—0 points. In order to better differentiate our players, we rewarded $\frac{3}{4}$ points for a material advantage (without mating the opponent).

The final score is the sum of all scores a player has received, divided by the number of games. This way, a player who always mates its opponent will receive a perfect score of 1. The score for a player that played against an opponent of comparable strength (where most games end in a draw), is $1/2$ on average.

4.1   Experiment 1: Competing Against a Human-Defined Strategy

As noted above, we developed most of our terminals by consulting several high-ranking chess players. In order to evaluate our system, we wished to test our evolved strategies against some of these players. Because we needed to play thousands of games in every run, these could not be conducted manually, but instead we programmed an optimal strategy, based on the guidance from the players we consulted. We wrote this evaluation program using the functions and terminals of our GP system.

During evolution, our chess programs competed against each other. However, every 10 generations the best individual was extracted and pitted in a 150-game tournament against the human-defined strategy. The results are depicted in Figure 1, showing runs for KRKR, KRRKRR and KQRKQR, respectively.

These figures clearly show that starting from a low level of performance, chess players evolve to play as good as high-ranking humans for all groups of endgames,

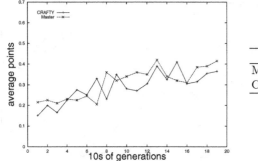

	%Wins	%Advs	%Draws
Master	6.00	2.00	68.00
CRAFTY	2.00	4.00	72.00

Fig. 2. Left: Results for multiple-endgame runs—wherein all endgames were used during evolution—against both CRAFTY and the Master-defined strategy. Each graph shows the average score over time of the best run of 20 runs carried out. A point represents the score of the best individual at that time, pitted in a 50-game tournament against CRAFTY, or a 150-game against the Master. Right: Percent of wins, advantages, and draws for best tournament of run (i.e., fitness peak of graph)

in one case even going beyond a draw to win (KQRKQR endgame, where a high score of 0.63 was attained). Improvement was rapid, typically requiring only a few dozens of generations (about 15 hours on a standard workstation).

4.2   Experiment 2: Competing Against a World-Class Chess Engine

Having attained good results against a human-defined strategy based on expert chess players, we went one step further and competed against a highly powerful chess engine. For this task, we used the CRAFTY engine (version 19.01) by Hyatt [3]. CRAFTY is a state-of-the-art chess engine, using a typical brute-force approach, with a fast evaluation function, NegaScout search, and all the standard enhancements [18]. CRAFTY finished *second* at the 12th World Computer Speed Chess Championship, held in Bar-Ilan University on July 2004. According to www.chessbase.com, CRAFTY has a rating of 2614 points, which places it at the human Grandmaster level. CRAFTY is thus, undoubtedly, a worthy opponent.

As expected, CRAFTY proved to be a formidable opponent, constantly mating the GP opponent at early generations. However, during the process of evolution, substantial improvement was observed to occur. As shown in Figure 1, our program managed to achieve near-draw scores, even for the complex KQRKQR endgame. Considering our evolved 2-lookahead programs' competing against a world-class chess player, our method seems quite viable and promising.

4.3   Experiment 3: Multiple-Endgame Runs

Aiming for general-purpose strategies, this third experiment involved the playing of one game *of each type* (rather than a single type)—both during evolution

---

[3] CRAFTY's source code is available at ftp://ftp.cis.uab.edu/pub/hyatt

and in the test tournaments. Evolved players were pitted against the Master-defined strategy and CRAFTY. As can be seen in Figure 2, near-draw scores were achieved under these conditions as well. We observed that performance kept improving and are confident that it would continue doing so with added computational resources.

5   Concluding Remarks and Future Work

We presented a method by which chess endgame players may be evolved to successfully hold their own against excellent opponents. One of the major prima facie problems with our scheme is its complexity, as evidenced by Tables 1, 2, and 3. In the time-honored tradition of computer science, we argue that this is not a bug but rather a feature—to be more precise, a somewhat overlooked feature of genetic programming.

We believe that GP represents a viable means to automatic programming, and perhaps more generally to machine intelligence, in no small part due to its being *cooperative with humans*. More than many other adaptive search techniques (e.g., genetic algorithms, artificial neural networks, ant algorithms), the GPer, owing to GP's representational affluence and openness, is better positioned to imbue the genomic language with his or her own intelligence. While artificial-intelligence (AI) purists may wrinkle their noses at this, taking the AI-should-emerge-from-scratch stance, we argue that a more practical path to AI involves man-machine cooperation. GP is a forerunning candidate for the 'machine' part.

We did not design our genome (Tables 1, 2, 3) in one fell swoop, but rather through an incremental, interactive process, whereby man (represented by the humble authors of this paper) and machine (represented by man's university's computers) worked hand-in-keyboard. To wit, we began our experimentation with small sets of functions and terminals, which were revised and added upon through our examination of evolved players and their performance and through consultation with high-ranking chess players. GP's design cooperativeness, often overlooked, is thus perhaps one of its major boons.

In addition, the number of terminals we used is small, compared to the number of patterns used by chess experts when evaluating a position: According to Simone and Gilmartin [19] this number is close to 100,000. Since most pattern-based programs nowadays are considered to be far from competitive (see [13]), the results we obtained may imply that we have made a step towards developing a program that has more in common with the way humans play chess.

In the future we aim to follow a number of paths: 1) improve the evolved programs' performance against the above and other endgames, 2) branch out beyond endgames, and 3) analyze the evolved cognition as to its resemblance and difference from human cognition.

References

1. Ferrer, G.J., Martin, W.N.: Using genetic programming to evolve board evaluation functions for a board game. In: 1995 IEEE Conference on Evolutionary Computation. Volume 2., Perth, Australia, IEEE Press (1995) 747–752
2. Bernstein, A., de V. Roberts, M.: Computer versus Chess-Player. Scientific American **198** (1958) 96–105
3. Russell, S.J., Norvig, P.: Artificial Intelligence: A Modern Approach. Prentice-Hall, Englewood Cliffs , NJ (1995)
4. DeCoste, D.: The Significance of Kasparov vs Deep Blue and the Future of Computer Chess. ICCA Journal **21** (1998) 33–43
5. Kendall, G., Whitwell, G.: An evolutionary approach for the tuning of a chess evaluation function using population dynamics. In: Proceedings of the 2001 Congress on Evolutionary Computation CEC2001, COEX, World Trade Center, 159 Samseong-dong, Gangnam-gu, Seoul, Korea, IEEE Press (2001) 995–1002
6. Chomsky, N.: Language and Thought. Moyer Bell, Wakefield, RI (1993)
7. Koza, J.R.: Genetic Programming: On the Programming of Computers by Means of Natural Selection. MIT Press, Cambridge, MA, USA (1992)
8. Koza, J.R., Keane, M.A., Streeter, M.J., Mydlowec, W., Yu, J., Lanza, G.: Genetic Programming IV: Routine Human-Competitive Machine Intelligence. Kluwer Academic Publishers, Norwell, MA (2003)
9. Gross, R., Albrecht, K., Kantschik, W., Banzhaf, W.: Evolving chess playing programs. In Langdon, W.B., Cantú-Paz, E., Mathias, K., Roy, R., Davis, D., Poli, R., Balakrishnan, K., Honavar, V., Rudolph, G., Wegener, J., Bull, L., Potter, M.A., Schultz, A.C., Miller, J.F., Burke, E., Jonoska, N., eds.: GECCO 2002: Proceedings of the Genetic and Evolutionary Computation Conference, New York, Morgan Kaufmann Publishers (2002) 740–747
10. Montana, D.J.: Strongly typed genetic programming. Evolutionary Computation **3** (1995) 199–230
11. Luke, S.: ECJ: A Java-based Evolutionary Computation and Genetic Programming Research System. (2000) http://www.cs.umd.edu/projects/plus/ec/ecj/.
12. Charness, N.: Expertise in chess: The balance between knowledge and search. In Ericsson, K.A., Smith, J., eds.: Toward a general theory of Expertise: Prospects and limits. Cambridge University Press, Cambridge (1991)
13. Fürnkranz, J.: Machine learning in computer chess: The next generation. International Computer Chess Association Journal **19** (1996) 147–161
14. Bonanno, G.: The logic of rational play in games of perfect information. Papers 347, California Davis - Institute of Governmental Affairs (1989) available at http://ideas.repec.org/p/fth/caldav/347.html.
15. Bain, M.: Learning Logical Exceptions in Chess. PhD thesis, University of Strathclyde, Glasgow, Scotland (1994)
16. Abelson, H., Sussman, G.J., with J. Sussman: Structure and Interpretation of Computer Programs. Second edn. The MIT-Press (1996)
17. Panait, L.A., Luke, S.: A comparison of two competitive fitness functions. In Langdon, W.B., Cantú-Paz, E., Mathias, K., Roy, R., Davis, D., Poli, R., Balakrishnan, K., Honavar, V., Rudolph, G., Wegener, J., Bull, L., Potter, M.A., Schultz, A.C., Miller, J.F., Burke, E., Jonoska, N., eds.: GECCO 2002: Proceedings of the Genetic and Evolutionary Computation Conference, New York, Morgan Kaufmann Publishers (2002) 503–511

18. Jiang, A.X., Buro, M.: First experimental results of ProbCut applied to chess. In: Proceedings of 10th Advances in Computer Games Conference, Kluwer Academic Publishers, Norwell, MA (2003) 19–32
19. Simon, H., Gilmartin, K.: A simulation of memory for chess positions. Cognitive Psychology **5** (1973) 29–46

Appendix: Brief Glossary of Basic Chess Terms

(More at www.arkangles.com)

Material value. Sum of all numerical values (see Point Count) for player's pieces (which are given positive values), and the opponent's (negative values).

Point count. Queen is worth 9 points, rooks – 5 points, bishops – 3 or 3.25 points, knights – 3 points, and pawns – 1 point. King is typically assigned an infinite value.

Advantage. When the current configuration of the game favors one side over another; includes: material advantage, permanent advantage, positional advantage, and temporary advantage.

Capture. Moving a piece to a square occupied by an enemy piece, thereby removing the enemy piece from the board.

Fork. A form of double attack where one piece threatens two enemy pieces at the same time. In a triple fork, three enemy pieces are threatened.

Endgame. The final phase of the game when there are few pieces left on the board. Endgame abbreviations are used to represent the remaining pieces (e.g., KRKR).

Ranking chess players. Both professional and amateur chess players may obtain a nationally (or internationally) recognized numerical rating (sometimes referred to as ELO). Independently, professional players may earn titles, gained in special official tournaments, in which title-holders must participate. A title, once earned, is the player's for life, while the point rating can oscillate. The lowest international title is Master (usually not gained before the player reaches ELO 2200). The highest titles are International Master (IM) and Grandmaster (GM). In 2003 there were only about 3000 IMs and GMs worldwide.

GP-Gammon: Using Genetic Programming to Evolve Backgammon Players

Yaniv Azaria and Moshe Sipper

Department of Computer Science, Ben-Gurion University, Israel
{azariaya, sipper}@cs.bgu.ac.il
www.moshesipper.com

Abstract. We apply genetic programming to the evolution of strategies for playing the game of backgammon. Pitted in a 1000-game tournament against a standard benchmark player—*Pubeval*—our best evolved program wins 58% of the games, the highest verifiable result to date. Moreover, several other evolved programs attain win percentages not far behind the champion, evidencing the repeatability of our approach.

1 Introduction

The majority of learning software for backgammon is based on artificial neural networks, which usually receive as input the board configuration and produce as output the suggested next best move. The main problem lies with the network's fixed topology: The designer must usually decide upon this *a priori*, whereupon only the internal synaptic weights change. (Nowadays, one sometimes uses evolutionary techniques to evolve the topology [1]).

The learning technique we have chosen to apply is *Genetic Programming* (GP), by which computer programs can be evolved [2]. A prime advantage of GP over artificial neural networks is the automatic development of structure, i.e., the program's "topology" need not be fixed in advance. In GP we start with an initial set of general- and domain-specific features, and then let evolution determine (evolve) the structure of the calculation (in our case, a backgammon-playing strategy). In addition, GP readily affords the easy addition of control structures such as conditional and loop statements, which may also evolve automatically.

This paper details the evolution of highly successful backgammon players via genetic programming. In the next section we present previous work on machine-learning approaches to backgammon. In Section 3 we present our algorithm for evolving backgammon-playing strategies using genetic programming. Section 4 presents results, followed by Section 5, wherein we conclude and describe future work.

2 Previous Work

The application of machine-learning techniques to obtain strong backgammon players has been done both in academia and industry. The best commercial

M. Keijzer et al. (Eds.): EuroGP 2005, LNCS 3447, pp. 132–142, 2005.

products to date are Jellyfish [3] and TD-Gammon [4]. Being commercial, with their innards unavailable for any scrutiny, we shall remain herein in the academic arena. Our benchmark competitor will thus be the freely available Pubeval—which has become a standard yardstick used by those applying AI techniques to backgammon. Pubeval is quite a strong machine player, trained on a database of expert preferences using comparison training [5].

Tesauro's approach is based on the Temporal Difference method, used to train a neural network through a self-playing model—i.e., learning is accomplished by programs playing against themselves and thus improving [4].

In 1997, Pollack, Blair, and Land [5] presented HC-Gammon, a much simpler Hill-Climbing algorithm that also uses neural networks. Under their model the current network is declared 'Champion', and by adding Gaussian noise to the biases of this champion network a 'Challenger' is created. The Champion and the Challenger then engage in a short tournament of backgammon; if the Challenger outperforms the Champion, small changes are made to the Champion biases in the direction of the Challenger biases.

Another interesting work is that of Sanner *et al.* [6], whose approach is based on cognition (specifically, on the ACT-R theory of cognition [7]). Rather than trying to analyze the exact board state, they defined a representational abstraction of the domain, consisting of general backgammon features such as blocking, exposing, and attacking. They maintain a database of feature neighborhoods, recording the statistics of winning and losing for each such neighborhood. All possible moves are encoded as sets of the above features; then, the move with the highest win probability (according to the record obtained so far) is selected.

Finally, Qi and Sun [8] presented a genetic algorithm-based multi-agent reinforcement learning bidding approach (GMARLB). The system comprises several evolving teams, each team composed of a number of agents. The agents learn through reinforcement using the Q-learning algorithm. Each agent has two modules, Q and CQ. At any given moment only one member of the team is in control—and chooses the next action for the whole team. The Q module selects the actions to be performed at each step, while the CQ module determines whether the agent should continue to be in or relinquish control. Once an agent relinquishes control, a new agent is selected through a bidding process, whereby the member who bids highest becomes the new member-in-control.

3   Evolving Backgammon-Playing Strategies Using Genetic Programming

We use Koza-style GP [2] to evolve backgammon strategies. In GP, a population of individuals evolves, where an individual is composed of LISP sub-expressions, each sub-expression being a LISP program constructed from *functions* and *terminals*. The functions are usually arithmetic and logical operators that receive a number of arguments as input and compute a result as output; the terminals are zero-argument functions that serve both as constants and as sensors. Sensors are a special type of function that query the domain environment (in our case, backgammon board configurations).

In order to improve the performance of the GP system, we used *Strongly Typed Genetic Programming* (STGP) [9], which allows to add data types and data-type constraints to the LISP programs, thereby affording the evolution of more powerful and useful programs.

In STGP, each function has a *return type* and *argument types* (if there are any arguments). In our implementation a type can be either an *atomic type*, which is a symbol, or a *set type*, which is a group of atomic types. A node n_1 can have a child node n_2 if and only if the return type of n_2 is compatible with the appropriate argument type of n_1. An atomic type is compatible with another atomic type if they are both identical, and a set type is compatible with another set type if they share at least one identical atomic type.

Note that the types are mere symbols and not real data types; their purpose is to force structural constraints on the LISP programs. The data passed between nodes consists only of real numbers.

Board Evaluation. Tesauro [4] noted that due to the presence of stochasticity in the form of dice, backgammon has a high branching factor, therefore rendering deep search strategies impractical. Thus, we opted for the use of a flat evaluator: after rolling the dice, generate all possible next-move boards, evaluate each one of them, and finally select the board with the highest score.

This approach has been used widely by neural network-based players and—as shown below—it can be used successfully with genetic programming. In our model, each individual is a LISP program that—using the sensors—receives a backgammon board configuration as input and returns a real number that represents the board score.

An artificial player is had by combining an (evolved) board evaluator with a program that generates all next-moves given the dice values.

Program Architecture. The game of backgammon can be observed to consist of two main stages: the 'contact' stage, where the two players can hit each other, and the 'race' stage, where there is no contact between the two players. During the contact stage, we expect a good strategy to block the opponent's progress and minimize the probably of getting hit. On the other hand, during the race stage, blocks and blots are of no import, rather, one aims to select moves that lead to the removal of a maximum number of pieces off the board.

This observation has directed us in designing the genomic structure of individuals in the population. Each individual contains a contact tree and a race tree. When a board is evaluated, the program checks whether there is any contact between the players and then evaluates the tree that is applicable to the current board state. The function set of the contact tree is richer and contains various general and specific board query functions. The function set of the race tree is much smaller and contains only functions that examine the checkers' positions. This is because at the race phase, the moves of each player are independent of the opponent's status, and thus are much simpler.

Functions and Terminals. Keeping in mind our use of STGP, we need to describe not only the functions and terminals but also their type constraints. We use two atomic types: *Float* and *Boolean*. We also use one set type—*Query*—that includes both atomic types.

The function set contains no domain-specific operators, but only arithmetic and logical ones, so we use the same function set for both contact and race trees. The function set is given in Table 1.

Table 1. Function set of the contact and race trees

F=Add(F, F)	Add two real numbers
F=Sub(F, F)	Subtract two real numbers
F=Mul(F, F)	Multiply two real numbers
F=If(B, F, F)	If first argument evaluates to a non-zero value, return value of second argument, else return value of third argument
B=Greater(F, F)	If first argument is greater than second, return 1, else return 0
B=Smaller(F, F)	If first argument is smaller than second, return 1, else return 0
B=And(B, B)	If both arguments evaluate to a non-zero value, return 1, else return 0
B=Or(B, B)	If at least one of the arguments evaluates to a non-zero value, return 1, else return 0
B=Not(B)	If argument evaluates to zero, return 1, else return 0

With terminals we use the ERC (*Ephemeral Random Constant*) mechanism, as described in Koza [2]. An ERC is a node that—when first initialized—is assigned a constant value from a given range; this value does not change during evolution, unless a mutation operator is applied.

The terminal set is specific to our domain (backgammon), and contains three types of functions:

1. The Float-ERC function calls upon ERC directly. When created, the terminal is assigned a constant, real-number value, which becomes the return value of the terminal.
2. The board-position query terminals use the ERC mechanism to query a specific location on the board. When initialized, a value between 0 and 25 is randomly chosen, where 0 specifies the bar location, 1-24 specify the inner board locations, and 25 specifies the off-board location (Figure 1).
   The term 'Player' refers to the contender whose turn it is, while 'Enemy' refers to the opponent. After completing the move, the contenders are swapped. When a board query terminal is evaluated, it refers to the board location that is associated with the terminal, from the player's point of view.
3. The last type of terminal is a function that provides general information about the board as a whole.

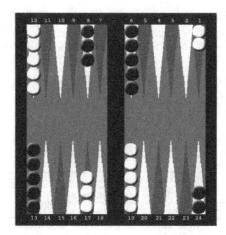

Fig. 1. Initial backgammon configuration. The White player's home positions are labeled 19-24, and the Black player's home positions are labeled 1-6

The terminal set for contact trees is given in Table 2 and that for race trees in Table 3.

Fitness Measure. The evolutionary process is internally driven, i.e., *the evolving strategies play against each other* (and not against an external opponent). As such, the fitness of an individual is relative to its cohorts. To avoid overly lengthy evaluation times, methods such as Round Robin—where each individual is pitted against all others—were avoided. Through experimentation we concluded that a good evaluation method is the Single Elimination Tournament: Start with a population of n individuals, n being a power of two. Then, divide the individuals into $\frac{n}{2}$ arbitrary pairs, and let each pair engage in a relatively short tournament of 50 games. Finally, set the fitness of the $\frac{n}{2}$ losers to $\frac{1}{n}$. The rest $\frac{n}{2}$ winners are divided into pairs again, engage in tournaments as before, and the losers are assigned fitness values of $\frac{1}{n/2}$. This process continues until one champion individual remains. Thus, the more tournaments an individual "survives," the higher its fitness.

Breeding Strategy. After the evaluation stage, we need to create the next generation of individuals from the current generation. This process involves two primary operators: *breeding* and *selection*. Of a finite set of breeding operators (described below), one is chosen probabilistically; then, one or two individuals (depending on the breeding operator) are selected from the current generation. Finally, the breeding operator is applied to the selected individual(s).

We use four breeding operators in our model, either unary (operating on one individual) or binary (operating on two individuals): *reproduction, sub-tree crossover, point mutation,* and *MutateERC*:

- The unary reproduction operator is the simplest one: copy one individual to the next generation with no modifications. The main purpose of this operator is to preserve a small number of good individuals.

Table 2. Terminal set of the contact tree. Note that zero-argument functions—which serve both as constants and as sensors—are considered as terminals

F=Float-ERC	ERC – random real constant in range [0,5]
Q=Player-Exposed(n)	If player has exactly one checker at location n, return 1, else return 0
Q=Player-Blocked(n)	If player has two or more checkers at location n, return 1, else return 0
Q=Player-Tower(n)	If player has h or more checkers at location n (where $h \geq 3$), return $h - 2$, else return 0
Q=Enemy-Exposed(n)	If enemy has exactly one checker at location n, return 1, else return 0
Q=Enemy-Blocked(n)	If enemy has two or more checkers at location n, return 1, else return 0
F=Player-Pip	Return player *pip-count* divided by 167 (*pip-count* is the number of steps a player needs to move in order to win the game. This value is normalized through division by 167—the *pip-count* at the beginning of the game)
F=Enemy-Pip	Return enemy *pip-count* divided by 167
F=Total-Hit-Prob	Return sum of hit probability over all exposed player checkers
F=Player-Escape	Measure the effectiveness of the enemy's barrier over his home positions. For each enemy home position that does not contain an enemy block, count the number of dice rolls that could potentially lead to the player's escape. This value is normalized through division by 131—the number of ways a player can escape when the enemy has no blocks
F=Enemy-Escape	Measure the effectiveness of the player's barrier over his home positions using the same method as above

Table 3. Terminal set of the race tree

F=Float-ERC	ERC – random real constant in range [0,5]
Q=Player-Position(n)	Return number of checkers at location n

- The binary crossover operator randomly selects an internal node in each of the two individuals (belonging to corresponding trees—either race or contact) and then swaps the sub-trees rooted at these nodes.
- The unary mutation operator randomly selects one node from one of the trees, deletes the subtree that is rooted at that node and grows a new subtree instead. (Crossover and mutation are described in detail in Koza [2].)
- The unary MutateERC operator selects one random node and then mutates every ERC within the sub-tree that is rooted at that node. The mutation operation we used is the addition of a small Gaussian noise to the ERC. We used this breeding operator to achieve two goals: first, this is a convenient

way to generate new constants as evolution progresses; and, second, it helps to balance the constants on good individuals. The MutateERC operation is described in [10].

We chose a selection method that supports relative fitness—*tournament selection*, as described in Koza [2]: randomly choose a small subset of individuals, and then select the one with the best fitness. This method is simple, respects the relative fitness scale, and also affords a fair chance of selecting low-fitness individuals in order to prevent early convergence.

4   Results

For benchmark purposes we used *Pubeval*—a free, public-domain board evaluation function written by Tesauro [11]. The program—which plays very well—seems to have become the *de facto* yardstick used by the growing community of backgammon-playing program developers. Several researchers in the field have pitted their own creations against Pubeval.

Our population consisted of 128 individuals, which evolved for 500 generations. We used the ECJ GP System of Luke [12]. We repeated the experiment 20 times and calculated the average, minimum, and maximum benchmark values every five generations. Figure 2 shows the benchmark curve of our individuals. Table 4 shows how our best evolved players fared against Pubeval, alongside the performance of the other approaches described in Section 2.

To get an idea of the human-competitiveness of our players we referred to the HC-gammon statistics (demo.cs.brandeis.edu/hcg/stats1.html), according

Table 4. Comparison of backgammon players. GP-Gammon-*i* designates the best GP strategy evolved at run *i*, which was tested in a tournament of 1000 games against Pubeval. Only the top 5 runs are shown (out of 20). For ACT-R-Gammon and HC-Gammon, the values cited are the best values obtained. For GMARLB-gammon, the authors cited a best value of 56%, apparently a fitness peak obtained during one evolutionary run, computed over **50 games.** This is too short a tournament and hence we cite their average value. Indeed, we were able to obtain win percentages of over 65% (!) for randomly selected strategies over 50-game tournaments, a result which dwindled to 40-45% when the tournament was extended to 1000 games

Player	% Wins vs. Pubeval
GP-Gammon-1	56.8
GP-Gammon-2	56.6
GP-Gammon-3	56.4
GP-Gammon-4	55.7
GP-Gammon-5	54.6
GMARLB-Gammon [8]	51.2
ACT-R-Gammon [6]	45.94
HC-Gammon [5]	40.00

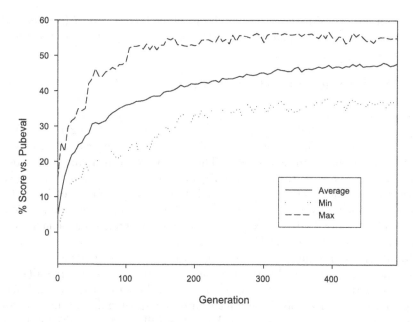

Fig. 2. Benchmark curve. The benchmark score of an individual equals the score it attained in a 1000-game tournament against Pubeval, divided by the sum of scores gained by both the individual and Pubeval

to which HC-Gammon wins 58% of the games when counting abounded games as wins, and 38% when not counting them. Considering that HC-Gammon wins 40% of the games versus Pubeval, we expect—by transitivity—that GP-gammon (with win percentage of 56% vs. Pubeval) is a very strong player in human terms.

On a standard workstation our system plays about 700–1,000 games a minute. As can be seen in Figure 2, to achieve good asymptotic performance our method requires on the order of 500,000–2,000,000 games (100–300 generations) per evolutionary run—about 2-3 days of computation. In comparison, GMARLB-Gammon required 400,000 games to learn, HC-Gammon – 100,000, and ACT-R-Gammon – 1000 games. The latter low figure is due to the explicit desire by ACT-R-Gammon's authors to model human cognition, their starting point being that a human can at best play 1,000 games a month (should he forego all other activities). Note that as opposed to the other individual-based methods herein discussed (e.g., employing one or a few neural networks), our approach is population based; the learning cost *per individual* is therefore on the order of a few thousand games.

Our primary goal herein has not been to reduce computational cost, but to attain the best machine player possible. As quipped by Milne Edwards (and quoted by Darwin in *Origin of Species*), "nature is prodigal in variety, but niggard in innovation." With this in mind, we did not mind having our processes run for a few days. After all, backgammon being a hard game to play expertly (our reason for choosing it), why should a machine learn rapidly? (see also [13]) Be that as it may, we do plan to tackle the optimization issue in the future.

5   Concluding Remarks and Future Work

As is often the case with genetic programming, evolved individuals are highly complex, especially when the problem is a hard one—e.g., backgammon. Much like a biologist examining naturally evolved genomes, one cannot divine the workings of the program at a glance. Thus, we have been unable—despite intense study—to derive a rigorous formulation concerning the structure and contribution of specific functions and terminals to the success of evolved individuals (this we leave for future work). Rigorousness aside, though, our examination of many evolved individuals has revealed a number of interesting behaviors and regularities, hereafter delineated.

Recall that our function set contains two types of board-query functions: those that perform specific board-position queries (e.g., Player-Exposed(n) and Player-Blocked(n)), and those that perform general board queries (e.g., Enemy-Escape and Total-Hit-Prob). These latter are more powerful, and, in fact, some of them can be used as stand-alone heuristics (albeit very weak) for playing backgammon.

We have observed that general query functions are more common than position-specific functions. Furthermore, GP-evolved strategies seem to "ignore" some board positions. This should come as no surprise: the general functions provide useful information during most of the game, thus inducing GP to make use of them often. In contrast, information pertaining to a specific board position has less effect on overall performance, and is relevant only at a few specific moves during the game.

We surmise that the general functions form the lion's share of an evolved backgammon strategy, with specific functions used to balance the strategy by catering for (infrequently encountered) situations. In some sense GP strategies are reminiscent of human game-playing: humans rely on general heuristics (e.g., avoid hits, build effective barriers), whereas local decisions are made only in specific cases. (As noted above, the issue of human cognition in backgammon was central to the paper by Sanner et al. [6].)

Our model divides the backgammon game into two main stages, thus entailing two types of trees. A natural question arising is that of refining this two-fold division into more sub-stages. The game dynamics may indeed call for such a refined division, with added functions and terminals specific to each game stage.

However, it is unclear how this refining is to be had: Any (human) suggestion beyond the obvious two-stage division is far from being obvious—or correct. One possible avenue of future research is simply to let GP handle this question altogether and evolve the stages themselves. For example, we can use a main tree to inspect the current board configuration and decide which tree should be used for the current move selection. These 'specific' trees would have their own separately evolving function and terminal sets. Automatically defined functions

(ADFs) [14] and architecture-altering operations [15] will most likely come in quite handy here.[1]

GP is known to be computer-intensive, being both memory- and time-avaricious. Witness Koza's use of a 1,000-Pentium cluster[2] and populations of up to 10,000,000 individuals [16]. Unfortunately, our own resources were limited to but a few workstations. We believe quite firmly that upping the resources will lead to the evolution of much better players. Part of our belief stems from a few multi-cpu experiments, which we performed on a cluster of workstations that were made available to us for a short period of time. Jumping at the occasion, we were able to attain a win percentage of 58% against Pubeval—the best known result to date. Hopefully, we will gain access to more resources in the future, thereby attempting to improve our players yet further.

Our application of an adaptive—so-called "intelligent"—search technique in the arena of games is epitomic of an ever-growing movement. Our evolved backgammon players are highly successful, able to beat previous automatically obtained strategies.

Acknowledgements

We are grateful to Assaf Zaritsky for helpful comments. Special thanks to Diti Levy for helping us with the drawing in Figure 1.

References

1. Yao, X.: Evolving artificial neural networks. Proceedings of the IEEE **87** (1999) 1423–1447
2. Koza, J.R.: Genetic programming: On the Programming of Computers by Means of Natural Selection. MIT Press, Cambridge, MA (1992)
3. Dahl, F.: JellyFish Backgammon. (1998-2004)
   http://www.jellyfish-backgammon.com.
4. Tesauro, G.: Temporal difference learning and TD-Gammon. Communications of the ACM **38** (1995) 58–68
5. Pollack, J.B., Blair, A.D., Land, M.: Coevolution of a backgammon player. In Langton, C.G., Shimohara, K., eds.: Artificial Life V: Proceedings of the Fifth International Workshop on the Synthesis and Simulation of Living Systems, Cambridge, MA, MIT Press (1997) 92–98
6. Sanner, S., Anderson, J.R., Lebiere, C., Lovett, M.: Achieving efficient and cognitively plausible learning in backgammon. In Langley, P., ed.: Proceedings of the 17th International Conference on Machine Learning (ICML-2000), Stanford, CA, Morgan Kaufmann (2000) 823–830

---

[1] Early experiments with ADFs in our current work produced lower results and—as non-ADF runs worked quite nicely—we decided to concentrate our efforts there. This does not preclude, however, the beneficial use of ADFs in the refinement of our methodology described in the paragraph.

[2] www.genetic-programming.com/machine1000.html

7. Anderson, J.R., Lebiere, C.: The Atomic Components of Thought. Lawrence Erlbaum Associates, Mahwah, NJ (1998)
8. Qi, D., Sun, R.: Integrating reinforcement learning, bidding and genetic algorithms. In: Proceedings of the International Conference on Intelligent Agent Technology (IAT-2003), IEEE Computer Society Press, Los Alamitos, CA (2003) 53–59
9. Montana, D.J.: Strongly typed genetic programming. Evolutionary Computation **3** (1995) 199–230
10. Chellapilla, K.: A preliminary investigation into evolving modular programs without subtree crossover. In Koza, J.R., Banzhaf, W., Chellapilla, K., Deb, K., Dorigo, M., Fogel, D.B., Garzon, M.H., Goldberg, D.E., Iba, H., Riolo, R., eds.: Genetic Programming 1998: Proceedings of the Third Annual Conference, University of Wisconsin, Madison, Wisconsin, USA, Morgan Kaufmann (1998) 23–31
11. Tesauro, G.: Software–Source Code Benchmark player "pubeval.c". (1993) http://www.bkgm.com/rgb/rgb.cgi?view+610.
12. Luke, S.: ECJ: A Java-based Evolutionary Computation and Genetic Programming Research System. (2000)
http://www.cs.umd.edu/projects/plus/ec/ecj/.
13. Sipper, M.: A success story or an old wives' tale? On judging experiments in evolutionary computation. Complexity **5** (2000) 31–33
14. Koza, J.R.: Genetic Programming II: Automatic Discovery of Reusable Programs. MIT Press, Cambridge, Massachusetts (1994)
15. Koza, J.R., Bennett III, F.H., Andre, D., Keane, M.A.: Genetic Programming III: Darwinian Invention and Problem Solving. Morgan Kaufmann, San Francisco, California (1999)
16. Koza, J.R., Keane, M.A., Streeter, M.J., Mydlowec, W., Yu, J., Lanza, G.: Genetic Programming IV: Routine Human-Competitive Machine Intelligence. Kluwer Academic Publishers, Norwell, MA (2003)

GP-Robocode: Using Genetic Programming to Evolve Robocode Players

Yehonatan Shichel, Eran Ziserman, and Moshe Sipper

Department of Computer Science, Ben-Gurion University, Israel
{shichel, eranz, sipper}@cs.bgu.ac.il
www.moshesipper.com

Abstract. This paper describes the first attempt to introduce evolution-arily designed players into the international Robocode league, a simulation-based game wherein robotic tanks fight to destruction in a closed arena. Using genetic programming to evolve tank strategies for this highly ac-tive forum, we were able to rank third out of twenty-seven players in the category of HaikuBots. Our GPBot was the only entry not written by a human.

"I wonder how long handcoded algorithms will remain on top."
Developer's comment at a Robocode discussion group,
robowiki.net/cgi-bin/robowiki?GeneticProgramming

1   Introduction

The strife between humans and machines in the arena of intelligence has fertil-ized the imagination of many an artificial-intelligence (AI) researcher, as well as numerous science fiction novelists. Since the very early days of AI, the domain of games has been considered as epitomic and prototypical of this struggle. Design-ing a machine capable of defeating human players is a prime goal in this area: From board games, such as chess and checkers, through card games, to computer adventure games and 3D shooters, AI plays a central role in the attempt to see machine beat man at his own game—literally.

Program-based games are a subset of the domain of games in which the human player has no direct influence on the course of the game; rather, the actions during the game are controlled by programs that were written by the (usually human) programmer. The program responds to the current game en-vironment, as captured by its percepts, in order to act within the simulated game world. The winner of such a game is the programmer who has provided the best program; hence, the programming of game strategies is often used to measure the performance of AI algorithms and methodologies. Some famous ex-amples of program-based games are *RoboCup* (www.robocup.org), the robotic soccer world championship, and *CoreWars* (corewars.sourceforge.net), in which assembly-like programs struggle for limited computer resources.

M. Keijzer et al. (Eds.): EuroGP 2005, LNCS 3447, pp. 143–154, 2005.

While the majority of the programmers actually write the code for their players, some of them choose to use machine-learning methods instead. These methods involve a process of constant code modifications, according to the nature of the problem, in order to achieve as best a program as possible. If the traditional programming methods focus on the ways to solve the problem (the 'how'), machine-learning methods focus on the problem itself (the 'what')—to evaluate the program and constantly improve the solution.

We have chosen the game of *Robocode* (`robocode.alphaworks.ibm.com`), a simulation-based game in which robotic tanks fight to destruction in a closed arena. The programmers implement their robots in the Java programming language, and can test their creations either by using a graphical environment in which battles are held, or by submitting them to a central web site where online tournaments regularly take place; this latter enables the assignment of a relative ranking by an absolute yardstick, as is done, e.g., by the Chess Federation. The game has attracted hundreds of human programmers and their submitted strategies show much originality, diversity, and ingenuity.

One of our major objectives is to attain what Koza and his colleagues have recently termed *human-competitive machine intelligence* [1]. According to Koza *et al.* [1] an automatically created result is human-competitive if it satisfies one or more of eight criteria (p. 4; ibid), the one of interest to us here being:

H. The result holds its own or wins a regulated competition involving human contestants (in the form of either live human players or human-written computer programs).

Since the vast majority of Robocode strategies submitted to the league were coded by hand, this game is ideally suited to attain the goal of human-competitiveness.

The machine-learning method we have chosen to use is *Genetic Programming* (GP), in which the code for the player is created through evolution [2]. The code produced by GP consists of a tree-like structure (similar to a LISP program), which is highly flexible, as opposed to other machine-learning techniques (e.g., neural networks).

This paper is organized as follows: Section 2 describes previous work. Section 3 delineates the Robocode rules and Section 4 presents our GP-based method for evolving Robocode strategies, followed by results in Section 5. Finally, we present concluding remarks and future work in Section 6.

2   Previous Work

In a paper published in 2003, Eisenstein described the evolution of Robocode players using a fixed-length genome to represent networks of interconnected computational units, which perform simple arithmetic operations [3]. Each element takes its input either from the robot's sensors or from another computational unit. Eisenstein was able to evolve Robocode players, each able to defeat a single opponent, but was not able to generalize his method to create players that could beat numerous adversaries and thus hold their own in the international league. This latter failure may be due either to problems with the methodology or to lack of computational resources—no conclusions were provided.

Eisenstein's work is the only recorded attempt to create Robocode players using GP-like evolution. The number of works that have applied machine-learning techniques to design Robocode players is meager, mostly ANN-based (Artificial Neural Network), and produced non-top-ranked strategies. In most cases the ANN controls only part of the robot's functionality, mainly the targeting systems. We found no reports of substantive success of ANNs over hand-coded robots. Applications of GP in robotics have been studied by several researchers, dating back to one of Koza's original experiments—evolving wall-following robots [4] (a full review of GP works in the field of robotics is beyond the scope of this paper).

3   Robocode Rules

A Robocode player is written as an event-driven Java program. A main loop controls the tank activities, which can be interrupted on various occasions, called *events*. Whenever an event takes place, a special code segment is activated, according to the given event. For example, when a tank bumps into a wall, the *HitWallEvent* will be handled, activating a function named *onHitWall()*. Other events include: hitting another tank, spotting another tank, and getting hit by an enemy shell.

There are five actuators controlling the tank: movement actuator (forward and backward), tank-body rotation actuator, gun-rotation actuator, radar-rotation actuator, and fire actuator (which acts as both trigger and firepower controller).

As the round begins, each tank of the several placed in the arena is assigned a fixed value of energy. When the energy meter drops to zero, the tank is disabled, and—if hit—is immediately destroyed. During the course of the match, energy levels may increase or decrease: a tank gains energy by firing shells and hitting other tanks, and loses energy by getting hit by shells, other tanks, or walls. Firing shells costs energy. The energy lost when firing a shell, or gained, in case of a successful hit, is proportional to the firepower of the fired shell.

The round ends when only one tank remains in the battlefield (or no tanks at all), whereupon the participants are assigned scores that reflect their performance during the round. A battle lasts a fixed number of rounds.

In order to test our evolved Robocode players and compare them to human-written strategies, we had to submit them to the international league. The league comprises a number of divisions, classified mainly according to allowed code size. Specifically, we aimed for the *one-on-one HaikuBot challenge*, in which the players play in duels, and their code is limited to four instances of a semicolon (four lines), with no further restriction on code size (`robocode.yajags.com`). Since GP naturally produces long lines of code, this league seemed most appropriate for our research. Moreover, a code size-limited league places GP at a disadvantage, since, ceteris paribus, GP produces longer programs due to much junk "DNA" (which a human programmer does not produce—usually).

4   Evolving Robocode Strategies Using Genetic Programming

We used Koza-style GP [2], in which a population of individuals evolves. An individual is represented by an ensemble of LISP expressions, each composed of functions and terminals. The functions we used are mainly arithmetic and logical ones, which receive several arguments and produce a numeric result. Terminals are zero-argument functions, which produce a numerical value without receiving any input. The terminal set is composed of zero-argument mathematical functions, robot perceptual data, and numeric constants. The list of functions and terminals is given in Table 1, and will be described below.

As part of our research we examined many different configurations for the various GP characteristics and parameters. We have tried, for instance, to use Strongly Typed Genetic Programming (STGP) [5], in which functions and terminals differ in types and are restricted to the use of specific types of inputs; another technique that we inspected was the use of Automatically Define Functions (ADFs) [6], which enables the evolution of subroutines. These techniques and a number of others proved not to be useful for the game of Robocode, and we concentrate below on a description of our winning strategy.

Program Architecture. We decided to use GP to evolve numerical expressions that will be given as arguments to the player's actuators. As mentioned above, our players consist of only four lines of code (each ending with a semicolon). However, there is much variability in the layout of the code: we had to decide which events we wished to implement, and which actuators would be used for these events.

To obtain the strict code-line limit, we had to make the following adjustments:

- Omit the radar rotation command. The radar, mounted on the gun, was instructed to turn using the gun-rotation actuator.
- Implement the fire actuator as a numerical constant which can appear at any point within the evolved code sequence (see Table 1).

The main loop contains one line of code that directs the robot to start turning the gun (and the mounted radar) to the right. This insures that within the first gun cycle, an enemy tank will be spotted by the radar, triggering a *ScannedRobotEvent*. Within the code for this event, three additional lines of code were added, each controlling a single actuator, and using a single numerical input that was evolved using GP. The first line instructs the tank to move to a distance specified by the first evolved argument. The second line

Table 1. GP Robocode system: Functions and terminals

Game-status indicators	
Energy()	Returns the remaining energy of the player
Heading()	Returns the current heading of the player
X()	Returns the current horizontal position of the player
Y()	Returns the current vertical position of the player
MaxX()	Returns the horizontal battlefield dimension
MaxY()	Returns the vertical battlefield dimension
EnemyBearing()	Returns the current enemy bearing, relative to the current player's heading
EnemyDistance()	Returns the current distance to the enemy
EnemyVelocity()	Returns the current enemy's velocity
EnemyHeading()	Returns the current enemy heading, relative to the current player's heading
EnemyEnergy()	Returns the remaining energy of the enemy
Numerical constants	
Constant()	An ERC in the range [-1, 1]
Random()	Returns a random real number in the range [-1, 1]
Zero()	Returns the constant 0
Arithmetic and logical functions	
Add(x, y)	Adds x and y
Sub(x, y)	Subtracts y from x
Mul(x, y)	Multiplies x by y
Div(x, y)	Divides x by y, if y is nonzero; otherwise returns 0
Abs(x)	Returns the absolute value of x
Neg(x)	Returns the negative value of x
Sin(x)	Returns the function $\sin(x)$
Cos(x)	Returns the function $\cos(x)$
ArcSin(x)	Returns the function $\arcsin(x)$
ArcCos(x)	Returns the function $\arccos(x)$
IfGreater(x, y, exp_1, exp_2)	If x is greater than y returns the expression exp_1, otherwise returns the expression exp_2
IfPositive(x, exp_1, exp_2)	If x is positive, returns the expression exp_1, otherwise returns the expression exp_2
Fire command	
Fire(x)	If x is positive, executes a fire command with x being the firepower, and returns 1; otherwise, does nothing and returns 0

instructs the tank to turn to an azimuth specified by the second evolved argument. The third line instructs the gun (and radar) to turn to an azimuth specified by the third evolved argument (Figure 1).

Functions and Terminals. Since terminals are actually zero-argument functions, we found the difference between functions and terminals to be of little

```
┌──────────┤ Robocode Player's Code Layout ├──────────┐
│ while (true)                                          │
│     TurnGunRight(INFINITY); //main code loop          │
│ ...                                                   │
│ OnScannedRobot() {                                    │
│     MoveTank(<GP#1>);                                 │
│     TurnTankRight(<GP#2>);                            │
│     TurnGunRight(<GP#3>);                             │
│ }                                                     │
└───────────────────────────────────────────────────────┘
```

Fig. 1. Robocode player's code layout

importance. Instead, we divided the terminals and functions into four groups according to their functionality:

1. *Game-status indicators:* A set of terminals that provide real-time information on the game status, such as last enemy azimuth, current tank position, and energy levels.
2. *Numerical constants:* Two terminals, one providing the constant 0, the other being an ERC (Ephemeral Random Constant), as described by Koza [2]. This latter terminal is initialized to a random real numerical value in the range [-1, 1], and does not change during evolution.
3. *Arithmetic and logical functions:* A set of zero- to four-argument functions, as specified in Table 1.
4. *Fire command:* This special function is used to preserve one line of code by not implementing the fire actuator in a dedicated line. The exact functionality of this function is described in Table 1.

Fitness Measure. The fitness measure should reflect the individual's quality according to the problem at hand. When choosing a fitness measure for our Robocode players, we had two main considerations in mind: the opponents and the calculation of the fitness value itself.

Selection of Opponents and Number of Battle Rounds: A good Robocode player should be able to beat many different adversaries. Since the players in the online league differ in behavior, it is generally unwise to assign a fitness value according to a single-adversary match. On the other hand, it is unrealistic to do battle with the entire player set—not only is this a time-consuming task, but new adversaries enter the tournaments regularly. We tested several opponent set sizes, including from one to five adversaries. Some of the tested evolutionary configurations involved random selection of adversaries per individual or per generation, while other configurations consisted of a fixed group of adversaries. The configuration we ultimately chose to use involved a set of three adversaries—fixed throughout the evolutionary run—with unique behavioral patterns, which we downloaded from the top of the HaikuBot league. Since the game is nondeterministic, a total of three rounds were played versus each adversary to reduce the randomness factor of the results.

Calculation of the Fitness Value: Since fitness is crucial in determining the trajectory of the evolutionary process, it is essential to find a way to translate battle results into an appropriate fitness value. Our goal was to excel in the online tournaments; hence, we adopted the scoring algorithms used in these leagues. The basic scoring measure is the fractional score F, which is computed using the score gained by the player S_P and the score gained by its adversary S_A:

$$F = \frac{S_P}{S_P + S_A}$$

This method reflects the player's skills in relation to its opponent. It encourages the player not only to maximize its own score, but to do so at the expense of its adversary's. We observed that in early stages of evolution, most players attained a fitness of zero, because they could not gain a single point in the course of the battle. To boost population variance at early stages, we then devised a modified fitness measure $\widetilde{F}$:

$$\widetilde{F} = \frac{\epsilon + S_P}{\epsilon + S_P + S_A},$$

where ϵ is a fixed small real constant.

This measure is similar to the fractional-score measure, with one exception: when two evolved players obtain no points at all (most common during the first few generations), a higher fitness value will be assigned to the one which avoided its adversary best (i.e., lower S_A). This proved sufficient in enhancing population diversity during the initial phase of evolution.

When facing multiple adversaries, we simply used the *average* modified fractional score, over the battles against each adversary.

Evolutionary Parameters:

- *Population size:* 256 individuals. Though some GP researchers, such as Koza, use much larger populations (up 10,000,000 individuals [1]), we had limited computational resources. Through experimentation we arrived at 256.
- *Termination criterion and generation count:* We did not set a limit for the generation count in our evolutionary runs. Instead, we simply stopped the run manually when the fitness value stopped improving for several generations.
- *Creation of initial population:* We used Koza's ramped-half-and-half method [2], in which a number d, between *mindepth* (set to 4) and *maxdepth* (set to 6) is chosen randomly for each individual. The genome trees of half of the individuals are then grown using the Grow method, which generates trees of any depth between 1 and d, and the other half is grown using the Full method, which generates trees of depth d exactly. All trees are generated randomly, by selection of appropriate functions and terminals in accordance with the growth method.

- *Breeding operators:* Creating a new generation from the current one involves the application of "genetic" operators (namely, crossover and mutation) on the individuals of the extant generation. We used two such operators:
  - *Mutation (unary):* randomly selects one tree node (with probability 0.9) or leaf (with probability 0.1), deletes the subtree rooted at that node and grows a new subtree instead, using the Grow method. Bloat control is achieved by setting a maxdepth parameter (set to 10), and invoking the growth method with this limit.
  - *Crossover (binary):* randomly selects a node (with probability 0.9) or a leaf (with probability 0.1) from each tree, and switches the subtrees rooted at these nodes. Bloat control is achieved using Langdon's method [7], which ensures that the resulting trees do not exceed the maxdepth parameter (set to 10).

  The breeding process starts with a random selection of genetic operator: a probability of 0.95 of selecting the crossover operator, and 0.05 of selecting the mutation operator. Then, a selection of individuals is performed (as described in the next paragraph): one individual for mutation, or two for crossover. The resulting individuals are then passed on to the next generation.

- *Selection method:* We used tournament selection, in which a group of individuals of size k (set to 5) is randomly chosen. The individual with the highest fitness value is then selected.

  In addition, we added elitism to the breeding mechanism: The two highest-fitness individuals were passed to the next generation with no modifications.

- *Extraction of best individual:* When an evolutionary run ends, we should determine which of the evolved individuals can be considered the best. Since the game is highly nondeterministic, the fitness measure does not explicitly reflect the quality of the individual: a "lucky" individual might attain a higher fitness value than better overall individuals. In order to obtain a more accurate measure for the players evolved in the last generation, we let each of them do battle for 100 rounds against 12 different adversaries (one at a time). The results were used to extract the optimal player—to be submitted to the league.

On Execution Time and the Environment. Genetic programming is known to be time consuming, mainly due to fitness calculation. We can estimate the time required for one run using this simple equation:

$$ExceutionTime = RoundTime \times NumRounds \times$$
$$NumAdversaries \times PopulationSize \times NumGenerations$$

A typical run involved 256 individuals, each battle carried out for 3 rounds against 3 different adversaries. A single round lasted about one second, and our best evolutionary run took approximately 400 generations, so the resulting total run time was:

$$ExecutionTime = 1 \times 3 \times 3 \times 256 \times 400 \approx 9.2 \times 10^5 \; seconds = 256 \; hours,$$

or about 10 days. In order to overcome the computational obstacle, we distributed the fitness calculation process over up to 20 computers. Needless to say, with additional computational resources run time can be yet further improved upon.

We used Luke's *ECJ11* system, a Java-based evolutionary computation and genetic programming research system (cs.gmu.edu/~eclab/projects/ecj/).

5 Results

We performed multiple evolutionary runs against three leading opponents, as described in Section 4. The progression of the best run is shown in Figure 2.

Due to the nondeterministic nature of the Robocode game, and the relatively small number of rounds played by each individual, the average fitness is worthy of attention, in addition to the best fitness. The first observation to be made is that the average fractional score converged to a value equaling 0.5, meaning that the average Robocode player was able to hold its own against its adversaries. When examining the average fitness, one should consider the variance: A player might defeat one opponent with relatively high score, while losing to the two others.

Though an average fitness of 0.5 might not seem impressive, two comments are in order:

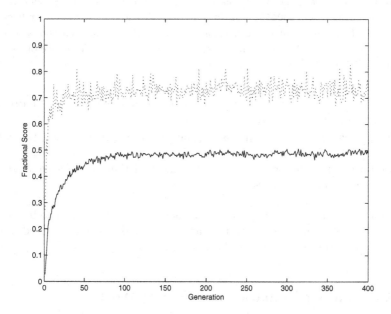

Fig. 2. Modified fractional score (Section 4) averaged over three different adversaries, versus time (generations). Top (dotted) curve: best individual, bottom (solid) curve: population average

Rank	Rating	Robot	Total Score	Survival	Last surv.	Bullet dmg.	Bonus	Ram dmg.	Bonus	1sts	2nds	3rds
1	588.87	ms.AresHaiku 0.4	36006	15400	3080	14346	2801	346	28	308	12	0
2	209.60	kawigi.haiku.HaikuTrogdor 1.1	36355	11950	2390	18692	3047	156	112	241	81	0
3	188.50	geep.haiku.GPBotC 1.0	36477	10750	2150	17901	2966	2275	424	223	99	0
4	179.57	pez.femto.HaikuPoet 0.2	27379	10250	2050	12987	1997	89	0	211	111	0
5	177.87	kawigi.femto.FemtoTrogdor 1.0	37213	11250	2250	17913	2785	2663	340	226	95	0
6	166.37	mz.HaikuGod 1.01	35602	12600	2520	15169	2259	2532	513	293	68	0
7	120.97	kawigi.haiku.HaikuCircleBot 1.0	32786	10700	2140	16942	2836	160	0	217	106	0
8	117.65	pez.femto.WallsPoetHaiku 0.1	29520	8950	1790	14476	2224	1867	202	181	141	0
9	111.24	cx.haiku.Escape 1.0	32086	12300	2460	14353	2509	408	49	248	74	0
10	91.26	cx.haiku.Xaxa 1.1	32396	10800	2160	16555	2614	209	51	219	104	0
11	76.56	ms.ChaosHaiku 0.1	32865	8900	1780	17012	2573	2223	367	186	138	0
12	54.34	kawigi.haiku.HaikuLinearAimer 1.0	18911	4850	970	11765	1239	79	0	99	221	0
13	35.87	soup.haiku.RammerHK 1.0	31684	4300	860	16968	1530	7262	751	92	230	0
14	21.70	cx.haiku.MeleeXaxa 1.0	32843	8950	1790	18100	1958	1329	708	183	141	0
15	-1.07	soup.haiku.MirrorHK 1.0	30132	7150	1430	18959	2312	273	0	143	177	0
16	-5.88	kawigi.haiku.HaikuSillyBot 1.2	18110	6150	1230	9818	854	50	0	140	187	0
17	-49.65	shf.HaikuAndrew .1	23614	7550	1510	12779	1609	158	0	154	168	0
18	-65.13	kawigi.haiku.HaikuChicken 1.0	20870	5600	1120	12622	1446	77	0	116	204	0
19	-77.20	soup.haiku.CutoffHK 1.0	34705	6650	1330	19957	2277	3818	662	137	183	0
20	-109.83	davidalves.net.PhoenixHaiku 1.0	23609	6300	1260	14341	1669	33	0	126	194	0
21	-145.71	cx.haiku.Smoku 1.1	24434	7450	1490	13535	1587	224	139	150	170	0
22	-174.44	dummy.haiku.Disoriented 1.0	18624	6700	1340	9696	785	98	0	143	181	0
23	-210.68	klo.haikuBounC 1.0	22201	5400	1080	13907	1468	283	58	112	209	0
24	-223.57	soup.haiku.RandomHK 1.0	17140	4000	800	11126	1129	80	0	86	235	0
25	-273.43	tango.haiku.HaikuTango 1.0	15977	4300	860	9508	907	365	28	87	233	0
26	-347.32	soup.haiku.DodgeHK 1.0	13235	2350	470	9579	518	312	0	47	273	0
27	-408.43	soup.haiku.WallDroidHK 1.0	8135	2950	590	3991	135	391	71	65	255	0

Fig. 3. Best GPBot takes third place at HaikuBot league on October 9, 2004 (`robocode.yajags.com/20041009/haiku-1v1.html`). The table's columns reflect various aspects of robotic behavior, such as *survival* and *bullet damage* measures. The final rank is determined by the *rating* measure, which reflects the performance of the robot in combat with randomly chosen adversaries

- This value reflects the average fitness of the population; some individuals attain much higher fitness.
- The adversaries used for fitness evaluation were excellent ones, taken from the top of the HaikuBot league. In the "real world," our evolved players faced a greater number of adversaries, most of them inferior to those used in the evolutionary process.

To join the HaikuBot challenge, we extracted what we deemed to be the best individual of all runs. Its first attempt at the HaikuBot league resulted in third place out of 27 contestants (Figure 3).

6   Concluding Remarks and Future Work

As noted in Section 1, Koza *et al.* [1] delineated eight criteria for an automatically created result to be considered human-competitive, the one of interest to us herein being:

H. The result holds its own or wins a regulated competition involving human contestants (in the form of either live human players or human-written computer programs).

Currently, all players in the HaikuBot league but GPBot are human-written computer programs. We believe that our attaining third place fulfills the eighth criterion: GPBots are human competitive.

In addition, the complexity of the problem should be taken under consideration: The game of Robocode, being nondeterministic, continuous, and highly diverse (due to the unique nature of each contestant), induces a virtually infinite search space, making it an extremely complex (and thus interesting) challenge for the GPer.

Generalization. When performing an evolutionary run against a single adversary, winning strategies were always evolved. However, these strategies were specialized for the given adversary: When playing against other opponents (even relatively inferior ones), the evolved players were usually beaten. Trying to avoid this obstacle, our evolutionary runs included multiple adversaries, resulting in better generalization, as evidenced by our league results (where our players encountered previously unseen opponents). Nevertheless, there is still room for improvement where generalization is concerned. A simple (yet highly effective, in our experience) enhancement booster would be the increase of computational resources, allowing more adversaries to enter into the fitness function.

Coevolution. One of the evolutionary methods that was evaluated and abandoned is *coevolution*. In this method, the individuals in the population are evaluated against each other, and not by referring to an external opponent. Coevolution has a prima facie better chance of attaining superior generalization, due to the diversity of opponents encountered during evolution. However, we found that the evolved players presented primitive behavior, and were easily defeated by human-written programs. Eisenstein [3] described the same phenomenon, and has suggested that the problem lies with the initial generation: The best strategies that appear early on in evolution involve idleness—i.e., no moving nor firing—since these two actions are more likely to cause loss of energy. Breeding such players usually results in losing the genes responsible for movement and firing, hence the poor performance of the latter generations. We believe that coevolution can be fruitful if carefully planned, using a two-phase evolutionary process. During the first stage, the initial population will be evolved using one or more human-written adversaries as fitness measure; this phase will last a relatively short period of time, until basic behavioral patterns emerge. The second stage will involve coevolution over the population of individuals that was evolved in the first stage. This two-phase approach we leave for future work.

Exploring Other Robocode Divisions. There are a number of other divisions apart from HaikuBot in which GP-evolved players might compete in the future. Among these is the *MegaBot challenge*, in which no code-size restrictions hold. Some players in this category employ a unique strategy for each adversary, using a predefined database. Since GP-evolved players are good at specializing, we might try to defeat some of the league's leading players, ultimately creating an overall top player by piecing together a collection of evolved strategies.

Other Robocode battle divisions are yet to be explored: melee games—in which a player faces multiple adversaries simultaneously, and team games—in which a player is composed of several robots that act as a team.

Acknowledgements

We thank Jacob Eisenstein for helpful discussions and for the genuine interest he took in our project.

References

1. Koza, J.R., Keane, M.A., Streeter, M.J., Mydlowec, W., Yu, J., Lanza, G.: Genetic Programming IV: Routine Human-Competitive Machine Intelligence. Kluwer Academic Publishers, Norwell, MA (2003)
2. Koza, J.R.: Genetic Programming: On the Programming of Computers by Means of Natural Selection. The MIT Press, Cambridge, Massachusetts (1992)
3. Eisenstein, J.: Evolving robocode tank fighters. Technical Report AIM-2003-023, AI Lab, Massachusetts Institute Of Technology (2003) citeseer.ist.psu.edu/647963.html.
4. Koza, J.R.: Evolution of subsumption using genetic programming. In Varela, F.J., Bourgine, P., eds.: Proceedings of the First European Conference on Artificial Life. Towards a Practice of Autonomous Systems, Paris, France, MIT Press (1992) 110–119
5. Montana, D.J.: Strongly typed genetic programming. Evolutionary Computation **3** (1995) 199–230
6. Koza, J.R.: Genetic Programming II: Automatic Discovery of Reusable Programs. MIT Press, Cambridge Massachusetts (1994)
7. Langdon, W.B.: Size fair and homologous tree genetic programming crossovers. Genetic Programming and Evolvable Machines **1** (2000) 95–119

Incorporating Learning Probabilistic Context-Sensitive Grammar in Genetic Programming for Efficient Evolution and Adaptation of Snakebot

Ivan Tanev

Department of Information Systems Design, Doshisha University,
1-3 Miyakodani, Tatara, Kyotanabe, Kyoto 610-0321, Japan
ATR Network Informatics Laboratories,
2-2-2 Hikaridai, "Keihanna Science City", Kyoto 619-0288, Japan
itanev@mail.doshisha.ac.jp
http://isd-si.doshisha.ac.jp/itanev/

Abstract. In this work we propose an approach of incorporating learning probabilistic context-sensitive grammar (LPCSG) in genetic programming (GP) employed for evolution and adaptation of locomotion gaits of simulated snake-like robot (Snakebot). In our approach LPCSG is derived from the originally defined context-free grammar, which usually expresses the syntax of genetic programs in canonical GP. During the especially introduced steered mutation the probabilities of applying each of particular production rules with multiple right-hand side alternatives in LPCSG depend on the context, and these probabilities are learned from the aggregated reward values obtained from the evolved best-of-generation Snakebots. Empirically obtained results verify that employing LPCSG contributes to the improvement of computational effort of both (i) the evolution of the fastest possible locomotion gaits for various fitness conditions and (ii) adaptation of these locomotion gaits to challenging environment and degraded mechanical abilities of Snakebot. In all of the cases considered in this study, the locomotion gaits, evolved and adapted employing GP with LPCSG feature higher velocity and are obtained faster than with canonical GP.

Keywords: Snakebot, adaptation, locomotion, genetic programming, grammar.

1 Introduction

Wheelless, limbless snake-like robots (Snakebots) feature potential robustness characteristics beyond the capabilities of most wheeled and legged vehicles – ability to traverse terrain that would pose problems for traditional wheeled or legged robots, and insignificant performance degradation when partial damage is inflicted. Some useful features of Snakebots include smaller size of the cross-sectional areas, stability, ability to operate in difficult terrain, good traction, high redundancy, and complete sealing of the internal mechanisms [3, 4]. Robots with these properties open up several critical applications in exploration, reconnaissance, medicine and inspection. However, compared to the wheeled and legged vehicles, Snakebots feature (i) more difficult control of

M. Keijzer et al. (Eds.): EuroGP 2005, LNCS 3447, pp. 155–166, 2005.

locomotion gaits and (ii) inferior speed characteristics. In this work we intend to address the following challenge: how to develop control sequences of Snakebot's actuators, which allow for achieving the fastest possible speed of locomotion.

For many tasks and robot morphologies, it might be seen as a natural approach to handcraft the locomotion control code by applying various theoretical approaches [12, 14]. However, handcrafting might not be feasible for developing the control code of real Snakebot due to its morphological complexity and the need of prompt adaptation under degraded mechanical abilities and/or unanticipated environmental conditions. The automated mechanisms for prompt generation of near-optimal solutions to complex, ill-posed problems are usually based on various models of learning or evolution of species in the Nature [5, 7, 13]. The proposed approach of employing genetic programming (GP) implies that the code, which controls the locomotion of Snakebot is automatically designed by a computer system via simulated evolution through selection and survival of the fittest in a way similar to the natural evolution of species [6].

The *objectives* of our work are (i) to explore the feasibility of applying GP for automatic design of the fastest possible locomotion of realistically simulated Snakebot and (ii) to investigate the adaptation of such locomotion to challenging environment and degraded abilities (due to partial damage) of simulated Snakebot. We are especially interested in the implications of incorporating a learning probabilistic context-sensitive grammar (LPCSG) in GP on the efficiency of evolution and adaptation of Snakebot. Presented approach of employing LPCSG is related to the approach of grammatical evolution [8] and to the use of probability distribution algorithms in evolutionary computations, mostly – in genetic algorithms [9, 10]. In neither of these methods however the incorporation of LPCSG in GP has been explored and our interest in the feasibility of such approach additionally motivated us in this work.

The remainder of this document is organized as follows. Section 2 emphasizes the main features of GP proposed for evolution of locomotion of Snakebot. Section 3 introduces the proposed approach of incorporating LPCSG in GP and discusses the empirically obtained results of efficiency of evolution and adaptation of Snakebot to challenging environment and partial damage. Section 4 draws a conclusion.

2   GP for Automatic Design of Locomotion Gaits of Snakebot

2.1   Representation of Snakebot

Snakebot is simulated as a set of identical spherical morphological segments ("vertebrae"), linked together via universal joints. All joints feature identical (finite) angle limits and each joint has two attached actuators ("muscles"). In the initial, standstill position of Snakebot the rotation axes of the actuators are oriented vertically (vertical actuator) and horizontally (horizontal actuator) and perform rotation of the joint in the horizontal and vertical planes respectively (Figure 1). Considering the representation of Snakebot, the task of designing the fastest locomotion can be rephrased as developing temporal patterns of desired turning angles of horizontal and vertical actuators of each segment, that result in fastest overall locomotion of Snakebot. The proposed representation of Snakebot as a homogeneous system comprising identical morphological segments is intended to significantly reduce the size of the search space of the

GP. Moreover, because the size of the search space does not necessarily increase with the increase of the complexity of Snakebot (i.e. the number of morphological segment), the proposed approach allows achievement of favorable scalability characteristics of GP.

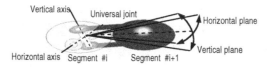

Fig. 1. Morphological segments of Snakebot linked via universal joint. Horizontal and vertical actuators attached to the joint perform rotation of the segment #i+1 in vertical and horizontal planes respectively

2.2 Algorithmic Paradigm

GP. GP [6] is a domain-independent problem-solving approach in which a population of computer programs (individuals' genotypes) is evolved to solve problems. The simulated evolution in GP is based on the Darwinian principle of reproduction and survival of the fittest. The fitness of each individual is based on the quality with which the phenotype of the simulated individual is performing in a given environment.

Function Set and Terminal Set. In applying GP to evolution of Snakebot, the genotype is associated with two algebraic expressions, which represent the temporal patterns of desired turning angles of both the horizontal and vertical actuators of each morphological segment. Because locomotion gaits, by definition, are periodical, we include the functions sin and cos in the GP function set in addition to the basic algebraic functions. Terminal symbols include the variables time, index of the segment of Snakebot, and two constants: Pi, and random constant within the range [0, 2]. The main parameters of the GP are summarized in Table 1.

Context-Free Grammar for Canonical GP. The context-free grammar (CFG) *G*, usually employed to define the allowed syntax of individuals in GP consists of (*N*, , *P*, *S*) where *N* is a finite set of nonterminal symbols,   is a finite set of terminal symbols that is disjoint from *N*, *S* is a symbol in *N* that is indicated as the start symbol, and *P* is a set of production rules, where a rule is of the form

```
V -> w
```

where *V* is a non-terminal symbol and *w* is a string consisting of terminals and/or non-terminals. The term "context-free" comes from the feature that the variable *V* can always be replaced by *w*, in no matter what context it occurs. The set of non-terminal symbols of *G* of GP, is employed to develop the temporal patterns of desired turning angles of horizontal and vertical actuators of segments, that result in fastest overall locomotion of Snakebot, is defined as follows:

```
N = {GP, STM, STM1, STM2, VAR, CONST_x10, CONST_PI, OP1, OP2}
```

where STM is a generic algebraic statement, STM1 – a generic unary (e.g., sin, cos)nop) algebraic statement, STM2 – a generic binary (dyadic, e.g. +, -, *, and /) algebraic statement, VAR – a variable, OP1 – an unary operation, OP2 – a binary (dyadic) operation, CONST_x10 is a random constant within the range [0..20], and CONST_PI equals either 3.1416 or 1.5708. The set of terminal symbols is defined as:

Σ = {sin, cos, nop, +, -, *, /, time, segment_id}

where sin, cos, nop, +, -, * and / are terminals which specify the functions in the generic algebraic statements. The start symbol is GP, and the set of production rules expressed in Backus-Naur form (BNF) is as shown in Figure 2.

Table 1. Main parameters of GP

Category	Value
Function set	{sin, cos, nop, +, -, *, /}
Terminal set	{time, segment_ID, Pi, random constant}
Population size	200 individuals
Selection	Binary tournament, selection ratio 0.1, reproduction ratio 0.9
Elitism	Best 4 individuals
Mutation	Random subtree mutation, ratio 0.01
Fitness	Velocity of simulated Snakebot during the trial
Trial interval	180 time steps, each time step account for 50ms of "real" time
Termination criterion	(Fitness >100) *or* (Generations>40) *or* (no improvement of fitness for 16 generations)

```
(1)          GP  ⟶  STM
(2.1-2.5)    STM  ⟶  STM1|STM2|VAR|CONST_x10|CONST_PI
(3)          STM1  ⟶  OP1 STM
(4.1-4.6)    OP1  ⟶  sin|cos|nop|-|sqr|sqrt
(5)          STM2  ⟶  OP2 STM STM
(6.1-6.4)    OP2  ⟶  +|-|*|/
(7.1-7.2)    VAR  ⟶  time|segment_id
(8)          CONST_x10  ⟶  0..20
(9.1-9.2)    CONST_PI  ⟶  3.1416|1.5708
```

Fig. 2. BNF of production rules of the context free grammar G of GP, employed for automatic design of locomotion gaits of Snakebot. The following abbreviations are used: STM – generic algebraic statement, STM1 – unary algebraic statement, STM2 – binary (dyadic) algebraic statement, VAR – variable, OP1 – unary operation, and OP2 – binary operation

GP uses the defined production rules of *G* to create the initial population and to mutate genetic programs. Production rules with multiple alternative right-hand sides (such as rules 2, 4, 6, 7 and 9) are usually chosen randomly in these operations.

Fitness Evaluation. The fitness function is based on the velocity of Snakebot, estimated from the distance, which the center of the mass of Snakebot travels during the trial. Fitness of 100 (the one of termination criteria shown in Table 1) is equivalent to a velocity, which displaced Snakebot a distance equal to twice its length.

Representation of Genotype. Inspired by its flexibility, and the recently emerged widespread adoption of document object model (DOM) and extensible markup language (XML), we represent evolved genotypes of simulated Snakebot as DOM-parse trees featuring equivalent flat XML-text. Both (i) the calculation of the desired turning angles during fitness evaluation and (ii) the genetic operations are performed on DOM-parse trees using API of off-the shelf DOM-parser.

Genetic Operations. Selection is a binary tournament. Crossover is defined in a strongly typed way in that only the DOM-nodes (and corresponding DOM-subtrees) of the same data type (i.e. labeled with the same tag) from parents can be swapped. The sub-tree mutation is allowed in strongly typed way in that a random node in genetic program is replaced by syntactically correct sub-tree. The mutation routine refers to the data type of currently altered node and applies randomly chosen rule from the set of applicable rewriting rules as defined in the grammar of GP.

ODE. We have chosen Open Dynamics Engine (ODE) [12] to provide a realistic simulation of physics in applying forces to phenotypic segments of Snakebot. ODE is a free, industrial quality software library for simulating articulated rigid body dynamics. It is fast, flexible and robust, and it has built-in collision detection.

3 Incorporating Learning Context-Sensitive Grammar in GP

3.1 Learning Probabilistic Context-Sensitive Grammar of Strongly-Typed GP

In the proposed approach, the learning probabilistic context-sensitive grammar (LPCSG) $G*$ is introduced as a set of the same attributes $(N*, \sum*, P*, S*)$ as CSG G defined in Section 2.2. The attributes $N*$, $\sum*$, and $S*$ are identical to the corresponding attributes N, $\sum$, and S of G. The set of production rules $P*$ of $G*$ are derived from P of G as follows: (i) Production rules of P_S $(P_S \subset P)$ of G which have a single right-hand side are defined in the same way in $P*$ as in P, while (ii) the production rules in P_M $(P_M \subset P)$ of G, which feature multiple right-hand side alternatives $V \rightarrow w_1 | w_2 | \ldots | w_N$ are re-defined for each instance i of the context as follows:

$$context_i \; V \rightarrow context_i \; w_1 \; (p^i_{1})$$
$$context_i \; V \rightarrow context_i \; w_2 \; (p^i_{2})$$
$$\ldots$$
$$context_i \; V \rightarrow context_i \; w_N \; (p^i_{N})$$

where p^i_1, p^i_2, $\ldots p^i_N$ ($\sum p^i_n = 1$, $n=1,2\ldots N$.) are the probabilities of applying each alternative rule with the left-hand side non-terminal V for the given $context_i$.

The proposed approach is based on the idea of introducing bias in applying the most preferable rule from rules with multiple, alternative right-hand sides. We assume that the preferences of applying certain production rules depend on the surrounding grammatical context, defining which rules have been applied before. The initial probability distributions (PD) p^i_1, p^i_2, $\ldots p^i_N$ for each $context_i$ is even and then learned (adjusted) incrementally at each generation from the best performing Snakebots. The learned PD is then used as a bias to steer the mutation of Snakebots. A sample of biased application of production rules of G^* according to the learned PD for the considered context is shown in Figure 3.

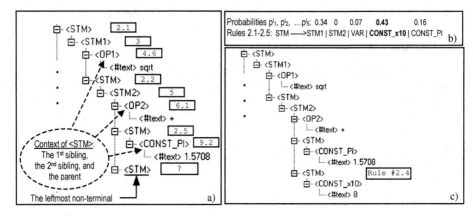

Fig. 3. Sample of biased application of production rules of G^*: the current leftmost non-terminal, as shown in (a) is STM, which requires applying one of the production rules 2.1-2.5 (refer to Figure 2). For the considered context (a), the "learned" preferences (b) of applying rules 2.1-2.5 indicate highest probability of production rule 2.4. Consequently, the rule 2.4 will be most likely applied producing the genetic program as shown in (c)

3.2   Algorithm of GP Incorporating LPCSG

The principal steps of algorithm of GP incorporating LPCSG are shown in Figure 4. As figure illustrates, additional Steps 6 and 9 are introduced in the canonical algorithm of GP. The PD is updated on Step 6, and the new offsprings, created applying the proposed "steered" mutation via PD on Step 9 are inserted into already reproduced via canonical crossover (Step 7) and mutation (Step 8), growing new population of Snakebots. The ratio K_{PD} of number of offsprings #N_{PD} created via "steered" mutation using PD and the number of offsprings #N_{co} created via canonical crossover is kept within the range [0, 5]. K_{PD} is dynamically tuned on Step 6 based on the stagnation counter C_S, which maintains the number of most recent generations without improvement of the fitness. K_{PD} is defined according to the following rule:

K_{PD} = 5 - smaller_of(5,C_S)

Lower values of K_{PD} in stagnated population (i.e., for $C_S>0$) favor the reproduction via canonical random genetic operations over the reproduction using steered mutation via PD. As we empirically investigated, low K_{PD} facilitates avoiding premature convergence of population by increasing the diversity of population and concequently, accelerating the escape from the (most likely) local optimal solutions, discovered by the steering bias of the current PD. Conversely, replacing the usually random genetic operations of canonical GP with the proposed steered mutation when K_{PD} is close to its maximum value (i.e., for $C_S=0$) can be viewed as a mechanism for growing and preserving the proven to be beneficial building blocks in evolved solutions rather than destroying them by usually random crossover and mutation.

Obtaining and applying PD during the steered mutation (Figure 4, Step 9) implies maintaining a PD table. Each entry in the table stores the context, the left-hand side non-terminal, the list of right-hand side symbols, the aggregated reward values and the calculated probability of applying the given production rule for the given context. A new entry is added or the aggregated reward value of existing entry is updated by extracting the syntactic features of the genetic programs comprising the mating pool of current generation.

```
Step 0: Creating Initial Population and Clearing PDD;
Step 1: While (true) do begin
Step 2:    Evaluating Population;
Step 3:    Ranking Population;
Step 4:    if TerminationCriteria then Go to Step 9
Step 5:    Selecting the Mating Pool;
Step 6:    Updating PDD and K_{PD};
Step 7:    Creating #N_{co} offsprings via canonical crossover;
Step 8:    Mutating current population via canonical mutation;
Step 9:    Creating #N_{PD} offsprings via mutation of mating pool using PD;
Step10: end;
```

Fig. 4. Algorithm of GP incorporating LPCSG. Steps 6 and 9 are specific for the proposed approach. Steps 0, 2-5, 7 and 8 are common principal steps of canonical GP

3.3   Empirical Results

This section discusses empirically obtained results verifying the effects of incorporating LPCSG on the efficiency of GP applied for the following two tasks: (i) *evolution* of the fastest possible locomotion gaits of Snakebot for various fitness conditions and (ii) *adaptation* of these locomotion gaits to challenging environment and degraded mechanical abilities of Snakebot. These tasks, considered as relevant for successful accomplishment of anticipated exploration, reconnaissance, medicine or inspection missions, feature different fitness landscapes. Therefore, the experiments discussed in this section are intended to verify the versatility and the scope of applicability of the proposed approach.

In all of the cases considered, the fitness of Snakebot reflects the low-level objective (i.e. *what* is required to be achieved) of Snakebot in these missions, namely, to be able to move fast regardless of environmental challenges or degraded abilities. The experiments discussed illustrate the ability of the evolving Snakebot to learn *how* (e.g. by discovering beneficial locomotion traits) to accomplish the required objective without being explicitly taught about the means to do so. Such *know-how* acquired by Snakebot automatically and autonomously can be viewed as a demonstration of emergent intelligence [1], in that the task-specific knowledge of *how* to accomplish the task emerges in the Snakebot from the interaction of the problem solver and the fitness function.

Evolution of Fastest Locomotion Gaits. Figure 6 shows the results of evolution of locomotion gaits for cases where fitness is measured as velocity in any direction. Despite the fact that fitness is unconstrained and measured as velocity in *any* direction, *sidewinding* locomotion (defined as locomotion predominantly perpendicular to the long axis of Snakebot) emerged in all 10 independent runs of GP, suggesting that it provides superior speed characteristics for considered morphology of Snakebot. As Figure 6c illustrates, incorporating LPCSG in GP is associated with computational effort (required to achieve probability of success 0.9) of about 20 generations, which is about 1.6 times faster that canonical GP with CFG. Sample snapshots of evolved best-of-run sidewinding locomotion gaits are shown in Figures 6d and 6e.

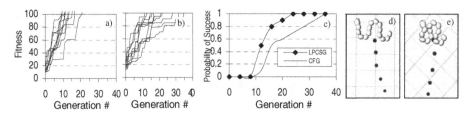

Fig. 6. Evolution of locomotion gaits for cases where fitness is measured as velocity in any direction: fitness convergence characteristics of 10 independent runs of GP with LPCSG (a), canonical GP with CFG (b), probability of success (c), and snapshots of sample evolved via GP with LPCSG best-of-run sidewinding Snakebots (d and e). The dark trailing circles in (d) and (e) depict the trajectory of the center of the mass of Snakebot

In order to verify the superiority of velocity characteristics of sidewinding we compared the fitness convergence characteristics of evolution in unconstrained environment for the following two cases: (i) unconstrained fitness measured as velocity in any direction (as discussed above and illustrated in Figure 6), and (ii) fitness, measured as velocity in forward direction only. The results of evolution of forward (rectilinear) locomotion, shown in Figure 7 indicate that non-sidewinding motion, compared to sidewinding, features much inferior velocity characteristics. The results also demonstrate that GP with LPCSG in average converges almost 4 times faster and to higher values than canonical GP. Snapshots taken during the motion of a sample evolved best-of-run sidewinding Snakebot are shown in Figures 7c and 7d.

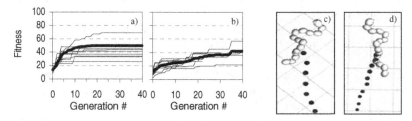

Fig. 7. Evolution of locomotion gaits for cases where fitness is measured as velocity in forward direction only. Fitness convergence characteristics of 10 independent runs of STGP with LPCSG (a), canonical STGP with CFG (b), and snapshots of sample evolved via STGP with LPCSG best-of-run forward locomotion (c and d)

The results of evolution of rectilinear locomotion of simulated Snakebot confined in narrow "tunnel" are shown in Figure 8. As the fitness convergence characteristics of 10 independent runs (Figure 8a and Figure 8b) illustrate, GP with LPCSG is almost twice faster than canonical GP. Compared to forward locomotion in unconstrained environment (Figure 7), the velocity in this experiment is superior, and even comparable to the velocity of sidewinding (Figure 6). This, seemingly anomalous phenomenon demonstrates an emergent intelligence – i.e. the ability of evolution to discover a way to utilize the walls of "tunnel" as (i) a source of extra grip and as (ii) an additional mechanical support for fast yet unbalanced locomotion gaits (e.g., vertical undulation) in an eventual unconstrained environment.

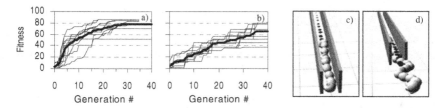

Fig. 8. Evolution of locomotion gaits of Snakebot is confined in narrow "tunnel": fitness convergence characteristics of 10 independent runs of STGP with LPCSG (a), canonical STGP with CFG (b), and snapshots of sample evolved best-of-run gaits at the intermediate (c) and final stages of the trial (d)

Adaptation of Sidewinding to Challenging Environment and Degraded Mechanical Abilities of Snakebot. Adaptation in Nature is viewed as an ability of species to discover the best phenotypic (i.e. pertaining to biochemistry, morphology, physiology, and behavior) traits for survival in continuously changing fitness landscape. The adaptive phenotypic traits are result of beneficial genetic changes occurred during the course of evolution (phylogenesis) and/or phenotypic plasticity (ontogenesis – learning, polymorphism, polyphenism, immune response, adaptive metabolism, etc.) occurring during the lifetime of the individuals. In our approach we employ GP with LPCSG for adaptation of Snakebot to changes in the fitness landscape caused by (i) challenging environment and (ii) partial damage to 1, 2, 4 and 8 (out of 15) morphological segments. In all of the cases of adaptation, GP is initialized with a population comprising 20 best-of-run genetic programs, obtained from 10 independent runs of evolution of Snakebot in unconstrained environment, plus additional 180 randomly created individuals.

The challenging environment is modeled by the introduction of immobile obstacles comprising 40 small, randomly scattered boxes, a wall with height equal to the 0.5 diameters of the cross-section of Snakebot, and a flight of 3 stairs, each with height equal to the 0.33 diameters of the cross-section of Snakebot. The results of adaptation of Snakebot, shown in Figure 9 demonstrate that the computational effort (required to reach fitness values of 100 with probability of success 0.9) of GP with LPCSG is about 20 generations. Conversely, only half of all runs of GP with CFG achieve the targeted fitness value, implying that the corresponding probability of success converges to the value of 0.5. Snapshots illustrating the performance of Snakebot initially evolved in unconstrained environment, before and after the adaptation (via GP with LPCSG) to challenging environment are shown in Figure 10. The additional elevation of the body, required to faster negotiate the obstacles represents the emergent knowhow in the adapting Snakebot. As Figure 11 illustrates, the trajectory of the central segment around the center of the mass of sample adapted Snakebot (Figure 11b) is twice higher than before the adaptation (Figure 11a).

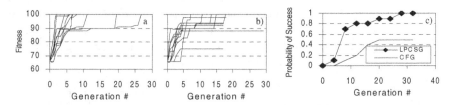

Fig. 9. Adaptation of sidewinding locomotion to challenging environment: fitness convergence characteristics of 10 independent runs of STGP with LPCSG (a), canonical STGP with CFG (b), and probability of success (c)

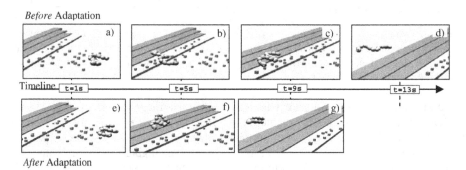

Fig. 10. Snapshots illustrating the sidewinding Snakebot, initially evolved in unconstrained environment, before the adaptation – initial (a), intermediate (b and c) and final stages of the trial (d), and after the adaptation to challenging environment via STGP with LPCSG - initial (e), intermediate (f) and final stages of the trial (g)

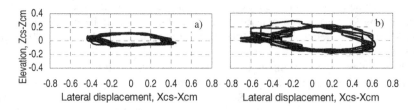

Fig. 11. Trajectory of the central segment (cs) around the center of mass (cm) of Snakebot for sample best-of-run sidewinding locomotion before (a) and after the adaptation (b) to challenging environment

The adaptation of sidewinding Snakebot to partial damage to 1, 2, 4 and 8 (out of 15) segments by gradually improving its velocity is shown in Figure 12. Demonstrated results are averaged over 10 independent runs for each case of partial damage to 1, 2, 4 and 8 segments. The damaged segments are evenly distributed along the body of Snakebot. Damage inflicted to a particular segment implies a complete loss of functionality of both horizontal and vertical actuators of the corresponding joint.

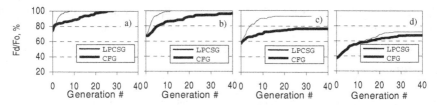

Fig. 12. Adaptation of Snakebot to damage of 1 (a), 2 (b), 4 (c) and 8 (d) segments

As Figure 12 depicts, Snakebot completely recovers from damage to single segment attaining its previous velocity in 25 generations with "canonical" GP employing CFG, and only in 7 generations with GP with LPCSG, resulting in a mean real-time of adaptation of a few hours of runtime on PC featuring Intel® 3GHz Pentium® 4 microprocessor and 2GB RAM under Microsoft Windows NT OS. Snakebots recovers to average of 94% (CFG) and 100% (LPCSG) of its previous velocity in the case where 2 segments are damaged. With 4 and 8 damaged segments the degree of recovery is 77% (CFG) and 92% (LPCSG), and 68% (CFG) and 72% (LPCSG) respectively. In all of the cases considered incorporating LPCSG contributes to faster adaptation of Snakebot, and in all cases the Snakebot recovers to higher values of velocity of locomotion.

4 Conclusion

In this work we propose an approach of incorporating LPCSG in GP employed for evolution and adaptation of locomotion gaits of simulated Snakebot. We introduced a "steered" mutation in which the probabilities of applying each of particular production rules with multiple right-hand side alternatives in LPCSG depend on the context, and these probabilities are "learned" from the aggregated reward values obtained from the evolved best-of-generation Snakebots. Empirically obtained results verify that employing LPCSG contributes to the improvement of computational effort of both (i) the evolution of the fastest possible locomotion gaits for various fitness conditions and (ii) adaptation of these locomotion gaits to challenging environment and degraded mechanical abilities of Snakebot.

Recent discoveries in molecular biology and genetics suggest that mutations do not happen randomly in the Nature. Instead, some fragments of DNA tend to repel the mutations away, while other fragments seem to attract it [2]. It is assumed that the former fragments are related to the very basics of life, and therefore, any mutation within them can be potentially fatal to the species. Biasing the mutation operation towards the proven to be beneficial genotypic structures in the proposed approach of steering of mutation via GP with LPCSG can be viewed as an attempt (i) to mimic the Natural mechanisms of genomic control over the mutation operations and (ii) to investigate the computational implication of these mechanisms on efficiency of simulated evolution and adaptation of engineering artifacts.

Acknowledgments

The author thanks Katsunori Shimohara, Thomas Ray and Andrzej Buller for their immense support of this work. The research was supported in part by the National Institute of Information and Communications Technology of Japan.

References

1. Angeline, P. J.: Genetic Programming and Emergent Intelligence. In Kinnear, K.E. Jr., editor, Advances in Genetic Programming, MIT Press (1994) 75-98
2. Caporale L. H., Darwin In the Genome: Molecular Strategies in Biological Evolution, 2002, McGraw-Hill/Contemporary Books
3. Dowling, K.: Limbless Locomotion: Learning to Crawl with a Snake Robot, doctoral dissertation, tech. report CMU-RI-TR-97-48, Robotics Institute, Carnegie Mellon University (1997)
4. Hirose, S.: Biologically Inspired Robots: Snake-like Locomotors and Manipulators, Oxford University Press (1993)
5. Kimura, H., Yamashita, T., and Kobayashi, S., Reinforcement Learning of Walking Behavior for a Four-Legged Robot, 40th IEEE Conference on Decision and Control (2001) 411-416
6. Koza, J. R., Genetic Programming: On the Programming of Computers by Means of Natural Selection, Cambridge, MA, MIT Press (1992)
7. Mahdavi, S., Bentley, P.J.: Evolving Motion of Robots with Muscles. In Proc. of Evo-ROB2003, the 2nd European Workshop on Evolutionary Robotics, EuroGP 2003 (2003) 655-664
8. O'Neill, M. and Ryan, C.: Grammatical Evolution: Evolutionary Automatic Programming in an Arbitrary Language, Series: Genetic Programming, Vol. 4, Springer (2003)
9. Pelikan M., Goldberg D. E., and Cantú-Paz, E.: BOA: The Bayesian optimization algorithm. Proceedings of the Genetic and Evolutionary Computation Conference (GECCO-99), I, (1999) 525-532
10. Salustowicz, R. and Schmidhuber, J.: Probabilistic incremental program evolution. Evolutionary Computation, Vol.5 No.2, (1997) 123-141
11. Smith, R.: Open Dynamics Engine (2001-2003) http://q12.org/ode/
12. Stoy, K., Shen W.-M., Will, P.M., A simple approach to the control of locomotion in self-reconfigurable robots, Robotics and Autonomous Systems, Vol.44 , No.3 (2003) 191-200
13. Takamura, S., Hornby, G. S., Yamamoto, T., Yokono, J. and Fujita, M., Evolution of Dynamic Gaits for a Robot, IEEE International Conference on Consumer Electronics (2000) 192-193
14. Zhang, Y., Yim, M. H., Eldershaw, C., Duff, D. G., and Roufas, K. D., Phase automata: a programming model of locomotion gaits for scalable chain-type modular robots, IEEE/RSJ International Conference on Intelligent Robots and Systems (IROS 2003), October 27 - 31; Las Vegas, NV (2003) 2442- 2447

Multi-logic-Unit Processor: A Combinational Logic Circuit Evaluation Engine for Genetic Parallel Programming

Lau Wai Shing[1], Li Gang[2], Lee Kin Hong[3], Leung Kwong Sak[4], and Cheang Sin Man[5]

[1-4] Department of Computer Science and Engineering,
The Chinese University of Hong Kong, Hong Kong
{wslau, gli, khlee, ksleung}@cse.cuhk.edu.hk
[5] Department of Computing,
The Hong Kong Institute of Vocational Education (Kwai Chung), Hong Kong
smcheang@vtc.edu.hk

Abstract. Genetic Parallel Programming (GPP) is a novel Genetic Programming paradigm. GPP Logic Circuit Synthesizer (GPPLCS) is a combinational logic circuit learning system based on GPP. The GPPLCS comprises a Multi-Logic-Unit Processor (MLP) which is a hardware processor built on a Field Programmable Gate Array (FPGA). The MLP is designed to speed up the evaluation of genetic parallel programs that represent combinational logic circuits. Four combinational logic circuit problems are presented to show the performance of the hardware-assisted GPPLCS. Experimental results show that the hardware MLP speeds up evolutions over 10 times. For difficult problems such as the 6-bit priority selector and the 6-bit comparator, the speedup ratio can be up to 22.

1 Introduction

Genetic Programming (GP) [1] is a robust method in Evolutionary Computation. There are two main streams in GP, standard GP [2] and linear-structured GP (linear GP) [3]. In standard GP, a genetic program is represented in a tree structure. In linear GP, a genetic program is represented in a linear list of machine code instructions or high-level language statements. A linear genetic program can be run on a target machine directly without performing any translation process. The Genetic Parallel Programming (GPP) [4] paradigm is developed on the basis of linear GP. In GPP, a genetic parallel program consists of a sequence of parallel-instructions. A parallel-instruction comprises multiple sub-instructions that can perform multiple operations simultaneously in an execution step. Cheang *et al.* [5] have demonstrated that GPP can be used to evolve combinational logic circuits with 2-input logic units.

In the last decade, advances in Field Programmable Gate Array (FPGA) [6] have made efficient Evolvable Hardware (EHW) [7] possible. EHW uses Evolutionary Algorithms to evolve hardware architecture extrinsically or intrinsically. One of the major usages of EHW is to design combinational logic circuits [8,9,10]. Moreover,

M. Keijzer et al. (Eds.): EuroGP 2005, LNCS 3447, pp. 167 – 177, 2005.

FPGA has been adopted to speed up Genetic Algorithms (GAs) and GP systems [11,12,13,14]. The basic idea is to put the whole or a part of a GA or GP system in hardware so as to solve problems in a shorter time than a pure software system.

In this paper, we propose a hardware-assisted combinational logic circuit learning system, GPP Logic Circuit Synthesizer (GPPLCS). The GPPLCS consists of a software Evolution Engine (EE) and an FPGA-based logic circuit evaluation engine, a Multi-Logic-Unit Processor (MLP) (see Fig. 1). Combinational logic circuits are represented in genetic parallel programs which can be run on the MLP.

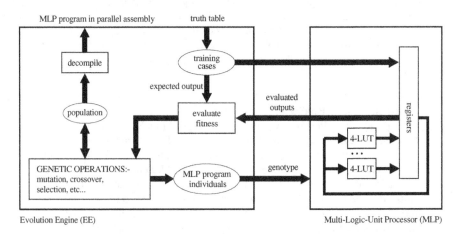

Fig. 1. The system block diagram of the GPPLCS

The main purpose of this paper is to show the performance of the GPPLCS with the hardware MLP. We performed experiments on four combinational logic circuit problems (i.e. a 6-bit multiplexer, a 2-bit full-adder, a 3-bit comparator and a 6-bit priority selector). Experimental results show that the hardware MLP speeds up the evolution by at least 10 times even for the easier problems which are less computationally intensive.

The rest of the paper is organized as follows: Section 2 contains descriptions of the GPPLCS; Section 3 presents the implementation details of the hardware MLP; Section 4 describes details of experiments and experimental settings; Section 5 presents results; and finally, Section 6 concludes our work.

2   Genetic Parallel Programming Logic Circuit Synthesizer

Genetic Parallel Programming (GPP) is a linear GP paradigm that evolves parallel programs based on a Multi-ALU Processor (MAP). GPP has been used to evolve compact parallel programs for different problems, such as numeric function regression [4] and data classification problems [15]. The GPP accelerating phenomenon [16] revealed that parallel programs can be evolved with less computational efforts relative to its sequential counterpart. This phenomenon allows a new two-step approach: 1) evolves a solution in a highly parallel program format; and 2) serializes the parallel program to a functionally equivalent sequential program. Based on GPP, a combinational logic circuit design system, GPP+MLP [5], was developed. Different combinational logic circuits were evolved with 2-input lookup-tables (2-LUTs). The core of the GPP+MLP system consists of an Evolution Engine (EE) and a Multi-Logic-Unit Processor (MLP) (see Fig.1 above). The EE manipulates the genetic parallel programs and performs genetic operations. The MLP evaluates the genetic parallel programs to determine their fitness.

In conventional combinational logic circuit design, only a few standard logic gate types, such as AND, NAND, OR, NOR and NOT, are used. The final circuit is a hardwired gate network with fixed types of logic gates. Thus, all conventional combinational logic circuit design techniques are based on some restricted assumptions such as the gate types and the gate network structures. The invention of FPGAs completely redefines the rules. FPGA is a novel type of re-programmable VLSIs. FPGAs have been developed rapidly and used widely in prototyping circuits or products with higher costs but lower volume productions, e.g. reconfigurable hardware, networking devices, etc. One of the main differences between FPGAs and the conventional programmable logic devices is that the former use k-input lookup-tables (k-LUTs) while the latter are based on a fixed AND-OR (or OR-AND) matrix to implement Boolean functions. A k-LUT is a 2^k-memory-location, single-bit memory module, which uses the address lines as the inputs and returns the contents of the addressed location as the output of a Boolean function. However, it is not a trivial task to determine an optimal LUT network directly from a truth table. It usually goes through several stages such as synthesis, mapping, placement and routing in the translation of design into a bitstream ready for FPGA. The major enhancement of the GPPLCS is that we adopt a 4-LUT (instead of 2-LUT in the GPP+MLP) as the basis functional unit. The evolved 4-LUT network can be directly implemented on most commercial FPGAs, e.g. Xilinx Virtex-E Series FPGAs [6].

2.1 The Multi-logic-Unit Processor

The 4-LUT MLP used in the GPPLCS is a general-purpose, tightly coupled processor. As shown in Fig. 2, the MLP consists of 16 4-LUTs (L0-L15) that perform logic operations; 16 variable registers (R0-R15) that store intermediate values and program outputs; and 16 read-only registers (R16-R31) that store program inputs and logic constants. A variable register can only be modified by a dedicated 4-LUT (e.g. L0 can write to R0 only). The EE will preload the program inputs and the constants into the read-only registers before a parallel program is executed.

In each processor clock cycle, each 4-LUT takes four input values from any four registers, then performs a logic operation and finally, writes a single-bit result to its corresponding output register. For the MLP shown in Fig. 2, in each clock cycle, at most 16 operations can be performed simultaneously and 16 intermediate results can be carried forward to the subsequent parallel-instructions through variable registers. The function of a 4-LUT can be changed by modifying the content of the corresponding 4-LUT. With the advances in semiconductor technologies, multiple MLPs can be placed in a high capacity FPGA. Multiple MLPs in an FPGA can be driven by a single MLP program to evaluate multiple training cases in parallel. It will further reduce the fitness evaluation time.

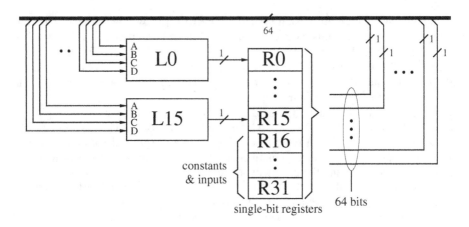

Fig. 2. The 4-LUT MLP used by the GPPLCS

2.2   The Evolved Parallel Programs

Any combinational logic circuit can be represented by an MLP program. An MLP program represents a combinational logic circuit. The MLP is designed to accept any bit pattern (genotype) as a valid program without causing processor fatal errors such as invalid opcodes. The genotype of a genetic parallel program is loaded and executed in the MLP directly without pre-evaluation correction. This is possible because all Boolean function outputs are Boolean variables. This closure property is especially important for the GPPLCS because of its random nature based on evolutionary technique. The phenotype of an MLP program can be expressed as a parallel assembly program (see Program 1).

```
#data
CONSTANTS: (r16-r21)=0,(r22-r26)=1
INPUTS:    (r27,r28,r29,r30,r31)<=(Cin,A1,A0,B1,B0)
OUTPUTS:   (r00,r01,r02)=>(Cout,S1,S0)
#program
0: bF6E0 r31 r27 r08 r29 r00
1: b3AA4 r00 r28 r06 r30 r00, bCB9E r00 r28 r30 r21 r01,
   b849E r31 r27 r31 r29 r02
```

Program 1. A 2-bit full-adder evolved by the GPPLCS

As shown in the program, an MLP program consists of two sections, the *#data* and the *#program* sections. The *#data* section defines constants, inputs and outputs. Before starting an execution, the MLP always initializes all variable registers (R0-R15) to logic '0'. The *CONSTANTS* line initializes read-only registers R16-R21 to logic '0' and R22-R26 to logic '1'. The *INPUTS* line defines input variables (*Cin, A1, A0, B1* and *B0*) and assigns them to read-only registers (R27-R31). The *OUTPUTS* line defines output variables (*Cout, S1* and *S0*) and assigns them to variable registers (R0-R2). The *#program* section contains parallel-instructions that perform Boolean operations. For example, the numbered lines in the *#program* section in Program 1 list out two parallel-instructions. For easy interpretation, all *nop* (no operations) sub-instructions are removed from the original program. Each sub-instruction consists of three parts: 1) a function name (*b0000-bFFFF* or *nop*); 2) four input registers; and 3) an output register. The Boolean function of each sub-instruction is denoted by a four-digit hexadecimal number which represents the 16-bit memory contents of the 4-LUT. For example, the *bF6E0 r31 r27 r08 r29 r00* sub-instruction in the parallel-instruction "0:" is implemented by loading "1111 0110 1110 0000" to the corresponding 4-LUT. The corresponding 4-LUT network of the program is shown in Fig. 3.

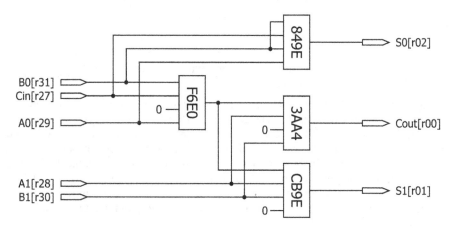

Fig. 3. The 2-bit full-adder (four 4-LUTs in two levels) shown in Program 1

3   Hardware Design and Implementation

This section presents the design and implementation details of the MLP. Fig. 4 shows the architecture of the core part of the MLP. The 16 sub-instruction registers (SIR0-SIR15) store the individual sub-instructions in the current parallel-instructions. The 16 processing elements (PE0-PE15) run sub-instructions and store results to their corresponding variable registers. The Control Unit (CU) decodes parallel-instructions and gives control signals to all MLP components. Due to the limited size of the interface bus between the CU and the host (64-bit only), more than one bus cycle is needed to transfer evaluation results of all rows in a truth table to the host.

In most cases, the GPPLCS only uses the first eight variable registers (R0-R7) to store program outputs. Thus, the MLP only needs to transfer the first eight variable registers to the host. In order to maximize the usage of the 64-bit interface bus, the MLP is designed to buffer eight sets of program outputs (of eight training cases). In this way, the evaluation results of the entire truth table are passed to the host in burst mode. For example, if there are N rows in a truth table, it takes $N/8$ clock cycles to transfer all program outputs to the host.

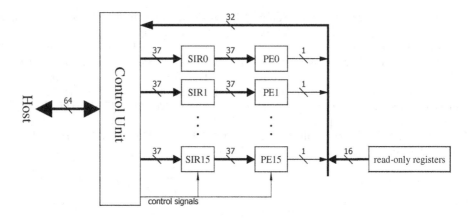

Fig. 4. The architecture of the MLP core

Fig. 5 shows a PE (PE*i*) which receives a sub-instruction from SIR*i*. It stores the result in the variable register R*i*. The core of the PE is a 4-LUT. It takes two processor clock cycles for the PE to execute one sub-instruction. In the first cycle, four input registers are selected by four multiplexers (M1-M4), and their values are then latched into an Internal Operand Register (IOR). In the second cycle, the 4-LUT uses the four latched operands to look up one bit and stores the result into R*i*. The IOR is used to pipeline the operations, i.e. selecting operands and looking up results, and to balance the long delay time on the route from the registers' outputs to the multiplexers' inputs.

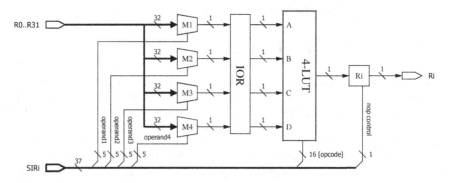

Fig. 5. A Processing Element

The MLP is implemented on a Pilchard board [17,18] which is a high performance reconfigurable computing development environment employing an FPGA. The Pilchard board is plugged into a 133 MHz synchronous dynamic RAM Dual In-line Memory Modules (DIMMs) slot of a PC. The Pilchard board can achieve a very high data transfer rate by making use of the DIMM RAM interface of the PC. Its efficient interface and low cost make it suitable for implementing the MLP. Here are some major features of the Pilchard board:

- Host interface: DIMM interface (a 64-bit data bus and a 14-bit address bus)
- Operating frequency: 100 MHz
- FPGA device: XCV1000E-HQ240-6
- OS supported: GNU/Linux

The FPGA used in the Pilchard board belongs to the Virtex-E series. The MLP uses only 2,515 slices. It is about 20% of the 12,288 slices available in the FPGA. Moreover, only one (out of 96) BlockRAM is used by the MLP. The critical path delay of the MLP is 9.965ns. Hence, it can operate at 100 MHz.

The MLP is coded in Very High Speed Integrated Circuit Hardware Description Language (VHDL) [19] which is a standard language for describing the structure and function of integrated circuits (ICs).

4   Experiments and Settings

To investigate the performance of the GPPLCS, we have used the system to evolve 4-LUT networks for four combinational logic circuit problems (see Table 1). Note that the 6-bit priority selector is to show the position of value '1' which first appears starting from the least significant bit in the 6-bit input. If none of the bits is set to value '1', an extra output bit which shows the case of all zero value is responsible for this special case. Since we have got six input bits (Input5 – Input0), we need extra three bits to indicate the position. Therefore, there are 4-bit outputs.

Table 1. Four combinational logic circuit problems. The N_{input} and N_{output} denote the numbers of inputs and outputs respectively. The N_{row} denotes the number of rows in the truth tables. The N_{case} denotes the total number of training cases

name	description	N_{input}	N_{output}	$N_{row}(=2^{N_{input}})$	$N_{case}(=N_{row}×N_{output})$
MUX	6-bit multiplexer	6	1	64	64
ADD	2-bit full-adder	5	3	32	96
CMP	3-bit comparator	6	3	64	192
PRI	6-bit priority selector	6	4	64	256

4.1   Experimental Settings

The main purpose of this paper is to show that the performance of the GPPLCS can be enhanced by the hardware MLP. All experimental settings are listed out in Table 2 below. In order to have a fair comparison in the performance between hardware-assisted GPPLCS and the pure software counterpart, evolutions of combinational logic circuits for the four combinational logic circuit problems are run on the same host (i.e. the PC where a Pilchard board locates). The host in which the Pilchard board locates is a Pentium III 800 MHz PC with ASUS CUSL2-C motherboard. The Pilchard board relies on the PC to communicate. User can transfer data to the Pilchard board via the DIMM slot in the host PC. The PC host is chosen because of the low level control required to mange the Pilchard board.

We test the problems with both the hardware-assisted GPPLCS and the pure software GPPLCS. The time for each tournament is recorded for comparison.

Table 2. Experimental settings

maximum program length	25 parallel-instructions
initialization	bit random, average 12.5 parallel-instructions
selection method	tournament (size=10)
4-LUT function set	b0000, ... , bFFFF, nop
inputs	$R_{32-Ninput} \dots R_{31}$
outputs	outputs: $R_0 \dots R_{Noutput-1}$
constants	logic 0, logic 1
crossover Prob.	0.1
bit-mutation Prob.	0.002
population size	2000
raw fitness (f)	the ratio of unsolved training cases
success predicate	all training cases are solved (f=0)

5   Results and Evaluations

Promising results are obtained for all the four combinational logic circuit problems. Table 3 summarizes the total elapsed times for the GPPLCS to evolve complete correct solutions with a pure software MLP and a hardware MLP. The t_H and t_S columns list out the execution times of the hardware-assisted GPPLCS and the pure software GPPLCS respectively.

Table 3. Summary of experimental results

problems	t_H (in sec)	t_S (in sec)	speedup ratio (t_S / t_H)
MUX	68	689	10.13
ADD	346	3824	11.05
CMP	1,575	34,760	22.07
PRI	720	14,651	20.35

It can be seen that the speedup of hardware over software is significant. For the ADD and MUX problems, the speedups are more than 10 times. For the CMP and PRI problems, the speedups are more than 20 times. The CMP problem takes nearly 10 hours to complete with the pure software GPPCLS, but it only takes less than half an hour with the hardware-assisted GPPLCS. Thus, problems of different levels of difficulties gain different speedups. This is easily recognized because the more difficult the problems, the more tournaments (computational effort) are taken to complete. Fig. 6 shows the speedup curves for the four tested problems. In these figures, the X-axis is the number of tournaments taken while the Y-axis is the speedup ratio (t_S/t_H).

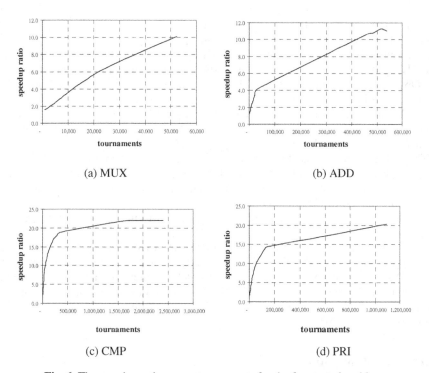

Fig. 6. The speedup ratio versus tournaments for the four tested problem

Figures 6(a) and (b) show that the speedup ratios for the MUX and ADD problems increase steadily to around 10. These two problems are relatively simple. Thus, the required computational efforts to evolve solutions for them are not so large. Conversely, in Figures 6(c) and (d), the speedup ratios are less than five initially when the evolution takes only a few thousand tournaments. As the evolution completes more tournaments, the speedup ratio increases rapidly to 22 times.

It is found that the speedup ratio increases with the number of tournaments taken in the evolution. It is obvious since execution time of each hardware evaluation is faster than that of each software evaluation by a certain theoretical limit. Therefore, it is expected that the speedup ratio is higher in those problems which have a larger number of tournaments taken. For example, in the MUX problem, only 10-time speedup is obtained due to the small number of tournaments taken (52,286). However, 22-time speedup is found in the CMP problem which takes 2,398,865 tournaments.

6   Conclusions and Further Work

In this paper, we have presented the design and implementation of a hardware-assisted GPP Logic Circuit Synthesizer (GPPLCS) prototype which uses a 4-LUT Multi-Logic-Unit Processor (MLP). The MLP uses a generic register machine architecture which can represent any combinational logic circuits. Moreover, the architecture of the MLP is so simple that multiple MLPs can be placed in an FPGA.

The hardware-assisted GPPLCS shows promising results in the speedup. With the help of hardware, GPPLCS achieves a 22-time speedup at most in our tested problems. Furthermore, the speedup ratio increases with the number of tournament taken in solving the problems. It is particularly suitable for solving difficult problems.

Obviously, the evaluations of individual rows on a truth table are independent. Thus, multiple rows of a truth table can be evaluated simultaneously. If there are enough hardware resources to implement multiple MLPs, multiple rows of a truth table can be evaluated in parallel. Based on the hardware resources usage of the current implementation of the single MLP (i.e. 20% hardware resources of the Xilinx XCV1000E FPGA on the Pilchard board), we can place four MLPs in the current Pilchard board. With this highly parallelized fitness evaluation engine, the evolution will be further sped up. The speedup can further be increased to 4 times as the current one. Hopefully, it can be up to about 90 times.

In the near future, we have planned to implement a full-scale GPPLCS system on the Virtex-II Pro ML310 System [20] which comprises a Virtex-II Pro XC2VP30 FPGA. We shall design and implement a hardware evolution engine to perform genetic operators. Since the XC2VP30 is directly connected to the system bus of the ML310 system, the bottle-neck of the host-Pilchard interface can also be removed. We envisage that the full-scale GPPLCS will speed up the evolution significantly so that more complex combinational logic circuits, e.g. 4-bit multiplier, can be evolved in shorter evolution time.

Acknowledgements

The work described in this paper was partially supported by two grants from the Research Grants Council of the Hong Kong Special Administrative Region, China (Project No. CUHK4192/03E and CUHK4127/04E).

References

1. Banzhaf, W., Koza, J.R., Ryan, C., Spector, L., Jocob, C.: Genetic Programming. IEEE Intelligent Systems Journal, Vol.17, No.3 (2000) 74-84
2. Koza, J.R.: Genetic Programming: On the Programming of Computers by Means of Natural Selection. MIT Press (1992)
3. Banzhaf, W., Nordin, P., Keller, R.E., Francone, F.D.: Generic Programming: An Introduction on the Automatic Evolution of Computer Programs and its Applications. Morgan Kaufmann (1998)
4. Leung, K.S., Lee, K.H., Cheang, S.M.: Evolving Parallel Machine Programs for a Multi-ALU Processor. Proceedings of IEEE Congress on Evolutionary Computation – CEC'02 (2002) 1703-1708
5. Cheang, S.M., Lee, K.H., Leung, K.S.: Designing Optimal Combinational Digital Circuits Using a Multiple Logic Unit Processor. Keijzer M., O'Reilly U, Lucas, S.M., Costa E., Soule T., Eds., Proceedings of the 7th European Conference on Genetic Programming – EuroGP'2004, LNCS 3003 (2004) 23-34
6. Virtex™ E Platform FPGAs: Introduction and Overview. Xilinx, Inc. (2002)
7. Yao, X., Higuchi, T.: Promises and Challenges of Evolvable Hardware. IEEE Transactions on Systems, Man, and Cybernetics – Part C, Vol.29, No.1 (1999) 87-97

8. Kalganova, T.: An Extrinsic Function-Level Evolvable Hardware Approach. Poli R., Banzhaf W., Langdon W.B., Miller J., Nordin P., Fogarty T.C., Eds., Proceedings of the 3rd European Conference on Genetic Programming – EuroGP'2000, LNCS 1802 (2000) 60-75

9. Miller, J.F., Job, D., Vassilev, V.K.: Principles in the Evolutionary Design of Digital Circuits – Part I. Genetic Programming and Evolvable Machines, Vol.1, No.1 (2000) 7-35

10. Coello, C.A., Luna, E.H., Aguirre, A.H.: Use of Particle Swarm Optimization to Design Combinational Logic Circuits. Tyrrell A.M., Haddow P.C., Torresen J., Eds., Proceedings of the 5th International Conference on Evolvable Systems: From Biology to Hardware – ICES'2003, LNCS 2606 (2003) 398-409

11. Koza J.R., Bennett III, F.H., Hutchings, J.L., Bade, S.L., Keane, M.A., Andre, D.: Rapidly Reconfigurable Field-Programmable Gate Arrays for Accelerating Fitness Evaluation in Genetic Programming. Proceedings of the 6th International Symposium on Field Programmable Gate Arrays – FPGA'98 (1998) 209-219

12. Heywood, M.I., Zincir-Heywood, A.N.: Register Based Genetic Programming on FPGA Computing Platforms. Poli R., Banzhaf W., Langdon W.B., Miller J., Nordin P., Fogarty T.C., Eds, Proceedings of the 3rd European Conference on Genetic Programming – EuroGP'2000, LNCS 1802, (2000) 44-59

13. Shackleford, B., Snider, G., Carter, R.J., Okushi, E., Yasuda, M., Seo, K., Yasuura, H.: A High-Performance, Pipelined, FPGA-Based Genetic Algorithm Machine. Genetic Programming and Evolvable Machine, Vol.2, No.1 (2001) 33-60

14. Martin P.: A Pipelined Hardware Implementation of Genetic Programming Using FPGAs and Handel-C. Foster J.A., Lutton E., Miller J., Ryan C. Tettamanzi A.G.B., Eds, Proceedings of the 5th European Conference on Genetic Programming – EuroGP'2002, LNCS 2278 (2002) 1-12

15. Cheang, S.M., Lee, K.H., Leung, K.S.: Evolving Data Classification Programs using Genetic Parallel Programs. Proceedings of IEEE Congress on Evolutionary Computation – CEC'03 (2003) 248-255

16. Leung, K.S., Lee, K.H., Cheang, S.M.: Parallel Programs are More Evolvable than Sequential Programming. Ryan C., Soule T., Keijzer M., Tsang E., Poli R., Costa E., Eds., Proceedings of the 6th European Conference on Genetic Programming – EuroGP'2003, LNCS 2610 (2003) 107-118

17. Leong, P.H.W., Leong, M.P., Cheung, O.Y.H., Tung, T., Kwok, C.M., Wong, M.Y. and Lee, K.H.: Pilchard – A reconfigurable computing platform with memory slot interface. Proc. of the 8th Annual IEEE Symposium on Field Programmable Custom Computing Machines (2001)

18. Tsoi, K.H.: Pilchard User Reference (v1.0). Department of Computer Science and Engineering, The Chinese University of Hong Kong (2004) [online:http://appsrv.cse.cuhk.edu.hk/~ceg5010/iftest/doc/ref.ps]

19. Shahill, K.: VHDL for Programmable Logic. Addison Wesley (1998)

20. Virtex-II Pro ML310 Embedded Development Platform – User Manual. Xilinx, Inc. (2004)

Operator-Based Distance for Genetic Programming: Subtree Crossover Distance

Steven Gustafson[1] and Leonardo Vanneschi[2]

[1] School of Computer Science & IT, University of Nottingham,
Jubilee Campus, Wollaton Rd. Nottingham, NG81BB, United Kingdom
smg@cs.nott.ac.uk
[2] Dipartimento di Informatica, Sistemistica e Comunicazione (DISCo),
University of Milano-Bicocca 20126 Milano, Italy
vanneschi@disco.unimib.it

Abstract. This paper explores distance measures based on genetic operators for genetic programming using tree structures. The consistency between genetic operators and distance measures is a crucial point for analytical measures of problem difficulty, such as fitness distance correlation, and for measures of population diversity, such as entropy or variance. The contribution of this paper is the exploration of possible definitions and approximations of operator-based edit distance measures. In particular, we focus on the subtree crossover operator. An empirical study is presented to illustrate the features of an operator-based distance. This paper makes progress toward improved algorithmic analysis by using appropriate measures of distance and similarity.

1 Introduction

In canonical tree-based genetic programming (GP), part of the search process is carried out using transformation operators on tree structures [1]. From a topological point of view, these operators can be thought of as defining the neighbourhood of these trees. To analyse various properties of the search process, it is often useful to know the distance between two trees. For example, if we wish to calculate a well-known measure of problem hardness, such as fitness distance correlation [2, 3, 4, 5], we have to calculate the distance of a sample of trees from a particular global optimum. As trees become closer to the optimum, we would like the improvement in fitness to be positively correlated with the decrease of distance, making the search more predictable. Furthermore, when studying diversity, the distance between two trees is required in order to find the average pair-wise degree of similarity of the trees in a population. When the average pair-wise distance of the population approaches 0, the population is converging and we can expect the search to become stuck in a local optimum. The use of tree structures and multi-node altering transformation operators (e.g. subtree crossover), typical of GP, makes, among other things, defining operator-based

M. Keijzer et al. (Eds.): EuroGP 2005, LNCS 3447, pp. 178–189, 2005.

distance measures complex. Thus, the study of fitness distance correlation has largely progressed using systems with single-node altering transformation operators like single-node mutation [4, 5]. This allows appropriate conclusions to be drawn from empirical results for correlating the improvement of fitness with the change in distance to an optimum. Likewise, most research for diversity methods and measures in turn rely on using distance measures based on single-node differences between trees [6, 7, 8, 9]. The most common are variations on the Levenshtein edit distance.

Defining a distance measure, or measure of similarity, that is, in some senses "bound" to (or "consistent" with) the genetic operators being used informally means that if two trees are *close* to each other, or similar, one can be transformed into the other in a few applications of the operator(s). The complexity involved in using operator-based distance measures is two-fold: firstly, distance needs to be re-defined for the specific operators being used. Thus, if we add a new mutation operator or a variation of subtree crossover, we will need to reconsider the distance definition. This results in a large design complexity. Secondly, actually computing an operator-based distance can be much more computationally expensive than the more straightforward edit distance. For example, complexity is increased for operator-based measures as typically the distance between two trees depends on the current population. The computational complexity involved in operator-based distance measures encourages the use of approximations such as edit distance.

The complexity of the edit distance between two trees is in $O(k)$, i.e. dependent upon the number of k nodes in the trees. Computing the pair-wise distance between every tree in the population has complexity $O(M^2 \times k)$, where M is the size of the population and k is the average size of the trees. If the edit distance measure defines a metric space, symmetry only requires $\frac{M(M-1)}{2}$ comparisons. For fixed-length bit-string genetic algorithms, Wineberg and Oppacher showed how this can be done in $O(M \times k)$ with preprocessing of the population[10]. This method would become more complex to design with a variable size and shape representation like GP trees.

The contribution of this paper is the exploration of defining operator-based distances and a discussion of the approximation of such distances. A distance measure based on subtree crossover is defined for a constructed problem, and an empirical study demonstrates its features. Section 2 introduces the concept of distance based on a genetic operator and its applications to a canonical GP system. Section 3 defines a new distance measure bound to standard subtree crossover. Since the calculation of this distance measure may require a lot of computing resources, some techniques to approximate this measure, thus reducing computational complexity without compromising its efficacy, are presented. Section 4 presents some experimental results showing the suitability of this distance measure, compared to a well known tree distance measure.

2   Defining Operator-Based Distance

Initially, we explore what operator distance means in a typical system. In this section, we consider a simple steady-state GP system using syntax trees to represent individuals, and subtree crossover for variation. Subtree crossover proceeds by selecting a subtree in a parent tree and a subtree in a donor tree, where any node in either tree can be selected. Next, subtree crossover replaces the parent's subtree with the donor's subtree. New trees replace the parent trees. Concerning distance, the question we want to answer is: what is the distance between two trees contained in the population according to a given operator, such as subtree crossover? In other words, we would like to know the *algorithmic* distance between two solutions according to a particular representation, operators, and fitness measure. First, we define some notation:

- P is the population, containing M trees,
- T_1 is the tree we want to compute a distance from, or the parent tree,
- T_2 is the tree which we would like to transform T_1 into,
- T_1/T_2 is the difference of the two trees. This operator produces a tuple (s_{T_1}, s_{T_2}), i.e. a pair of subtrees, where subtree $s_{T_2} \in T_2$ must replace $s_{T_1} \in T_1$ to make $T_1 = T_2$.

Supposing that $T_1 \in P$, the crossover distance between T_1 and T_2 depends on the ability to select s_{T_2} from some tree in P. Thus, the crossover distance[1] between T_1 and T_2 also depends on the population P: if $T_2 \in P$, then s_{T_2} will also be in P. In the case where $T_2 \in P$, we could state that the distance is equal to 1, since it is possible to transform T_1 into T_2 in just one crossover application. On the other hand, if $s_{T_2} \notin P$ then it will require more than one application of subtree crossover to make $T_1 = T_2$. Following this idea, calculating a distance value in the light of multiple applications of an operator would need to consider if any applications of subtree crossover to T_1 would result in a new tree T_1' that required subtree s_{T_2}' to transform T_1' into T_2. If $s_{T_2}' \in P$, the distance is 2, since it is possible to transform T_1 into T_2 with 2 crossover applications. To find distances greater than 2, we need to continue this process. This definition of operator-distance is an accurate reflection of the subtree crossover operator for a steady-state model with offspring-parent replacement. However, there are obvious problems with calculating this distance. Let us assume the average size of the M trees in P is k, and that crossover can choose any node in the population as the root of the new subtree. The number of potential intermediate trees T_1' to consider for distances greater than 1 is $M \times k$. The distance calculation needs to be carried out for each of these trees to decide if the distance is 2.

Now, if we consider a generational model, which seems to be largely used in the GP community, an operator defined in terms of the population makes defining distance using multiple operator applications even more difficult. For

---

[1] The generic term "metric" would probably be more suitable than the term "distance" here; anyway, we go on using the term "distance" for simplicity.

example, we may consider the new tree T_1' after an operator application on T_1 using the population P. In a generational algorithm, we will then need to consider the next population P' when thinking about another operator application on T_1'. We can either create all the possible next populations or we can approximate them. Creating the future populations presents computational limitations. Similarly, we might create the future expected populations using calculations similar to the ones found in the schema theorems for GP [11]. However, finding the future expected populations is also costly, essentially requiring a similar amount of computation as actually running the GP algorithm.

Furthermore, let us assume that we calculate the distance between T_1 and T_2 as 1, meaning that one application of our operator to T_1 can build T_2. However, when we actually execute our algorithm, it is not certain that this particular application will occur. If we calculate the distance between T_1 and T_2 as 3, which depends on two intermediate trees, we know with less confidence if either T_1' or T_1'' will actually be produced. The decreasing amount of confidence we will have in the accuracy of distance values based on future generations is likely to make this type of distance measure less useful. Therefore, in practical applications of operator-based distance measures, it may only be useful to know the likelihood of creating a particular tree T_2 in the next generation.

To overcome the difficulty in defining a multiple operator distance, and to incorporate the stochastic properties of the algorithm, we can consider operator-distance in terms of the probability of correctly applying the operator once. That is, if one tree is in the neighbourhood of another, how likely is it that this neighbour will be found. Since we know (or we can easily calculate) the values of parameters like the selection probability of trees and the frequency of all subtrees in the current population, we could assign a probability to the selection of all subtrees in the next population. If we know what subtree is required to make two trees equal, then we may approximate distance in terms of the probability of selecting this subtree. While this measure would only consider one operator application, it will do so with a higher confidence and at significantly less computational effort then considering the future expected populations. However, we could attempt to approximate the creation and selection of subtrees in future populations using a similar method. We now look at this new idea of distance measure more closely.

The new operator-based distance can now be formulated as: given an operator V, trees T_1 and T_2, and a population of trees P, can V be applied such that $V(T_1, P) = T_2$? That is, can an operator V, that uses the genetic material in P, be applied once to T_1 to produce T_2? If the answer is 'yes', we would not say that distance is 1, but we would bind this distance value to the probability of generating T_2 from the application of V to T_1. Instead of asking for the required number of edit operations to transform a tree T_1 into another tree T_2, we ask *how probable* is it that an operator will transform T_1 into T_2. In other words, if the required genetic material to transform T_1 into T_2 is present in the population, how probable is it that our operator V selects it? In case T_1 and T_2 share no common material, this will be the probability that subtree crossover selects the root of T_1 and a subtree equal to T_2. Note that using this type of distance measure reporting a probability may pose a problem doing fitness-distance correlation studies. We are currently looking at ways to overcome these problems by accurately approximating the likely construction of missing subtrees. We will now describe this probability-based operator distance measure for subtree crossover.

3   Subtree Crossover Distance

We will consider a subtree crossover distance between two trees in a population in context of the genetic material contained in the population. Given the subtree crossover operator V_{SC}, a distance function can be defined by the following pseudo-code:

```
func distance(T₁, T₂, V_SC, P ){
    (s_T₁, s_T₂) = T₁/T₂
    ps1 = probSelecting(s_T₁, T₁ )
    ps2 = probCreating(s_T₂, P )
    return  ps1 * ps2
}
```

Given the subtree s_{T_2} that needs to replace $s_{T_1} \in T_1$, the distance is defined in terms of the probability of selecting s_{T_1} in T_1 and the probability of creating (or selecting) s_{T_2} from P. Both functions, probSelecting() and probCreating(), require knowledge of the selection probabilities used in the algorithm. Finding s_{T_1} and s_{T_2} and determining the probability of selecting $s_{T_1} \in T_1$ can be done in linear time in the size of T_1 and T_2. The crux of the subtree crossover operator-based distance is finding the probability of generating the subtree s_{T_2}. As T_2 will be in the current population, the function will report a non-zero value (this is the case if selecting the root-node of T_2 is possible).

The *probSelecting()* function can be defined for subtree crossover based on the node selection probability. Given uniform node selection, selecting the subtree $s_{T_1} \in T_1$ has the probability of $\frac{1}{|T_1|}$. The *probCreating()* function for subtree crossover can be defined to consider all the occurrences of the subtree s_{T_2} in the population and their probability of selection. That is, for a tree that contains s_{T_2}, we may want to know how likely that tree will be selected by a selection method. We will then want to know the probability of selecting s_{T_2}. Determining the number and selection probability of each occurrence of $s_{T_2} \in P$ is in $O(M \times k)$, where k refers to the average size of an individual in the population. To carry out this search for each pair-wise distance computation would have a complexity in $O(M^3 \times k^2)$. Preprocessing the population prior to carrying out the pair-wise distance calculation can reduce this complexity.

So far we have limited our distance measure to the single application case, which would seem appropriate for standard GP as only one application of an operator is typically used to generate a new individual. We have also considered the whole population as a source of potential subtrees. However, as we know,

evolutionary algorithms use fitness-based selection to implement solution competition. Therefore, not all trees have the same likelihood of being selected as donor
trees. We can use this fact to provide an effective way of reducing complexity of
this operator distance while preserving the utility of the measure.

A way to reduce the complexity of the above subtree crossover-based definition, which can also be applied for other operator-based definitions, is to only
consider those trees and their subtrees that are *likely* to be selected. We can
define threshold values α for tree selection and β for subtree selection. Then,
we can produce the set of fit (or likely to be selected) trees F according to the
following definition:

$$F = \{\forall i \in P \mid \text{Better\_Than}(i, \alpha)\},$$

where the predicate $\text{Better\_Than}(i, \alpha)$ is True if individual i has better fitness
than more than α individuals in the population. The value of α should reflect the
behaviour of the selection method being used, e.g. in terms of the tournament
size in tournament selection. We can find the set F in $O(M)$ time and are only
required to do so once for the whole population (if we are calculating the pair-
wise measure). Now, for subtrees s and w in an individual $i \in F$, we can define
the set R of the likely-to-be selected subtrees according to:

$$R = \{(\forall s \in i) \wedge (i \in F) \mid \frac{|\{\forall w \in i \mid |s| = |w|\}|}{|i|} > \beta\},$$

That is, a subtree s in individual i (from F) will be in R if the number of subtrees
in i, with the same size as s, divided by the the total size of i is greater that β. In
effect, we are only considering those subtrees of a given size that are likely to be
selected by a uniform node selection probability. For example, in a full tree with
7 nodes (depth 3), and a uniform node selection probability, selecting a specific
subtree with 3 nodes has probability of $\frac{1}{7}$. Selecting any (of the two) subtrees of
size 3 has a probability of $\frac{2}{7}$, and selecting a subtree of size 7 has a probability
of $\frac{1}{7}$. However, selecting a subtree of size 1 has probability of $\frac{4}{7}$. Thus, while it
may not be the exact probability of selecting a specific subtree with 3 nodes, we
can focus our attention only on those subtrees that are more likely to be selected
due to their size. So, if subtree crossover needs to select a very large subtree to
transform one tree into another, we might assume the likelihood of doing this
will be so small that it is effectively 0. Additionally, if the missing subtree is only
in a tree that has a very low chance of being selected, due to its poor fitness,
again it may be reasonable to report a probability of selection as 0.

In summary, two ways to reduce the complexity of a subtree crossover-based
distance are: (1) only consider the trees in the population that are likely to be
selected, and (2) in those trees, only consider the subtrees that are likely to be
selected. Tuning the parameters α and β will allow us to reduce the complexity
of the computing the population pair-wise distance. However, a potential disadvantage of this approximation, particularly in discrete fitness spaces, is the fact
that the number of individuals with fitness better than α individuals in the population can vary. Thus, if the population has no duplicating fitness values, this

approximation scheme may reflect reality. However, if the population contains only one unique fitness value, then all the individuals have the same probability of being selected. Preprocessing the population prior to computing the pair-wise edit distances can also consider α and β to reduce the memory demands of the distance calculation. We now complement the above discussion of operator-based distance measures with an empirical study of one such operator-based measure.

4   Experimental Results

In this section, we investigate the practical side of implementing an operator-based edit distance measure for the subtree crossover operator. To reduce the complexity of this study and its presentation, we use a constructed problem similar to the problems that emphasise solution structure, or tree shape, like the Lid [12] problem.

4.1   The GP System and Problem

The GP system is the same as above, but a new tree produced by subtree crossover replaces the worst fit one in the population (instead of the parent tree). Again, only subtree crossover is used, where node selection for subtree crossover is uniform. Tournament selection is of size 3, and a population of 20 trees is used. The functions are two-argument nodes that have no meaning. Primitives are empty, null nodes. Thus, our trees are binary, where internal nodes always have two child nodes. We use a maximum depth of 7 during subtree crossover, and initialise the population with full trees of depth 3, where these tree shapes are all identical.

A random tree shape is generated to define the goal state, or instance. A tree shape is generated by randomly picking an odd number between 16, to ensure trees are not too small, and 31, the size of a full tree of depth (7-2). We then randomly, with uniform probability, assign two child nodes to each available leaf node, starting with a root node. All leaf nodes that have depth less than the maximum are deemed available. This tree shape defines the instance. This construction method is similar to one found in [12].

The tree shape, defining an instance, is abstracted by representing it by the number of nodes present at each depth. Fitness is the absolute difference between the number of nodes at each depth in a candidate tree and the target random tree shape. For example, Figure 1(a) shows a random tree that defines an instance, and Figure 1(b) shows a perfect solution with the same number of nodes at each depth. Note that the shape of the solution is not identical and allows flexibility, as well as possible deception, in the search process.

We generate 30 random instances, and collect 30 random runs of the system for each instance, with a maximum of 500 generations (subtree crossover applications). The GP algorithm found an optimum solution 614 out of the 900 runs, the earliest of these at generation 16 and the latest at generation 499. The average generation where an optimum was found was 291, with a standard deviation of 117 generations.

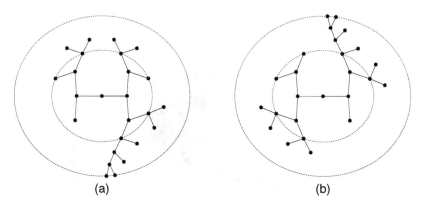

Fig. 1. An instance is defined by a randomly generated tree shape (a) and can be solved perfectly by a candidate tree shape that has the same number of nodes at each depth (b)

4.2   Operator-Based Distance

In this study, we will use a simpler version of the operator distance defined in Section 3. Specifically, we will only consider the frequency of subtrees required to make two trees equal. We define subtree crossover distance D_{SC} between two trees T_1 and T_2, given the population P, as:

$$D_{SC}(T_1, T_2, P) = \frac{occur(s_{T_2}, P)}{\#subtrees(P)}, \qquad (1)$$

where $subtrees(P)$ returns the set of all subtrees in a population, and $occur(s, P)$ counts the number of occurrences of a subtree s in a population P. Earlier we defined the difference of two trees as $T_1/T_2 = (s_{T_1}, s_{T_2})$, where the resulting tuple defined the subtree in T_1 that needed to be replaced by s_{T_2} to make $T_1 = T_2$. We will define our operator-based distance for an average pair-wise distance measure in a population, thus $s_{T_2} \in P$. However, it is possible that s_{T_2} will only be represented by the tree $T2$ itself, requiring that $s_{T_1} = T_1$.

4.3   Complexity

To calculate the average pair-wise distance using D_{SC} requires $M^2 - M$ distance calculations, where M is the number of trees in the population P. An edit distance pair-wise calculation would have an average complexity, assuming k is the average size of a tree in the population, of $O(k \times M^2)$. However, and of particular interest here, our operator-based distance D_{SC}, while having the same worst case bound, is likely to be less as the entire trees do not need to be explored. As soon as two subtrees that do not match are encountered, D_{SC} computes the frequency of the missing subtree and returns. That is, the complexity of D_{SC} is in $O(m \times M^2)$, where $m \leq k$. The function $subtrees$ can incur a memory cost as it needs to store all unique subtrees in P and their frequency. However,

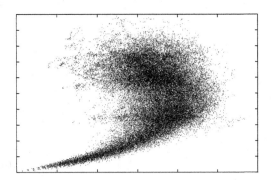

Fig. 2. A scatter plot showing the correlation between the edit distance and the operator-based distance, average pair-wise distance in each population

this function only needs to be carried out once prior to each average pair-wise distance calculation, and requires linear time to visit all subtrees.

4.4   Pair-Wise Distance

Measures of distance or similarity can be useful for a variety of analysis. For example, the average pair-wise distance between all the trees in the population can indicate the amount of genetic material remaining in the population, as well as the likely behaviour of the operators. In both cases, it is important that the distance measure reflects the behaviour of the operator for meaningful results. We carried out two average pair-wise distance calculations using a standard edit distance and our subtree crossover operator-based distance. The edit distance measure counts the number of non-identical nodes between two overlapped trees. We normalise the average pair-wise distance of a population by dividing it by the average tree size (number of nodes) in that population. The operator-based distance measure divides the number of occurrences of the missing subtree by the total number of subtrees in the population, indicating the likelihood of selecting this missing subtree (but not with respect to fitness). We subtract this number from 1 to produce a measure of dissimilarity similar to the edit distance, where values close to 0 indicate high similarity, and values close to 1 indicate high dissimilarity.

Figure 2 shows the correlation between the two above measures. As the operator-distance is based on the frequency of missing subtrees in the populations, which tend to contain more and more subtrees in subsequent populations, the range of these frequency values will also vary from population to population. Therefore, to visualise the frequencies with varying ranges more effectively, we multiply a population's average pair-wise operator distance by the population's average tree size. This value is then scaled for the [0, 1] range. We can see that for

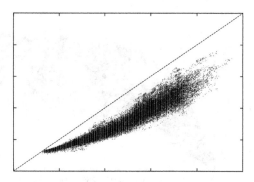

Fig. 3. A scatter plot showing the correlation between the edit distance and the operator-based distance complexities, or the average number of nodes visiting during the calculation of each pair-wise distance measure

low values of dissimilarity ($\sim$ 0 to 0.4), there is the expected positive correlation. However, as the edit distance dissimilarity increases, the operator-based dissimilarity takes on a wide range of values, and vice versa. Even in our simulations using a simplified problem of only tree shapes with no node contents and a fairly simple measure of operator distance, the disparity between the operator-based distance and the common edit distance is clear. That is, two trees that appear to be similar according to edit distance are not necessarily similar in terms of our operators.

4.5   Complexity Reductions

As mentioned earlier, there may be some cases where an operator-based distance measure can reduce the complexity of measuring distance. For example, using a basic string edit distance generally requires that all nodes of each tree need to be checked. However, if we are using only the subtree crossover operator, if the root of a subtree does not match with the root of another subtree in a second tree, we do not need to continue the comparison of these subtrees. That is, we know that the whole subtree will need to be replaced using subtree crossover to make the two trees equal.

Figure 3 shows the correlation between the complexities of each pair-wise distance measure. We can see that the edit distance complexity, approximated here as the average tree size in the population, grows at a faster rate than the operator-based distance, where the former is a function of size and the latter a function of dissimilarity. However, this should be taken in the light of the higher cost of preprocessing the population prior to the operator-based distance calculation as well as the memory constraints involved in storing and looking up the frequencies of various subtrees. For example, Figure 4 shows the evolution

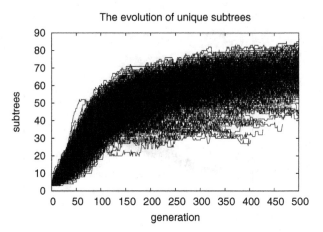

Fig. 4. The number of unique subtrees in each generation for all runs. Initially, with a population of full trees of depth 3, there are only three unique subtrees

of the number of unique subtrees in each population during the runs. For a population of size 20, with a depth limit of 7, there averages around 65 or so unique subtrees that need to be stored. While the representation used here is quite simple (binary trees with no node content), a memory requirement (number of unique subtrees) for computing operator distance that is within a constant multiple of the population size is promising.

5   Conclusions

This paper represents a first step in the study of the issues concerning operator-based distance measures for genetic programming. The variable shaped and sized solutions in the tree representation can make defining operator-based distance measures difficult. Therefore, it has become very common to use edit distance measures instead. Distance measures that do not capture the operator behaviour, however, are not always applicable for analytical studies like fitness distance correlation. Also, these measures need to be used carefully when studying population diversity. Thus, we aim to examine the practical difficulties in measuring distance using the canonical operator, subtree crossover. A series of possible definitions of operator-based distance were defined in this paper. Importantly, we discussed ways of reducing the complexity of such measures by means of various approximations. An empirical study showed how an edit distance measure and an operator-based distance measure can fail to correlate. We showed in the simulations that operator-based distance can result in a reduction of complexity over an edit distance measure, where complexity is the number of nodes examined during a pair-wise distance calculation. Our future work is exploring other definitions of operator-based measures, for subtree crossover and other operators, and the tradeoffs involved with reducing their complexity.

References

1. J.R. Koza. *Genetic Programming: On the Programming of Computers by Means of Natural Selection*. MIT Press, Cambridge, MA, USA, 1992.
2. T. Jones and S. Forrest. Fitness distance correlation as a measure of problem difficulty for genetic algorithms. In L. Eshelman, editor, *Proceedings of the Sixth International Conference on Genetic Algorithms*, pages 184–192, San Francisco, CA, 1995. Morgan Kaufmann.
3. P. Collard, M. Clergue, M. Tomassini, and L. Vanneschi, A study of fitness distance correlation as a difficulty measure in genetic programming. *Evolutionary Computation*, in press.
4. L. Vanneschi, M. Tomassini, P. Collard, and M. Clergue. Fitness distance correlation in structural mutation genetic programming. In C. Ryan et al., editors, *Genetic Programming, Proceedings of the European Conference*, volume 2610 of *LNCS*, pages 459–468, Essex, 14-16 April 2003. Springer-Verlag.
5. L. Vanneschi. *Theory and Practice for Efficient Genetic Programming*. Ph.D. thesis, University of Lausanne, Switzerland, 2004.
6. S. Gustafson, A. Ekárt, E.K. Burke, and G. Kendall. Problem difficulty and code growth in genetic programming. *Genetic Programming and Evolvable Hardware*, 5(3):271–290, 2004.
7. S. Gustafson. *An Analysis of Diversity in Genetic Programming*. PhD thesis, School of Computer Science and Information Technology, University of Nottingham, Nottingham, England, February 2004.
8. E.K. Burke, S. Gustafson, and G. Kendall. Diversity in genetic programming: An analysis of measures and correlation with fitness. *IEEE Transactions on Evolutionary Computation*, 8(1):47–62, 2004.
9. N.F. McPhee and N.J. Hopper. Analysis of genetic diversity through population history. In W. Banzhaf et al., editors, *Proceedings of the Genetic and Evolutionary Computation Conference*, pages 1112–1120, FL, USA, 1999. Morgan Kaufmann.
10. M. Wineberg and F. Oppacher. Distance between populations. In E. Cantú-Paz et al., editors, *Proceedings of the Genetic and Evolutionary Computation Conference*, volume 2724 of *LNCS*, pages 1481–1492, Chicago, USA, 2003. Springer-Verlag.
11. R. Poli and N.F. McPhee. General schema theory for genetic programming with subtree-swapping crossover: Part i. *Evolutionary Computation*, 11(1):53–66, 2003.
12. J.M. Daida, H. Li, R. Tang, and A.M. Hilss. What makes a problem GP-hard? validating a hypothesis of structural causes. In E. Cantú-Paz et al., editors, *Proceedings of the Genetic and Evolutionary Computation*, volume 2724 of *LNCS*, pages 1665–1677, Chicago, IL, USA, 12-16 July 2003. Springer-Verlag.

Repeated Patterns in
Tree Genetic Programming

William B. Langdon[1] and Wolfgang Banzhaf[2]

[1] Computer Science, University of Essex, UK
[2] Computer Science, Memorial University of Newfoundland, Canada

Abstract. We extend our analysis of repetitive patterns found in genetic programming genomes to tree based GP.

As in linear GP, repetitive patterns are present in large numbers. Size fair crossover limits bloat in automatic programming, preventing the evolution of recurring motifs. We examine these complex properties in detail: e.g. using depth v. size Catalan binary tree shape plots, subgraph and subtree matching, information entropy, syntactic and semantic fitness correlations and diffuse introns. We relate this emergent phenomenon to considerations about building blocks in GP and how GP works.

1   Introduction

Repeated sequences are commonplace in natural genomes. Biologists have discovered a vast amount of repetition in the DNA of microbes, plants and animals [1]. In fact it is now known that less than 3% of a human genome consists of protein-coding genes whereas around 50% of it consist of repetitive sequences [2; 3]. Biologists have recently turned their attention toward these patterned sequences [4; 5; 6] because the huge percentage of it indicates that these sequences play a major role in hereditary biology. The question we are asking is whether this emergent phenomenon might also be present in artificial genomes used for genetic programming.

Our initial search turned up repetitive sequences in linear GP genomes [7]. Here we turn to tree GP genomes. We find there are indeed, small and large repeated patterns in large trees which have been evolved by genetic programming. It can be observed that evolved trees are incrementally constructed from high fitness subtrees which are, however, not classic GP building blocks. Instead diffuse introns ensure that most code is robust to change.

We suggest that observations of this type can shed some new light on the old question of building blocks in GP [8]. Do they exist? If so, how does GP use them? If they do not, how does genetic search succeed?

Our route in this paper is roundabout: we start by following up on our work which suggests repeated patterns are prevalent in linear genetic programming [7] but now look at tree based GP. We use our time series modelling and Bioinformatics classification test problems (described in Section 2 and [7]) to show that, despite high mutation rates, multiple large syntactic and semantic repeated patterns can occur in standard subtree crossover as well (Section 3). We deepen

M. Keijzer et al. (Eds.): EuroGP 2005, LNCS 3447, pp. 190–202, 2005.

this analysis in Section 4: where we measure tree shape, entropy, sub-fitness and sensitivity within trees. This will lead us back to suggest (Section 5) at least in some simple modelling and prediction applications: 1) "introns" are somewhat diffused rather than discrete subtrees with a well defined root node that immediately nullifies their effect and 2) GP incrementally assembles solutions from large fit components, which are somewhat different from the classic "building block". Section 6 concludes.

2   Demonstration Problems

We have chosen two moderately difficult benchmark problems to represent typical modelling and prediction applications of genetic programming. Both were originally used as machine learning benchmarks. The Mackey-Glass chaotic time series has been used to demonstrate scientific, medical and financial modelling, e.g. [9]. The GP system is given historical data from which to predict a next value. We used the IEEE benchmark discretised into 8 bit unsigned integers, see Figure 1, left. All 1201 data points were used for training.

The second benchmark is a binary classification bioinformatics problem. Reinhardt and Hubbard [10] have shown that amino acids in a protein can be used to predict its location in the cell. They trained neural networks to distinguish between seven cellular locations in animals and microbes. We restrict ourselves to localising animal proteins (normally it is known if a protein is animal or bacterial) and a binary classification problem. To this end we evolve models which predict if an animal protein will be found in the cell nucleus or elsewhere. I.e. in the cell cytoplasm, in the mitochondria or outside the cell [10]. We used the same Swissprot data for 2427 proteins as used in [10]. There are 1097 nuclear (and 1330 non-nuclear) sequences of amino acids (see Figure 1, right). Data were split evenly into training and test sets.

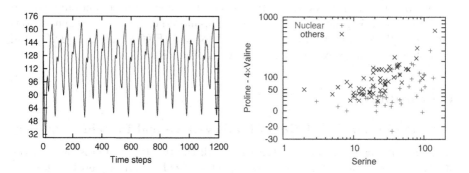

Fig. 1. Left: Mackey-Glass chaotic time series http://neural.cs.nthu.edu.tw/jang/benchmark/, $\tau = 17$. Right: Number of amino acids in nuclear and non-nuclear proteins. To reduce clutter only 5% of the proteins are plotted. The 3 (of 20) amino acids and function where suggested by sensitivity analysis of the smallest GP model

3   Genetic Programming Configuration

Even though we expect crossover [11] to be responsible for repeated sequences, we follow recent GP practise and use a high mutation rate and a mixture of different mutation operators. In some runs, to avoid bloat, we also used size fair crossover (FXO) [12]. See Table 1.

Table 1. GP Parameters for Mackey-Glass time series prediction. (Parameters for protein localisation, where different, are given in brackets [proteins:])

Function set:	MUL ADD DIV SUB operating on unsigned bytes [proteins: floats]
Terminal set:	Registers are initialised with historical values of time series. D128 128 time steps ago, D64 64, D32 32, D16 16, D8 8, D4 4, D2 2 and finally D1 with the previous value. Time points before the start of the series are set to zero. Constants 0..127.
	[Proteins: Number (integer) of each of the 20 amino acids in the protein. (Counts for code B were split evenly between aspartic acid D and asparagine N. Those for Z, between glutamic acid E and glutamine Q.) 100 unique constants randomly chosen from tangent distribution (50% between -10.0 and 10.0) [13]. (By chance none are integers.)]
Fitness:	RMS error [Proteins: $\frac{1}{2}$True Positive rate + $\frac{1}{2}$True Negative rate [14]]
Selection:	non elitist, tournament size 7. Pop Size 500 [proteins: 5000].
Initial pop:	ramped half-and-half (2:6) (50% of terminals are constants)
Parameters:	50% mutation (point 22.5%, constants 22.5%, shrink 2.5% subtree 2.5%). Max tree size 1000. Either 50% subtree crossover or 50% size fair crossover (90% on internal nodes), FXO fragments $\leq$ 30 [12]
Termination:	50 generations

Ten runs each with an initial population of 500, suggested this was too small for the protein localisation benchmark. There was a correlation (0.4 size fair and 0.2 two point (2XO) crossover) between the fitness of the best random tree and that of the best 50 generations later. So a population of 5000 and 50 generations was used. (The correlation co-efficient fell to 0.17 (FXO) and 0.12 (2XO) and mean holdout fitness rose 4% for both types of crossover.)

4   Results

4.1   Performance and Size of Mackey-Glass and Protein Programs

Table 2 summarises each of the ten runs with the two types of crossover on the Mackey-Glass modelling problem. As expected, size fair runs are both faster and evolve significantly smaller trees (Wilcoxon Two Sample Test p=0.007). Also as expected with standard GP, tree size increases up to the maximum size limit (1000) when evolution is continued to 500 generations. Figure 2 shows the fall in RMS error of the best individual in the population in each of the ten extended runs with standard crossover. It is the formation of repeated subtrees in

Table 2. Best Mackey-Glass prediction error after 50 generations of tree GP runs. Using size fair (FXO) and standard two point (2XO) crossover. Rows are RMS error and size of best of run tree and elapse time. Results after 500 generations (2XO only) show all runs improved fitness but trees increased enormously in size

											Mean
FXO error	4.42	4.38	4.85	4.89	4.01	4.92	3.84	4.65	3.66	4.80	4.44
size	33	53	81	39	55	25	15	13	69	27	41
secs	226	342	363	275	363	205	83	44	467	163	253
2XO error	3.82	3.59	3.81	4.27	4.28	2.20	2.78	4.16	2.38	3.47	3.48
size	59	45	143	117	47	87	91	43	123	145	90
secs	617	384	610	416	412	503	543	269	967	645	537
2XO error	3.74	1.51	1.18	3.66	3.41	1.09	2.78	3.78	1.08	1.85	2.41
500 size	793	705	669	957	963	883	847	923	957	467	816
gens secs	13200	12200	11400	16100	11900	14500	11000	14300	22300	9500	13600

Fig. 2. Evolution of smallest RMS error in ten 2XO M-G runs. Despite size and shape changing from one generation to the next, for many successive generations the best fitness is identical to that in the previous generation. (Initial fitness, not shown, of the ten runs varied from 5.5 to 18.3.)

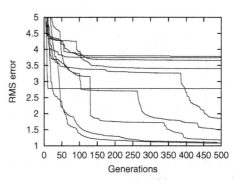

these runs (and similar protein prediction runs) that we shall concentrate upon. While at first sight progress appears continuous, note that there are many generations where the best fitness is identical to that in the previous generation even though the best individual in the population has been replaced (by crossover/ mutation).

Table 3 summarises the ten runs on the protein prediction problem with both types of crossover. Again size fair crossover produces small trees more quickly than standard GP. As with Mackey-Glass both tree GP approaches produce models with a similar performance to linear GP [7]. That is GP is comparable to the best neural network approaches given in [10].

To confirm our previous results on the evolution of tree shapes [15; 16] also hold on the two benchmarks, Figure 3 plots the size (total number of nodes) and (maximum) depth of trees at every 10 (left) or 100 (right) generations during each of the 2×ten standard GP runs. The cross hairs give the population mean and standard deviation. As expected, the GP runs do not converge, instead the populations contain trees of different sizes and depths. Figure 3 is plotted on top of statistics relating not to GP but to the underlying distribution of binary trees

Table 3. Holdout set fitness on Bioinformatics benchmark. (Fitness is mean accuracy over nuclear and non-nuclear animal proteins.) 10 tree GP runs with size fair (FXO) and 10 with standard two point (2XO) crossover with a population of 5000 and 50 gens. As with Mackey-Glass, size fair runs are both faster and evolve smaller trees

												Mean
FXO	percent	80	82	81	79	82	78	82	80	79	80	80
	size	57	77	43	47	69	77	85	59	53	41	61
	secs	1400	2300	1300	1200	2100	1700	1600	1700	1400	1400	1600
2XO	percent	81	82	80	82	83	82	83	83	82	81	82
	size	571	349	223	711	843	283	435	195	515	147	427
	secs	6100	5600	4200	6500	9600	4100	4500	4200	4800	3900	5400

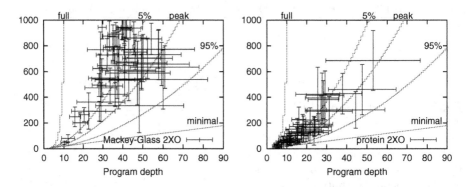

Fig. 3. Evolution of mean depth and size with mutation and standard crossover (2XO). 10 Mackey-Glass (left) and 10 protein runs (right). To reduce clutter standard deviations are only plotted every 100 generations (10 right). As expected [15], size increases until largest in population reach limit (1000) and much of the populations lie near the peak in the distribution of tree shapes

(labelled "full", "5%", "peak", "95%" and "minimal") [16]. Cf. the Catalan distribution of subtree sizes [17, p241–242]. While initial populations contain only small trees, Figure 3 shows they evolve into populations of trees whose shape lies near that of the most popular trees in the underlying distribution. Note Figure 3 shows: in radically different problems, similar shaped trees evolve.

4.2   Shape of Subtrees

The previous section has established that standard GP finds good models on both problems and programs' size and shape evolves as expected. This section starts to consider what is happening inside the trees. Figure 4 uses the same size-depth plots as Figure 3 to look at the evolved programs. Instead of one point per tree, there is a point for each node in each of the best trees in the last population of each run. Mostly subtrees lie between the 5% and 95% lines.

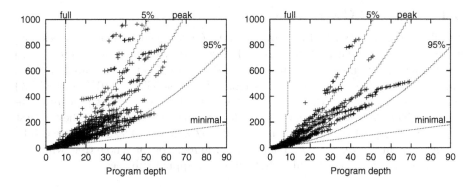

Fig. 4. Depth and size of every subtree in best of run trees (2XO). 10 Mackey-Glass (left) and 10 protein runs (right). Note the similarity with the shape of whole trees as they evolved, Figure 3. (Small amount of noise added to spread data that would otherwise be plotted directly on top of each other.)

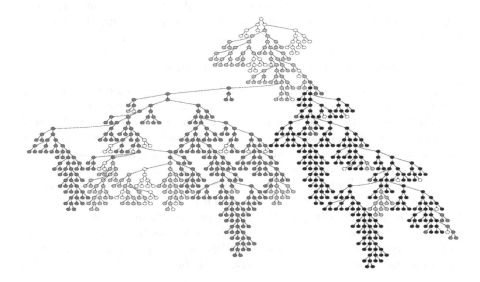

Fig. 5. Repeated patterns in the largest protein prediction program (2XO, 843 nodes). largest pattern (133 nodes) in black. Other nodes in repeated patterns are filled according to size of the repeated pattern (33–132 grey and 11–32 light grey). Unique nodes and nodes which are part of small patterns are not filled

This indicates that subtrees within the best program at the end of the runs have distributions of size and shape similar to that of the whole trees in previous generations. I.e. there is a strong tendency for trees to be composed of subtrees which are also randomly shaped. This fractal self similarity would be expected of random trees.

4.3   Repeated Code Fragments

In all cases using standard crossover (2XO), GP evolved best of run trees containing large repeated patterns. As with linear GP, this happens despite a high level of mutation and a size limit. Figure 5 shows the identical repeated patterns (allowing overlaps) for one evolved program. Between 56% and 91% (mean 71%) of the ten best of run Mackey-Glass (2XO) models are part of repeated subgraphs which are too big to have formed by chance. The figures for the ten best of run protein prediction programs are: 33%–92%, mean 74%. See Figure 6. The replications in Figures 5 and 6 refer to any fragment of the whole tree, while the rest of Section 4 considers only whole non-overlapping subtrees.

4.4   Syntactically Repeated Subtrees

Figure 7 shows the location and size of exactly repeated subtrees in the largest of the protein prediction trees. Figure 8 refers to the same twenty best of run programs as Figure 6, however it considers only exactly repeated subtrees (rather than any fragments). The requirement to include all the leafs in a repeated fragment tends to reduce their size but we see a similar picture: in every run repeated subtrees (too large to be due to chance) are evolved.

4.5   Semantically Repeated Subtree Outputs

The previous sections have only considered repeated code at the syntax level. Now we consider the semantics of the evolved programs. Since there are no side-effects, repeated subtrees must return exactly the same values. Figure 9 shows on the training examples the fraction of the program where the semantic value of the subtrees are the same, and where they are highly correlated. It shows semantic repetition is even higher than when just considering program syntax. Part of the difference is due to constants (which are always correlated with each other). However this is not enough to explain all of the difference, suggesting non-trivial syntactically different subtrees have been evolved which produce correlated answers. Part of the explanation may be symmetries, such as + and ×, whereby non-identical code calculates identical answers. Alternatively the monolithic fitness function may encourage redundant code.

4.6   Entropy of Subtrees

As might be expected, variation in values calculated by subtrees across the training set has a strong tendency to increase from the leafs to the root. This is also true of random programs. Figure 10 shows the variability within the largest protein location tree (2XO, 50 generations). We use information entropy [18] (calculated using signal value to 6 decimal places) as our measure of variation.

The protein location programs do not contain "classic" intron nodes. I.e. there are few places deep in the tree where information passes only from one input of a function to its output, totally ignoring the other input. The entropy, if any, of "classic" intron nodes would come from just one input. Thus the entropy of an "all or nothing" intron would be the same as that of its active argument.

Fig. 6. Same program as in Figure 5. Here whole subtrees are exactly repeated. Nodes are filled according to size of the repeated subtree. Unique nodes and nodes which are part of small patterns (3 nodes or less) are not filled. Two largest (59 nodes, right hand side) coloured red. Note these are partially repeated elsewhere in the tree (e.g. 55 node subtree shaded black)

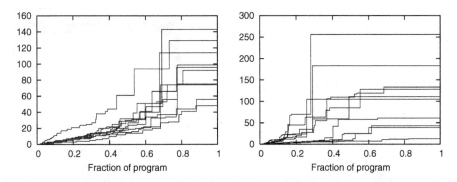

Fig. 7. Size of repeated pattern v. fraction of best of run trees (2XO). 10 Mackey-Glass (500 gens, left) and 10 protein runs (50 gens, right). In every run the largest repeated pattern is too big to arise by chance

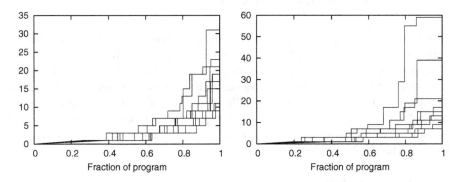

Fig. 8. Size of identical subtrees v. fraction of best of run trees (2XO). 10 Mackey-Glass (500 gens, left) and 10 protein runs (50 gens, right). In every run the largest repeated subtree is too big to arise by chance

Fig. 9. Upper curves show number of highly correlated subtrees v. fraction of the best largest protein prediction tree (2XO run 4, cf. Figures 5, 6 and 10–12). For comparison, the lower (solid) curve refers to syntactic repeats (rather than semantic). It is the top curve from Figure 8 (right). Many subtrees (22%) produce a constant. This gives rise to the sudden jump at 0.78 but only explains part of the difference between syntactic and semantic repeats

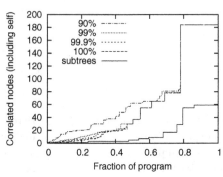

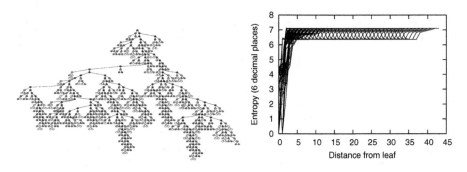

Fig. 10. Left: Entropy of each node in largest protein program (cf. Figures 5, 6 and 9–12.) Darker grey indicates more variation across the training set. At levels 7 and 9 there are two links where large subtrees pass through bottle necks. **Right:** Entropy on each of 422 paths from leaf to root. The bottle necks near the root show up as repeated dips in the tail. This structure is an artifact caused by paths passing through similar routes near the root but having different lengths

Sometimes entropy (i.e. variability) falls from the leaf towards the root are caused by a SUB subtree with both arguments referring to the same amino acid. This has no variation since it always yields zero, so the subtree has less entropy than either of its leafs. (Random programs also contain bottleneck nodes of low entropy.) Most cases where entropy falls are very close to a leaf. However a few of the largest protein location (2XO) programs do possess bottlenecks where entropy falls on the output of a large subtree. This means the subtree has less effect on the whole program.

4.7   Fitness of Subtrees

As might be expected, correlation or anti-correlation with training data tends to rise from the leafs to the root. Between 15 and 78 (depending on the run) subtrees in each best of run program exceed the performance of random search

Fig. 11. High fitness (or anti-fitness) subtrees as a fraction of the 10 best protein trees (2XO). Note range of horizontal axis. Since fitness is a very non-linear function, we define a normalised fitness as being, for each run, the generation in which a program of the corresponding fitness was first found. All runs exceeded the best fitness found in a million random trees programs by generation 8

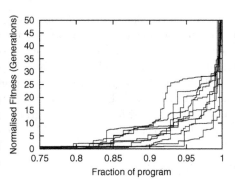

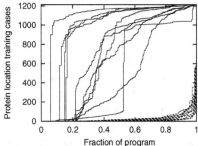

Fig. 12. Importance of nodes within protein prediction trees. **Left:** Largest protein prediction tree. The 125 (15%) subtrees which change more than 10 training cases are highlighted in black. (Same example as in Figures 5, 6–10.) Note several large repeated subtrees do not contribute to fitness. **Right:** Number of training cases which subtrees influences as a fraction of the 10 2XO best of run programs. Solid curves plot where impact is more than 0.005%. Dashed lines: node causes prediction to change

(10^6 ramped half-and-half trees). See Figure 11. Since fitness tends to fall away from the root, there are more lower fitness subtrees. Secondly, despite being non-elitist, fitness increases monotonically. Therefore the fitness distribution within the best subtrees can also be explained by saying: the longer evolution has had to work since a fitness level was reached the larger the number of subtrees exceeding that fitness there will be.

4.8   Importance of Subtrees (Sensitivity Analysis)

While the trees do not contain "classic introns", where one argument of a function has no impact on its output, some nodes do have much more impact than others. To see this, we replaced each subtree in turn by its median value and counted the number of training cases where this changed the output. The upper solid curves on the right of Figure 12 plot the number of fitness cases where the output was changed by more than 0.005%. While the lower dashed curves show the number of cases where subtrees contribute to fitness, i.e. the number of training cases where replacing it changed the program's prediction. Between 5% and 23% of nodes in protein prediction programs have less than 0.005% impact on all training cases. If we consider just fitness (lower dashed curves) this rises to between 7% and 57% of the program. I.e. on average 30% of subtrees can be replaced without changing any of the program's predictions.

5   Discussion

Sections 4.1 and 4.2 confirm (cf. [15]) trees have evolved to the same fractal shape as random trees but Sections 4.3–4.5 show repeated syntactic and semantic patterns which are far from random. Sections 4.6 and 4.7 suggest GP programs (with non-Boolean function sets without side effects) are composed of high fitness subtrees which mostly pass information upwards towards the root. That is, they are not dominated by classic "introns" (which ignore data from one or more subtrees). However the sensitivity analysis (Section 4.8) shows large parts of the tree, including repeated parts, can be replaced by a constant and have no or little effect on fitness.

We suggest the repeated patterns seen in GP used for modelling and prediction are not like classic GA "building blocks" [8]. They are not small. They have high fitness on the whole problem, rather than sub-components of it. It appears evolution is gradually, haphazardly, assembling a complete program by repeatedly reusing subtrees it has already discovered in ways allowing it to squeeze out marginal incremental improvements. In the process some components become of lesser importance in the final program.

6   Conclusions

Correlation between performance of initial and evolved populations suggests lack lustre initial random programs can have an impact on the final outcome. Correlation might be a useful population size analysis tool.

As expected, size fair crossover (FXO) [12] and a range of mutation operators controlled bloat [15]. In these experiments, the compact models were slightly worse than the much larger ones evolved with standard crossover and mutation.

Entropy and subtree fitness analysis suggest genetic programming (GP) succeeds in finding ways to put together moderately sized fit subtrees to yield larger trees containing few highly sensitive components with higher performance.

While it is always difficult to generalise from a limited number of examples, we have seen for two diverse non-trivial problems the spontaneous emergence of repeated patterns in both linear and tree based GP and with a variety genetic operations. This leads use to tentatively suggest on problems, without tight limits on tree size, depth, etc., where bloat is possible, GP will generally evolve programs containing copious repeated patterns. Although this work is far from complete, we suggest future analysis may: discover further spontaneous effects which arise from evolution rather than the programmer, cast light on the workings of GP and may lead to new automatic programming techniques.

Acknowledgements

This work was carried out at University College, London. WB acknowledges support from an NSERC discovery grant under RGPIN 283304-04.

Source Code

Code to generate Graphviz format dot files from GP programs can be found at `http://www.cs.ucl.ac.uk/staff/W.Langdon/lisp2dot.html`.

References

1. Britten, R.J., Kohnen, D.E.: Repeated sequences in DNA. Science **161** (1968) 529–540
2. Smit, A.F.A.: The origin of interspersed repeats in the human genome. Current Opinions in Genetics and Development **6** (1996) 743–748
3. Patience, C., Wilkinson, D.A., Weiss, R.A.: Our retroviral heritage. Trends in Genetics **13** (1997) 116–120
4. Lupski, J.R., Weinstock, G.M.: Short, interspersed repetitive DNA sequences in procaryotic genomes. Journal of Bacteriology **174** (1992) 4525–4529
5. Toth, G., Gaspari, Z., Jurka, J.: Microsatellites in different eukaryotic genomes: Survey and analysis. Genome Research **10** (2000) 967–981
6. Achaz, G., Rocha, E.P.C., Netter, P., Coissac, E.: Origin and fate of repeats in bacteria. Nucleic Acids Research **30** (2002) 2987–2994
7. Langdon, W.B., Banzhaf, W.: Repeated sequences in linear genetic programming genomes. Complex Systems (2005). In press.
8. O'Reilly, U.M., Oppacher, F.: The troubling aspects of a building block hypothesis for genetic programming. In Whitley, L.D., Vose, M.D., eds.: Foundations of Genetic Algorithms 3, Morgan Kaufmann (1995) 73–88
9. Oakley, H.: Two scientific applications of genetic programming: Stack filters and non-linear equation fitting to chaotic data. In Kinnear, Jr., K.E., ed.: Advances in Genetic Programming. MIT Press (1994) 369–389
10. Reinhardt, A., Hubbard, T.: Using neural networks for prediction of the subcellular location of proteins. Nucleic Acids Research **26**(9) (1998) 2230–2236
11. Koza, J.R.: Genetic Programming: On the Programming of Computers by Means of Natural Selection. MIT Press, Cambridge, MA, USA (1992)
12. Langdon, W.B.: Size fair and homologous tree genetic programming crossovers. Genetic Programming and Evolvable Machines **1**(1/2) (2000) 95–119
13. Langdon, W.B.: Genetic Programming and Data Structures. Kluwer (1998)
14. Langdon, W.B., Barrett, S.J.: Genetic programming in data mining for drug discovery. In Ghosh, A., Jain, L.C., eds.: Evolutionary Computing in Data Mining. Volume 163 of Studies in Fuzziness and Soft Computing. Springer (2004) 211–235

15. Langdon, W.B., Soule, T., Poli, R., Foster, J.A.: The evolution of size and shape. In Spector, L. *et at.*, eds.: Advances in GP 3. MIT Press (1999) 163–190
16. Langdon, W.B., Poli, R.: Foundations of Genetic Programming. Springer (2002)
17. Sedgewick, R., Flajolet, P.: An Introduction to the Analysis of Algorithms. Addison-Wesley (1996)
18. Shannon, C.E., Weaver, W.: The Mathematical Theory of Communication. The University of Illinois Press, Urbana (1964)

Tarpeian Bloat Control and Generalization Accuracy

Sébastien Mahler, Denis Robilliard, and Cyril Fonlupt

Laboratoire d'Informatique du Littoral,
Université du Littoral-Côte d'Opale,
BP719, 62228 Calais Cedex, FRANCE

Abstract. In this paper we focus on *machine-learning* issues solved with Genetic Programming (*GP*). Excessive code growth or *bloat* often happens in GP, greatly slowing down the evolution process. In [Pol03], Poli proposed the *Tarpeian Control* method to reduce bloat, but possible side-effects of this method on the generalization accuracy of GP hypotheses remained to be tested. In particular, since Tarpeian Control puts a brake on code growth, it could behave as a kind of *Occam's razor*, promoting shorter hypotheses more able to extend their knowledge to cases apart from any learning steps.

To answer this question, we experiment Tarpeian Control with symbolic regression. The results are contrasted, showing that it can either increase or reduce the generalization power of GP hypotheses, depending on the problem at hand. Experiments also confirm the decrease in size of programs. We conclude that Tarpeian Control might be useful if carefully tuned to the problem at hand.

1   Introduction

Genetic Programming (*GP*) [Koz92] is an automatic method for building programs. Usually a program is represented by a variable-length structure, and GP often produces a population of longer programs over generations but not always fitter. This leads to the so-called *bloat* phenomenon [BL02] [Luk00] when GP process drastically slows down during the evaluation step due to the increase of size of the programs. In order to reduce GP tendency to generate lengthy programs, Poli recently introduced the *Tarpeian Control* technique [Pol03], that is based on the general schema theory [Pol01].

In [Pol03], Tarpeian Control tackles the even 10 parity problem and symbolic regression with a 10-variate cubic polynomial target function. Fitness is proportional to the sum of errors made over some *training* examples and the author compares this average best fitness *versus* average mean size in the population. There is no analysis of generalization accuracy, although quality of an hypothesis is usually connected with the real error, made over all the problem instances, and not only over the samples set. Moreover there is a large agreement on considering that shorter hypotheses may have a stronger generalization accuracy.

M. Keijzer et al. (Eds.): EuroGP 2005, LNCS 3447, pp. 203–214, 2005.

Our work is an investigation to find whether or not Tarpeian Control has an effect on generalization. We use GP and Tarpeian Control over three symbolic regression benchmarks, to explore the potential benefits of the method. We approximate real error by testing hypotheses over a set of instances distinct from the learning set. In the first section, we detailed the Tarpeian Control method. Next, we focus on generalization accuracy and overfitting problem. The symbolic regression benchmarks and the experiments results are given in Sect. 4. Then we discuss in Sect. 5 about improvements added by Tarpeian Control to GP, before conclusion.

2   Tarpeian Control and Bloat

Bloat is an uncontrolled code growth that slows down the GP process and thus the fitness progress. In [Koz92], the author recommends limiting the depth of trees at both initializing and breeding steps. But definition of such a limit depends on the problem at hand. It should be based on empirical knowledge due to prior experiments. Even with depth limits, GP often increases the size of programs up to the upper limit. Luke studied in [Luk00] the complex dynamics of bloat. Parametric and non parametric methods like those of [LP02] are often justified empirically, but they may not be suitable for any given GP applications. On the other hand, Tarpeian Control is a method based on the schema theory established in [Pol01] and was recently proposed in [Pol03].

Tarpeian Control intends to slow down the growth of average size of programs in a population. Estimation of the average size for the next generation with the schema theory is closely linked to the selection probability of each program. Poli suggests to periodically reduce the selection probability of *some* longer-than-average programs. Since the selection probability directly depends on fitness value, zeroing the fitness of some programs will greatly reduce their chances of being selected for future generations. In this article, we denote as *Tarpeian target* the subset of above-average-sized individuals that will undergo the lost in fitness.

Following Poli's idea, we propose in Table 1 a possible pseudo-code for implementing Tarpeian Control. The *Target Ratio* parameter gives the percentage of over average sized programs that are targeted at every generation. The great

Table 1. A possible Tarpeian Wrapper

```
function evaluation(program)
  begin
    IF ( (size(program) > average_pop_size)
         AND (random() < target_ratio) )
      THEN return( very_low_fitness );
      ELSE return( fitness(program) );
  end;
```

advantage of such a code is that it does not require prior knowledge to estimate any limit in size or depth. Since Tarpeian Control acts on selection probability, no replacement step is required unlike the *Death-by-size* method proposed in [PL04]. Two remarks come about the code in Table 1:

1. Tarpeian Control behaves like a random selection operator for above-average-sized programs.
2. The number of programs in the Tarpeian target depends on the number of above-average-sized programs in the population, and thus it may vary from a generation to the next.

Notice that no rule is given in [Pol03] to choose a good value for Target Ratio, that turns out as a supplementary GP parameter that needs to be tuned. In Sect. 4, we experiment Tarpeian Control with five different ratios, to try to obtain some hints about satisfactory values.

Poli presented Tarpeian Control as a wrapper for the evaluation step. But wrapping often means recompiling the GP loop of an evolutionary system. We propose an implementation for the ECJ library [Luk03], such that the existing Java code is left unchanged. Details may be found in Appendix or at `http://www-lil.univ-littoral.fr/~mahler/TarpeianPipeline/`.

3 Generalization and Overfitting

Since GP tackles machine learning issues, the overfitting risk must be taken into account: it worsens accuracy of hypotheses. Definition 1 reminds what Mitchell calls overfitting in [Mit97].

Definition 1. *Given a hypothesis space H, a hypothesis $h \in H$ is said to overfit the training data if there exists some alternative hypothesis $h' \in H$, such that h has smaller error than h' over the training examples, but h' has a smaller error than h' over the entire distribution of instances.*

According to Def. 1, overfitting involves computing the error on the whole distribution of instances from a given problem, also called the real error **REr**. However in most real world cases, the real error can only be approximated: the available instances are often split into two sets, one for training, the other one –called the test set or validation set– for estimating **REr**.

Most often the result of GP is the best fitness program at the last generation. In the following sections, **REr**(*end*) denotes the generalization error of that particular program.

4 Experimental Procedure

We test GP over artificial benchmarks where samples are singletons $\{x, f(x)\}$ generated by a known formula $f : \mathbb{R}^2 \to \mathbb{R}$. Singletons $\{x, f(x)\}$ used as test cases have an x value taken from a regular mesh, while random x values are used to build the learning samples.

Three symbolic regression problems referred as $SR1$, $SR2$ and $SR3$ are defined in Table 2 and appeared in benchmarks in [Kei03], [SB01] and [TP01]. The function set is simpler than those in [Kei03] for an easier resolution of the problem, with less test cases. Table 3 shows GP settings of our experiments. We use Root-Mean-Squared error (RMS) defined in (1) as a standardized positive fitness –absolute best value is 0. We record the learning error –our fitness– of the best fitness program at every generation. We keep the generalization error **REr**(*end*) of the best fitness program (by computing the RMS error made over the test cases) and the average size of programs in the population, at the last generation. To compare GP and Tarpeian Control, we conduct six experiments for each problem: one with classical GP and five with Tarpeian Control, Target Ratio set up to values ranging from 10% to 50% by 10% steps.

After 35 runs per experiment, we observe values from the last generation: the average program size in the population, and the generalization error made by the best fitness individual. Results are shown in boxplots (Figures 1 to 3). We use statistical tests as explained in the next section. For a quick comparison, the evolution of best fitness values (average after 35 runs) is shown in Fig. 4.

Table 2. Symbolic regression settings

Problem name	$SR1$	$SR2$	$SR3$
Target: f(x,y)	x^y	$x*y + \sin((x-1)(y+1))$	$x^4 - x^3 + \frac{y^2}{2} - y$
Fitness Cases $\{(x,y), f(x,y)\}$	100 random points (x,y) in [0,1]x[0,1]	20 random points (x,y) in [-3,3]x[-3,3]	20 random points (x,y) in [-3,3]x[-3,3]
Test Cases	mesh with (x,y) in [0,1]x[0,1] step 0.01	mesh with (x,y) in [-3,3]x[-3,3] step 0.1	mesh with (x,y) in [-3,3]x[-3,3] step 0.1

$$RMS = \sqrt{\frac{1}{n} \times \sum (p(x_i, y_i) - f(x_i, y_i))^2} \qquad (1)$$

Root-Mean-Squared error of a program p with aim to fit a function $f : \mathbb{R}^2 \to \mathbb{R}$ over n samples.

Table 3. GP general parameters applied to GP alone, and to GP with Tarpeian Control set with 10%, 20%, 30%, 40% and 50% ratios

Population	500 programs
Evaluations	25000 evaluations
GP Nodes	Add, Sub, Mul, pDiv, ERC, X, Y (if $|den| < 10^{-6}$ then $pDiv(num, den) = 1.0$)
Initialization	ramped half & half
Operators	Mutation(5%), Crossover(85%), Reproduction(10%), Elitism
Max Depth	12

5   Results and Discussion

5.1   Statistics and Tests Preamble

We care about the generalization error in this article, and that error is not as regular as the learning error over generations and from one run to another. We can see some fitness –or learning error– improvements in Figure 4 whereas it is not so evident for generalization errors in Figures 1 to 3. Extreme generalization errors values ($\mathbf{REr}(end)$) have happened, especially for problem $SR\,3$. So only comparing average generalization values may lead to a wrong interpretation. We suspect the $\mathbf{REr}(end)$ variable does not follow a normal law. We then test over the 6 sets of 35 values of $\mathbf{REr}(end)$, the normality of that variable. The *Shapiro-Wilk* test for normality helps us to decide which one of these hypotheses is the most probable:

- *H0* = "With the 35 final generalization errors we have observed for an experiment, the $\mathbf{REr}(end)$ variable generally follows a normal law."
- *H1* = "With the 35 final generalization errors we have observed for an experiment, the $\mathbf{REr}(end)$ variable *does not* generally follow a normal law."

The result of a test is a probability (called *p-value*) of having taken the wrong decision when choosing H1 (the alternative hypothesis): this leads the user to *reject* the null hypothesis (H0) –or to say H1 is more probable, if the p-value if very low. Doing that choice we take a risk we choose in advance: the alpha

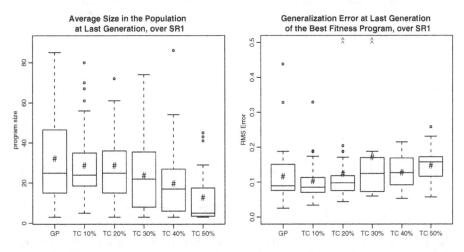

Fig. 1. Problem $SR\,1$: average size in the population and generalization error of the best fitness program, at last generation after 35 runs. Average value is the *sharp point*, and *hat points* ˆ represent extreme upper values. Median is the *horizontal line* in the middle of the box

quantity. For our work we choose $alpha = 0.05$, meaning that we can base decisions on a 95% confidence level.

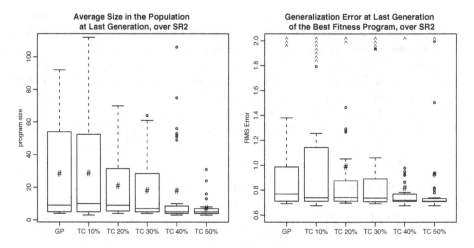

Fig. 2. Problem *SR*2: average size in the population and generalization error of the best fitness program, at last generation after 35 runs. Average value is the *sharp point*, out of the graph for GP generalization because of extreme upper values (*hat points* ^). Median is the *horizontal line* in the middle of the box

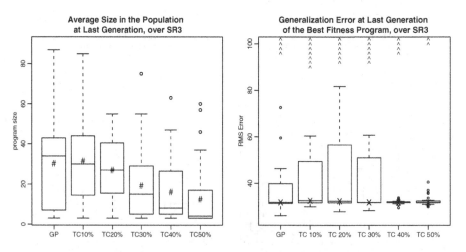

Fig. 3. Problem *SR*3: average size in the population and generalization error of the best fitness program taken at last generation after 35 runs. Average value is *sharp point*, often out of the graph because of extreme upper values (*hat points* ^), and *X points* show the median value

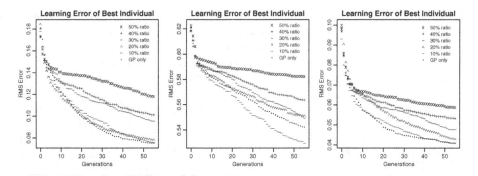

Fig. 4. Best Learning Error –or best fitness– evolving among generations (average value after 35 runs)

In Tables 4 to 6, we see the p-value for Shapiro-Wilk test is very low in most cases, which confirms the non-normality.

We then test GP *versus* Tarpeian Control wondering if the later is a brake on the generalization power or if it provides more easily generalizable programs. Thus we use the Wilcoxon test (that has no normality condition) to decide which one of two hypotheses is the most probable. The following hypotheses are tested:

Table 4. Does the generalization error of the best fitness program ($\mathbf{RER}(end)$) follow a normal law over Problem $SR1$? Is $\mathbf{RER}(end)$ of Tarpeian Control significantly better (or worse) than the $\mathbf{RER}(end)$ of GP ? After 35 runs per experiment, we use *Shapiro-Wilk* test for normality of $\mathbf{RER}(end)$. We reject the normality hypothesis in most cases (p-value $\leq$ 5%) with an alpha risk of 0.05. We compare GP results *versus* Tarpeian Control with the *Wilcoxon* two-samples test. Since Wilcoxon p-values are often high, Tarpeian Control with low ratios and GP $\mathbf{RER}(end)$ results are probably similar. We can assert with a 95% confidence level that $\mathbf{RER}(end)$ of Tarpeian Control with 40% and 50% ratios will be outperformed by that of GP, over $SR1$ problem

alpha 0.05 experiment	Shapiro p.value	Wilcoxon Test: H0 = $\mathbf{RER}(end)_{GP} \sim \mathbf{RER}(end)_{TC}$ over SR1		
		H1: GP *vs* TC	p.value	*test meaning* (at alpha risk)
GP	41.493%			
TC 10% ratio	9.307%	lower	70.758%	no difference
TC 20% ratio	$\leq 10^{-5}$	lower	46.022%	no difference
TC 30% ratio	$\leq 10^{-5}$	lower	15.055%	no difference
TC 40% ratio	0.002%	lower	2.662%	GP has a better $\mathbf{RER}(end)$ than TC
TC 50% ratio	$\leq 10^{-5}$	lower	0.048%	GP has a better $\mathbf{RER}(end)$ than TC
TC 10% ratio	9.307%	greater	29.647%	no difference
TC 20% ratio	$\leq 10^{-5}$	greater	54.444%	no difference
TC 30% ratio	$\leq 10^{-5}$	greater	85.218%	no difference
TC 40% ratio	0.002%	greater	97.410%	no difference
TC 50% ratio	$\leq 10^{-5}$	greater	99.954%	no difference

Table 5. Problem $SR2$: Generalization error ($\mathbf{REr}(end)$) normality test and comparison between $\mathbf{REr}(end)$ of Tarpeian Control and $\mathbf{REr}(end)$ of GP, after 35 runs per experiment. Tarpeian Control with low ratios and GP $\mathbf{REr}(end)$ results are similar. We can assert with a 95% confidence level that $\mathbf{REr}(end)$ of GP is outperformed by Tarpeian Control with 40% and 50% ratios, over $SR2$ problem

alpha 0.05 experiment	Shapiro p.value	Wilcoxon Test: H0 = $\mathbf{REr}(end)_{GP} \sim \mathbf{REr}(end)_{TC}$ over SR2 H1: GP vs TC	p.value	test meaning (at alpha risk)
GP	$\leq 10^{-5}$			
TC 10% ratio	$\leq 10^{-5}$	lower	57.917%	no difference
TC 20% ratio	$\leq 10^{-5}$	lower	68.917%	no difference
TC 30% ratio	$\leq 10^{-5}$	lower	88.232%	no difference
TC 40% ratio	$\leq 10^{-5}$	lower	97.896%	no difference
TC 50% ratio	$\leq 10^{-5}$	lower	99.570%	no difference
TC 10% ratio	$\leq 10^{-5}$	greater	42.543%	no difference
TC 20% ratio	$\leq 10^{-5}$	greater	31.499%	no difference
TC 30% ratio	$\leq 10^{-5}$	greater	12.001%	no difference
TC 40% ratio	$\leq 10^{-5}$	greater	2.164%	GP has a worse $\mathbf{REr}(end)$ than TC
TC 50% ratio	$\leq 10^{-5}$	greater	0.445%	GP has a worse $\mathbf{REr}(end)$ than TC

Table 6. Problem $SR3$: Generalization error ($\mathbf{REr}(end)$) normality test and comparison between $\mathbf{REr}(end)$ of Tarpeian Control and $\mathbf{REr}(end)$ of GP, after 35 runs per experiment. Tarpeian Control and GP $\mathbf{REr}(end)$ results are similar

alpha 0.05 experiment	Shapiro p.value	Wilcoxon Test. H0 = $\mathbf{REr}(end)_{GP} \sim \mathbf{REr}(end)_{TC}$ over SR3 H1: GP vs TC	p.value	test meaning (at alpha risk)
GP	$\leq 10^{-5}$			
TC 10% ratio	$\leq 10^{-5}$	lower	21.562%	no difference
TC 20% ratio	$\leq 10^{-5}$	lower	16.478%	no difference
TC 30% ratio	$\leq 10^{-5}$	lower	26.679%	no difference
TC 40% ratio	$\leq 10^{-5}$	lower	71.755%	no difference
TC 50% ratio	$\leq 10^{-5}$	lower	73.706%	no difference
TC 10% ratio	$\leq 10^{-5}$	greater	78.780%	no difference
TC 20% ratio	$\leq 10^{-5}$	greater	83.812%	no difference
TC 30% ratio	$\leq 10^{-5}$	greater	73.706%	no difference
TC 40% ratio	$\leq 10^{-5}$	greater	28.644%	no difference
TC 50% ratio	$\leq 10^{-5}$	greater	26.678%	no difference

- $H0$ = "GP and Tarpeian Control provides programs with similar $\mathbf{REr}(end)$."
- $H1$ = "GP provides programs making a better (lower) generalization error than Tarpeian Control." (H1 denoted *lower* in Tables 4 to 6)
- $H1'$ = "Tarpeian Control provides programs making a better (lower) generalization error than GP." (H1 denoted *greater* in Tables 4 to 6,)

In Tables 4 to 6 we test $H0$ *vs* $H1$ in the first five lines (with $H1$: lower), and then we test $H0$ *vs* $H1'$ in the last lines (with $H1$: greater).

5.2   Results

We first observe the average size improvement at last generation over the first problem $SR1$ in Figure 1: Tarpeian Control decreases the average program size. About generalization error, statistical tests in Table 4 show there must be no improvement with 10% to 30% Tarpeian Ratios. The bloat control method with 40% and 50% Tarpeian Ratios even builds programs with a worse error than that of GP programs.

When it comes to the second problem $SR2$, Tarpeian Control improves average size as seen in Figure 2. In Table 5, results of statistical tests lead us to conclude that Tarpeian Control with high ratios (40% and 50%) improves generalization error of the best fitness program when compared to GP programs. When lower ratios are used, the difference with GP generalization is not statistically significant.

Results over the third problem $SR3$ again show a decrease of the average program size in Figure 3. The generalization errors we observe do not reveal any statistical difference between generalization errors of programs produced with GP and with GP under Tarpeian Control, as seen in Table 6.

5.3   Overfitting

Generalization error over the first problem ($SR1$) does not have extreme values with GP, while some values above 0.5 appear with Tarpeian Control with 20% and 30% ratios. Moreover the extreme generalization error values observed with GP and Tarpeian Control, over problems $SR2$ and $SR3$ lead to high average values not shown in Figures 2 and 3. Keijzer pointed in [Kei03] that asymptotes and singular points may appear with the protected division. So RMS errors may have high values because of only one run giving a model with infinite asymptote close to one case in the test set. We suppose this "evident" overfitting phenomenon happens here. High ratios for Target Ratio do not prevent such a situation.

5.4   Tarpeian Control Effects

The Wilcoxon test shows there are some significant differences between GP alone and GP with Tarpeian Control. The high ratios may improve **REr**(end) over problem $SR2$. But the same ratios worsen the results over problem $SR1$. So Tarpeian Control does not seem to have a "silver bullet" ratio: the generalization benefits of Tarpeian Control depend on the application as well as on the Target Ratio value. While we do not mainly focus on size in this article, Figs. 1 to 3 show the average program size in the population at the last generation is reduced when Tarpeian Control is used. We can conclude Tarpeian Control reduces bloat. Since Tarpeian Control uses a random selection to keep out of reproduction the above-average-sized programs with no notice of their fitness, it looks the shorter hypotheses left in the population may not be the one with better generalization error. Despite its effectiveness against bloat, a blind use of Tarpeian Control is not recommended. However the fact that a too strong bloat control can degrade learning is not particular to Tarpeian Control: other bloat control schemes can also suffer from this.

6   Conclusion

We test the impact of Tarpeian Control over the generalization accuracy and overfitting threat for GP applied to symbolic regression problems. All problems are not equal towards Tarpeian Control. We suggest the difficulty of the problem and the Target Ratio value for Tarpeian Control both influence its action towards generalization accuracy.

Tarpeian Control might improve GP generalization accuracy as for problem $SR2$ with 40% and 50% ratios. It may also worsen that property as seen for problem $SR1$ with 40% and 50% ratios. As usual with bloat control methods, Tarpeian Control tries to keep a low average program size, and, if pushed too far, it may slow down the fitness progress too, because of the random "elimination" of longer-than-average programs that would have a good fitness.

A blind use of Tarpeian Control is no good help for GP: there is no wild-card setting for the Target Ratio parameter. However we have experimentaly demonstrated that a careful tuning can improve the generalization accuracy.

These experiments raise the question whether there could be an automated way of tuning the target ratio of Tarpeian Control, using some sort of cross validation measures with a test set during the run.

References

[BL02]   W. Banzhaf and W. B. Langdon. Some considerations on the reason for bloat. *Genetic Programming and Evolvable Machines*, 3(1):81–91, March 2002.

[DPB+04]   Kalyanmoy Deb, Riccardo Poli, Wolfgang Banzhaf, Hans-Georg Beyer, Edmund Burke, Paul Darwen, Dipankar Dasgupta, Dario Floreano, James Foster, Mark Harman, Owen Holland, Pier Luca Lanzi, Lee Spector, Andrea Tettamanzi, Dirk Thierens, and Andy Tyrrell, editors. *Proceedings of the Genetic and Evolutionary Computation Conference – GECCO-2004, Part II*, volume 3103 of *Lecture Notes in Computer Science*, Seattle, WA, USA, 26-30 June 2004. Springer-Verlag.

[Kei03]   Maarten Keijzer. Improving symbolic regression with interval arithmetic and linear scaling. In *[RSK+03]*, pages 71–83, Essex, 2003.

[Koz92]   John R. Koza. *Genetic Programming: On the Programming of Computers by Means of Natural Selection*. MIT Press, Cambridge, MA, USA, 1992.

[LP02]   Sean Luke and Liviu Panait. Lexicographic parsimony pressure. In W. B. Langdon, E. Cantú-Paz, K. Mathias, R. Roy, D. Davis, R. Poli, K. Balakrishnan, V. Honavar, G. Rudolph, J. Wegener, L. Bull, M. A. Potter, A. C. Schultz, J. F. Miller, E. Burke, and N. Jonoska, editors, *GECCO 2002: Proceedings of the Genetic and Evolutionary Computation Conference*, pages 829–836, New York, 9-13 July 2002. Morgan Kaufmann Publishers.

[Luk00]   Sean Luke. *Issues in Scaling Genetic Programming: Breeding Strategies, Tree Generation, and Code Bloat.* PhD thesis, Department of Computer Science, University of Maryland, A. V. Williams Building, University of Maryland, College Park, MD 20742 USA, 2000.

[Luk03]   Sean Luke. Ecj 10 : An evolutionnary computation research system in java, 2003.

[Mit97]   Tom M. Mitchell. *Machine Learning.* McGraw-Hill, 1997.

[PL04]   Liviu Panait and Sean Luke.   Alternative bloat control methods.   In *[DPB+ 04]*, pages 630–641, 2004.

[Pol01]   Riccardo Poli.   General schema theory for genetic programming with subtree-swapping crossover.   In Julian F. Miller, Marco Tomassini, Pier Luca Lanzi, Conor Ryan, Andrea G. B. Tettamanzi, and William B. Langdon, editors, *Genetic Programming, Proceedings of EuroGP'2001*, volume 2038 of *LNCS*, pages 143–159, Lake Como, Italy, 18-20 April 2001. Springer-Verlag.

[Pol03]   Riccardo Poli.   A simple but theoretically-motivated method to control bloat in genetic programming. In *[RSK+ 03]*, pages 200–210, 2003.

[RSK+03]   Conor Ryan, Terence Soule, Maarten Keikzer, Edward Tsang, Riccardo Poli, and Enersto Costa, editors. *6th European Conference, EuroGP 2003*, volume 2610 of *LNCS*, Colchester, UK, apr 2003. Springer.

[SB01]   Matthew Streeter and Lee A. Becker. Automated discovery of numerical approximation formulae via genetic programming.   In *[SGW+ 01]*, pages 147–154, 2001.

[SGW+01]   Lee Spector, Erik D. Goodman, Annie Wu, W. B. Langdon, Hans-Michael Voigt, Mitsuo Gen, Sandip Sen, Marco Dorigo, Shahram Pezeshk, Max H. Garzon, and Edmund Burke, editors. *Proceedings of the Genetic and Evolutionary Computation Conference (GECCO-2001)*, San Francisco, California, USA, 7-11 jul 2001. Morgan Kaufmann.

[TP01]   Alexander Topchy and William F. Punch.   Faster genetic programming based on local gradient search of numeric leaf values. In *[SGW+ 01]*, pages 155–162, 2001.

Appendix: Tarpeian Control Integrated in ECJ

With ECJ library [Luk03], we can modify the GP loop with the help of a parameter file, without recompiling the whole library. The *pipeline* class helps to setup any complex breeding steps: it connects selection and genetic operators together. GP individuals can be forwarded through several paths written in the parameter file and made of different operators as illustrated in Figure 5. ECJ evaluation uses a classical *evaluated flag*, so adding a new pipeline that turns some flags to on is similar to the wrapper action proposed by Poli in [Pol03]. We propose the following classes to implement Tarpeian Control:

- `TarpeianPipeline` class, that chooses a fraction of above-average-sized programs, assigns them a low fitness value and turns on their evaluated flag,
- `AverageSizeStatistics` class, that computes average size before breeding step.

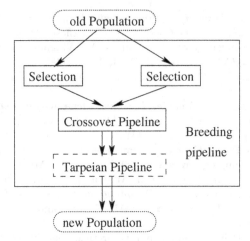

Fig. 5. A breeding pipeline example used in ECJ for Tarpeian Control integration

After insertion of two compiled class in the library, one needs to modify the parameter file to add the following settings:

- a new stage in the pipeline just before the evaluation step,
- a `target_ratio` value,
- a `very_low_fitness` value,
- a supplementary statistics class.

Source code and application example may be found at:
`http://www-lil.univ-littoral.fr/~mahler/TarpeianPipeline/`.

The Tree-String Problem: An Artificial Domain for Structure and Content Search

Steven Gustafson, Edmund K. Burke, and Natalio Krasnogor

School of Computer Science & IT, University of Nottingham,
Jubilee Campus, Wollaton Rd. Nottingham, NG81BB, United Kingdom
{smg, ekb, nxk}@cs.nott.ac.uk

Abstract. This paper introduces the Tree-String problem for genetic programming and related search and optimisation methods. To improve the understanding of optimisation and search methods, we aim to capture the complex dynamic created by the interdependencies of solution structure and content. Thus, we created an artificial domain that is amenable for analysis, yet representative of a wide-range of real-world applications. The Tree-String problem provides several benefits, including: the direct control of both structure and content objectives, the production of a rich and representative search space, the ability to create tunably difficult and random instances and the flexibility for specialisation.

1   Introduction

The behaviour of heuristic search algorithms in artificial intelligence domains (and other complex scenarios like operations research) is difficult to pin-down by conventional analytical methods. More specifically, as heuristic search algorithms are often stochastic in nature, they frequently result in incomplete searches, re-sample previously-visited states, oscillate between states and become trapped in local optimum. The fixed points of heuristics are usually hard to determine, making their run time average and worst case complexity difficult to assess [1]. Consequently, the design and application of heuristic methods for real-world problems typically proceeds by trial-and-error. However, artificial domains can provide insight into the search abilities of various algorithms, allowing future research to better apply these methods. Improving understanding of these methods is a step toward more general search and optimisation methods. This paper introduces a new artificial domain to improve the understanding of solution structure and content in heuristic methods.

Many real-world problems contain two key overlapping and often conflicting objectives: solutions must have a structure (e.g. topology), and the structure must be "filled" with the appropriate content. Examples of these objectives can be seen in planning, classification using decision trees and symbolic regression. Planning typically requires hierarchical solutions that encapsulate key low-level behaviours. An example in mobile robot planning is the issue of localisation [2]: robots have a difficult time maintaining a good approximation of their location

M. Keijzer et al. (Eds.): EuroGP 2005, LNCS 3447, pp. 215–226, 2005.

during moving and sensing. Localisation is a key low-level behaviour that needs to be carried out during a high-level strategy to allow for effective planning. The induction of decision trees for classification constructs solution structure simultaneously with data set features at each tree location. Higher level nodes typically encode more important features, while lower level nodes are used to make finer class distinctions. A classic example of such a method is Quinlan's ID3 algorithm [3]. Finally, structure and content issues can be found in the induction of mathematical expressions from data. In this case we look for both a functional form and its ideal operators and coefficients.

The above examples emphasise the interdependencies between solution structure and content. As these types of problems rarely contain features that can be optimised independent of the whole, artificial domains that allow direct manipulation of structure and content also need to ensure that the richness of the interdependencies is maintained. Genetic programming is the prototypical method that must deal with solution structure and content issues during its search for algorithmic solutions. Genetic programming handles structure and content issues implicitly in its search process. While genetic programming is shown to be competent in overcoming these conflicts in several real-world domains, it is not known whether the way it deals with structure and content is optimal or particularly good.

Genetic programming is an evolutionary algorithm that represents solutions as computer programs[4]. Artificial domains are frequently used as testbed problems: the most popular being the Artificial Ant problem, the family of Boolean problems (e.g. even-parity), and symbolic regression problems. These problems provide testbeds that represent problems such as planning, digital design, classification and mathematical regression. However, these domains typically lack random instance generation, the ability to easily create tunably difficult and large instances for studying asymptotic behaviour, and a clear distinction between the issues of solution structure and content conflicts.

Previous work has highlighted the desire of the community to address these issues. To improve solution generalisation, a random trail generator was created for the Artificial Ant problem to complement the existing use of the the Santa Fe trail [5]. While investigating *hardness* in genetic programming, tunably difficult instances of the Binomial-3 regression problem were found [6]. In this case, genetic programming was shown to have a harder time dealing with ill-suited constants. Also in the regression domain, tunably difficult random polynomials were created by considering the increased precision required by an approximation using the same search space (i.e. primitive constant ranges) [7]. This allowed the study of code growth under varied levels of difficulty for genetic programming. The aforementioned problems place emphasis on solution content, which is not independent from solution structure. The following problems direct attention back toward solution structure.

The Lid problem [8] focused only on the search for structure by using fixed arity primitives with no meaning themselves, other than for creating tree shapes. Instances in this problem, using a canonical representation and operator, were tunably difficult and allowed a more direct examination of structure mechanisms and representation issues during search. The Max problem [9, 10] and the Royal Tree problem [11] were created to contain a singular goal state to allow analysis of how structure acquires appropriate content. These problems define an ideal solution that requires specific primitives at specific structure locations. Although these problems do have intermediate reward states, they can appear to be like needle-in-the-haystack problems that may not accurately reflect real-world problems and are somewhat limited in their flexibility for producing random instances that are tunable. A more complex Royal-Tree-like problem was defined in [12] that consisted of finding the correct proportions of subprograms using multi-arity nodes. This problem, along with the ORDER and MAJORITY problems [13, 14], investigated the relationship between content and structure, where the latter two were mainly concerned with the occurrence and location of primitives in solutions. Again, while these problems address particular issues in understanding difficulty with the canonical representation and operators, it is less clear as to how they are representative of real-world problems.

The Tree-String problem attempts to bridge the gap between simple and highly-specific problems to real-world problems by providing instance tunability, random instance generation, and a rich and complex search space, while still being amenable to analysis. This last point, amenability to analysis, is gained from the use of simple and clear methods and the ability to use small population sizes while maintaining complex behaviour. The paper proceeds by first defining the Tree-String problem. We then provide an empirical study to further demonstrate the tunability of instances and the complex search space attained using the Tree-String problem.

2   The Tree-String Problem Definition

The Tree-String problem was originally intended to be an artificial domain for genetic programming, but the domain also has possible applications in other areas of artificial intelligence. The goal of the Tree-String problem is to derive specific structure and content elements simultaneously. Instances are defined using a target solution consisting of a tree shape and content. Candidate solutions are then measured for their similarity to the target solution with respect to both tree shape and content objectives.

The Tree-String problem is defined as a tuple Π:

$$\Pi = (\Psi, \Xi, t, \alpha, \gamma, \delta),$$

where an instance is represented by a target solution t, composed of content elements from the set Ψ and has a tree shape defined by elements from the set Ξ. For example, binary tree structures which have internal nodes n and leaf nodes l would have $\Xi = \{n, l\}$. The functions α and γ map the instance t to two linear string representations, such that:

For structure: $\alpha(t) \mapsto \Xi^*$,   and for content: $\gamma(t) \mapsto \Psi^*$.

Finally, the function δ provides a measure of similarity that will represent fitness objectives, i.e. similarity, between two strings representing tree shape and the similarity between two strings representing tree content. That is, given a candidate solution t_c and target solution t_t:

$$\delta(\alpha(t_t), \alpha(t_c)) \mapsto i \in \aleph, \quad \text{and} \quad \delta(\gamma(t_t), \gamma(t_c)) \mapsto j \in \aleph,$$

where i and j represent the heuristic solution quality of t_c compared to instance t_t. The fitness function of the genetic programming system, or other heuristic search method, can then use these quality measures in a multiobjective selection method or linear combination, where the former is used in this paper. While the implementations of α, γ and δ can vary, to represent more closely the particulars of a given problem domain, in this paper we propose to fix them as follows:

- α : depth-first, in-order, tree traversal for solution content,
- γ : breadth-first tree traversal for solution structure,
- δ : longest common substring (LCS).

To further illustrate the Tree-String problem, let us consider an example using binary trees $\Xi = \{n, l\}$ with content using two symbols, $\Psi = \{A, B\}$. Note that the symbol A can either be a node or a leaf. Next, consider an instance t_t which has the following properties:

- the γ function makes a breadth-first tree traversal over the shape elements in t_t to produce $\gamma(t_t) = nnnllnlll$, and
- the α function makes a depth-first tree traversal over the content of t_t to produce $\alpha(t_t) = AAAABBBBB$.

Now, let us imagine that a search method generated a candidate solution t_c such that $\gamma(t_c) = nnnnlllll$ and $\alpha(t_c) = BBBAABBAA$. We then compute the measure of solution quality using δ (i.e. the longest common substring between the components of t_t and t_c), where the common substrings are underlined:

- $\delta(\alpha(t_t), \alpha(t_c)) = \text{LCS}(\underline{nnnlln}lll, \underline{nnnnl}llll) = 5$ and
- $\delta(\gamma(t_t), \gamma(t_c)) = \text{LCS}(AA\underline{AABBBB}B, BBB\underline{AABBB}AA) = 4$.

The elements of the candidate solution t_c that contributed to solution quality are shown below. The tree on the left shows the target tree instance t_t. The tree on the right shows the candidate solution t_c. The **structure** components in t_c that contribute toward fitness are denoted in parentheses (e.g. (A)), and the **content** components that contribute toward fitness are emphasized in bold italics (e.g. *A*):

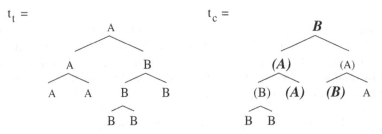

The above example demonstrates the conflicting nature of structure and content objectives, where the portion of the solution that contributes to the structure objective is different from the part that contributes toward the content objective. This property is likely to make it difficult for transformation operators to effect either content or structure objectives alone, making the two features interdependent.

The choices of breadth-first and depth-first traversals for γ and α was purposefully done to exploit the hierarchical nature of solution structure and element juxtaposition of solution content, respectively. These functions also allow the search to focus on key *features* of target solutions. By features, we refer to more general properties (e.g. for structure: balanced, sparse or bushy trees). While an instance of the Tree-String problem would use a pre-selected structure and content, these do not necessarily define one unique goal state that would achieve maximal fitness. This is different from other domains like the Royal Tree or Max problems. However, the use of the longest common substring measure guarantees that strings are compared with their order preserved. Other measures like edit distance would provide the same value if two strings match every-other symbol or the same number of consecutive symbols. The longest common substring function complements the flexibility in the depth/breadth-first traversals with the more strict requirement of contiguous matching elements. It is our goal that these definitions allow for suitably complex behaviour representing real-world domains, but that is well-defined and amenable to analysis.

To further illustrate the Tree-String problem, we report preliminary work toward furthering the understanding of problem difficulty in genetic programming.

3   A Preliminary Study of Difficulty

In [15], the Tree-String problem was used to represent key properties of other common testbed domains (Artificial Ant, Parity, Regression) to study dissimilarity. A single instance of the problem using a binary tree shape and four content symbols was randomly produced. The tree shape was selected from those found to be more easily encountered by genetic programming [8]. The use of four symbols was an approximation to the typical size of function and primitive sets used in other testbeds. Random trials were carried out on this instance. A subsequent analysis over the three common testbed problems suggested that their specific behaviours were captured by the Tree-String problem. That is, the instance of the Tree-String problem represented the general behaviour of the other problems, see Chapter 7 of [15]. However, the full potential of the Tree-String problem was not used in that study, which is now being extended using a range of tunable instances. We report on that progress next.

3.1   Experimental Methodology

The genetic programming algorithm is generational with a population size of 50. Two-parent subtree crossover is used to transform existing solutions into new

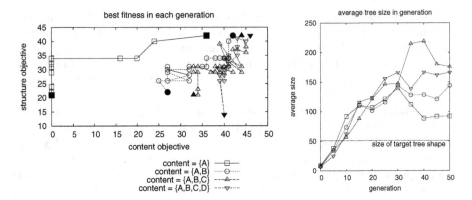

Fig. 1. The best fitness of four runs, using the same target tree shape and four different content strings, is plotted in the left graph. The first and last generation best fitness values are shown in solid symbols (■, •, ▲, ▼), where the first generation of each run is located in the upper right of the plot. The average tree size of each population is shown in the right graph

ones. Two parent crossover selects a subtree (where non-leaf nodes are selected 90% of the time) from each parent and swaps them. All children are valid provided they are within a predefined depth limit. To select parents for crossover, tournament selection with tournament size of 3 is used. The initial population is created by producing random trees using the Full and Grow methods equally between depths 2 and 4. A maximum depth for new trees is 17, and a stopping criterion of 50 generations is used. Ignoring the small population size, the system used here is a canonical system. A multiobjective pareto criterion is used for fitness evaluation with the objectives of structure and content, using the functions described in Section 2. A pareto optimal, or best fit solution is one which is better in at least one objective and no worse in the other compared to the rest of the population.

3.2   Single Structure, Multiple Content Behaviour

Initially, we look at the behaviour of four runs with the above system using one pre-selected tree shape (tree shape #2 in Figure 2 with depth 9 and 51 nodes) with four randomly created content strings (each with an increasing number of symbols, from 1 to 4). Figure 1 shows the evolution of the best fitness in each generation for each of the four runs. Here the fitness objectives report the size of the target strings (51 symbols) minus the longest common substring: a minimisation problem.

In Figure 1, the left graph shows that the more symbols in the content set Ψ, the more the search process optimises for tree shape. With one symbol in the content set, the search process can easily find a solution with the correct content (the size of the target tree in this case). However, as the number of content symbols is increased to four, the search process makes very little progress improving the content objective, but focuses instead on the tree shape. The right graph of Figure 1 shows the evolution of the average solution size in each of the four runs. We can see that at generation 10, the easiest instance ($\Psi = A$), had the largest average tree size. However, the average size in this instance also reduced the most toward the end of the run when the structure objective is being improved. However, in the harder instances (3 and 4 symbols in Ψ) a larger average tree size is produced at the end of the run. The behaviour of more difficult instances producing larger solutions is similar to previous results for tunably difficult instance in genetic programming [6, 7].

This typical instance demonstrates that the search for both structure and content are conflicting in the Tree-String problem. While a population of size of only 50 individuals was used, the problem induces a complex search space. The remainder of the paper describes a much larger study of *hardness* in genetic programming, which is the subject of our future research. We show the generation of tunable instances and how genetic programming has a more difficult time improving both objectives when either one becomes more difficult.

3.3   Tree-String Instances

We create instances in the Tree-String problem with an increasing number of nodes and increasing content alphabet size. These two features, tree size and content size, are likely to lead to increased difficulty for the genetic programming algorithm. To avoid the pitfall of selecting tree shapes which are in themselves difficult for genetic programming, and duplicating aspects of [8], we will use the method of creating tree shapes from [8] but select shapes that are the most commonly visited (also seen in other empirical studies in [10]). We are then ignoring two other ways of tuning instances: fixing content and tree size and choosing more difficult shapes – or – for a particular tree shape and content, using different generation of target content (e.g. non-random ways).

To create the set of tree shapes on which to place random content, forming an instance, we generate a tree shape using the iterative tree growth method from [8], similar to the hill-climbing method in [16]. The method iteratively adds two child nodes to a probabilistically chosen leaf node, starting with the root. The probability of selecting leaf nodes can be altered to restrict trees to be less than a particular depth. We produce 500 random trees with depths between 5 and 15, and with 15 and 272 nodes. We first randomly pick a tree size from the latter range. A tree shape is then grown with a limit of depth 15. Figure 2 shows the distribution of the depth and size of the 500 random trees. We select a tree shape from depths 7, 9, 11, and 13 that are close to the mean size for that depth, ensuring that tree shape alone will not effect difficulty. These trees are shown in Figure 2 using a circular lattice visualisation[1] [8]. The root node lies at the very center, and each two child nodes lie at the intersection of subsequent lines.

The second step to define our instances is selecting which symbols from Ψ to use. We will create four random strings. The first using one symbol from Ψ,

---

[1] Code to produce this visualisation is available at http://www.cs.nott.ac.uk/~ smg/

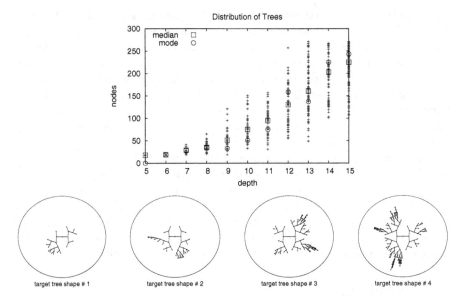

Fig. 2. Trees constructed by an iterative growth method, with bounds on maximum depth and randomly chosen size. The top graph shows the median and mode size for tree at each depth. A tree near to the median size was selected as a target tree from depths 7, 9, 11 and 13 in the empirical study. The bottom row of this Figure shows those four tree shapes

the second two, and so on. That is, one random string has $|\Psi| = 1$, another $|\Psi| = 2$, another $|\Psi| = 3$, and the last $|\Psi| = 4$. Each random string will be the same size as the tree shape under consideration - producing 4×4 instances. The genetic programming system will then use the same content set as used to create the current instance under consideration. That is, genetic programming will not need to address the additional potential problem of filtering out unnecessary elements from the content set.

The ability of genetic programming to search for tree content as well as tree shape can now be tested. By using tree shapes near the median of the distribution, we can assume with some confidence that they represent those shapes which genetic programming should be able to find more easily [8, 10]. However, by increasing the size of the instances, we hope to increase the difficulty of finding the correct tree shape. Also, by generating four random strings for each shape with an increasing content set size, we expect to control difficulty for finding correct content.

3.4   Experimental Study

The genetic programming method is run for 30 runs on each instance, creating 480 runs. We report the *improvement* of solution quality as the total size of the tree minus the longest common substring for structure and the total

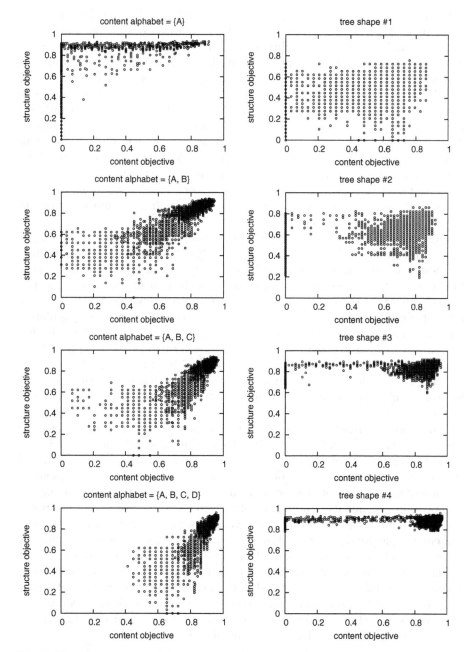

Fig. 3. The improvement of the content and structure objectives. Improvement is normalised by instance (target tree) size. The left column groups runs according to the size of the content set Ψ, increasing complexity from top to bottom. The right column groups runs according to the target tree shape, increasing size from top to bottom

size minus the longest common substring for content. These values are then normalised by dividing them by the target tree size. We report the best (pareto optimal) candidate solution quality found in each population during the run. A similar study using a linear combination of the structure and content objectives (instead of a pareto criterion) was not seen to be significantly different.

We first examine the fitness distributions for the runs with different sizes of the set Ψ. The left column of Figure 3 shows the best fitness in each generation of each of the runs. From top to bottom in the left column of Figure 3 the alphabet Ψ size is increased. We can see how it is initially very easy to find good content (top). However, as Ψ size increases to 4 (bottom), the search gradually shifts toward improving tree shape. Thus, over all the random instances consisting of different target tree sizes, shapes and depths, increasing alphabet size increases the difficulty in the algorithm, causing the search to shift from improving content toward improving structure.

We now examine the effect of target tree size and fitness improvement for the same experiments, but now the instances are grouped by target tree size from smallest to largest. Again, we normalise both fitness objectives by the size of the target tree. The right column of Figure 3 shows the best fitness improvement in both objectives as target tree size increases (from top to bottom). Note that the amount of computation given to all experiments is equal, i.e. the same population size and generations. With a content alphabet of size 1 (as seen in the left column of this figure), genetic programming is still capable of finding trees large enough to match the target content string, seen with all tree sizes. However, the algorithm is unable to find similar improvements with regard to structure. That is, overall improvement is not in proportion to size when content complexity is greater than one symbol.

The empirical results demonstrate the creation of random instances for the Tree-String problem that are tunably difficult. Instance difficulty was achieved by increasing either the content complexity or the size of the size complexity of the instance. When one aspect of the instance (content or size) is *easy enough* (i.e. a content alphabet of size one or small tree shape), genetic programming can improve solution quality. Adding complexity to one objective, however, greatly effects the ability to improve that objective, and sometimes both objectives. More content complexity (more symbols in the content set) makes it harder to improve the content objective (left column of Figure 3), and larger tree shapes make it harder to improve either objective in proportion to the size (right column of Figure 3). A similar behaviour was also seen in the context of the multiobjective optimisation of size and quality [17], where it was easier to reduce tree size than improve quality. Runs converged toward improving the easier objective of size rather than equally improving size and quality simultaneously.

4   Conclusions

Analytical work for genetic programming has always encountered difficulty due to large population sizes, variable sized solutions and expensive fitness evaluation. The Tree-String problem offers the ability to simulate complex solution behaviour (content and structure dependencies) using variable length strings. That is, we do not need to compile new individuals or use precompiled elements for calculating fitness. All the functions used to convert Tree-String elements to strings representing structural and content features are generic (i.e. breadth-first, depth-first tree traversals and longest common substring). We are currently developing a very simple genetic programming system for the Tree-String problem that incorporates efficiency improvements described in [18] for the iTree data structure. It is our goal that the Tree-String problem allows for efficient research to take place on a complex problem, ultimately making significant contributions to the scientific community. We feel that for such a problem to be useful, it must be relevant to realistic genetic programming applications. It is for this reason that the Tree-String problem requires explicit focus on solution structure and content.

Capturing elements of real-world problems in artificial domains can be difficult. Artificial domains are intended to allow precise and efficient analytical work but often focus on singular aspects of solutions (structure or content). Additionally, testbed domains typically handle properties of solution structure and content implicitly, making it difficult to glean their effects. The Tree-String problem is proposed to make a stronger bridge between testbed functions and real-world applications. Toward this goal, we have seen the following properties of the Tree-String problem:

1. Control over both structure and content issues,
2. Clear and simple methods defining fitness and representation,
3. A complex and behaviour-rich search space,
4. The ability to create tunably difficult and random instances,
5. Substantial room for specialisation toward specific research goals.

Our future work is examining problem hardness in genetic programming. We are also examining variants of the Tree-String problem to carry-out efficient algorithmic analysis with respect to other problems. While we have hypothesised that the Tree-String problem is representative of other genetic programming domains, our current work is attempting to create mappings between these domains or between important domain features.

Acknowledgements

This work was supported by EPSRC grant GR/S70197/01. SG thanks Jano van Hemert and the reviewers for their comments.

References

1. D.S. Johnson, C.H. Papadimitriou, and M. Yannakakis. How easy is local search? *Journal of Computer and System Sciences*, 37(1), August 1988.
2. D. Fox, W. Burgard, H. Kruppa, and S. Thrun. A probabilistic approach to collaborative multi-robot localization. *Autonomous Robots*, 8(3):325–344, 2000.

3. J.R. Quinlan. Induction of decision trees. *Machine Learning*, 1:81–106, 1986.
4. J.R. Koza et al. *Genetic Programming IV: Routine Human-Competitive Machine Intelligence*. Kluwer Academic Publishers, 2003.
5. I. Kushchu. An evaluation of evolutionary generalisation in genetic programming. *Artificial Intelligence Review*, 18(1):3–14, 2002.
6. J.M. Daida et al. What makes a problem GP-hard? analysis of a tunably difficult problem in genetic programming. *Genetic Programming and Evolvable Machines*, 2(2):165–191, June 2001.
7. S. Gustafson, A. Ekárt, E.K. Burke, and G. Kendall. Problem difficulty and code growth in genetic programming. *Genetic Programming and Evolvable Hardware*, 5(3):271–290, 2004.
8. J.M. Daida, H. Li, R. Tang, and A.M. Hilss. What makes a problem GP-hard? validating a hypothesis of structural causes. In E. Cantú-Paz et al., editors, *Proceedings of the Genetic and Evolutionary Computation*, volume 2724 of *LNCS*, pages 1665–1677, Chicago, IL, USA, 12-16 July 2003. Springer-Verlag.
9. C. Gathercole and P. Ross. An adverse interaction between crossover and restricted tree depth in genetic programming. In J.R. Koza et al., editors, *Genetic Programming 1996: Proceedings of the First Annual Conference*, pages 291–296, Stanford University, CA, USA, 28–31 July 1996. MIT Press.
10. W.B. Langdon and R. Poli. *Foundations of Genetic Programming*. Springer-Verlag, Berlin, 2002.
11. W.F. Punch, D. Zongker, and E.D. Goodman. The royal tree problem, a benchmark for single and multi-population genetic programming. In P.J. Angeline and K.E. Kinnear, Jr., editors, *Advances in Genetic Programming 2*, chapter 15, pages 299–316. The MIT Press, Cambridge, MA, USA, 1996.
12. U.-M. O'Reilly. The impact of external dependency in genetic programming primitives. In *Proceedings of the IEEE World Congress on Computational Intelligence*, pages 306–311, Anchorage, AL, USA, 5-9 May 1998. IEEE Press.
13. U.-M. O'Reilly and D.E. Goldberg. How fitness structure affects subsolution acquisition in genetic programming. In J.R. Koza et al., editors, *Proceedings of the Third Annual Genetic Programming Conference*, pages 269–277, Madison, WI, USA, 22-25 July 1998. Morgan Kaufmann.
14. D.E. Goldberg and U.-M. O'Reilly. Where does the good stuff go, and why? how contextual semantics influence program structure in simple genetic programming. In W. Banzhaf et al., editors, *Genetic Programming, Proceedings of the First European Workshop*, volume 1391 of *LNCS*, pages 16–36, Paris, 1998. Springer-Verlag.
15. S. Gustafson. *An Analysis of Diversity in Genetic Programming*. PhD thesis, School of Computer Science and Information Technology, University of Nottingham, Nottingham, England, February 2004.
16. A. Juels and M. Wattenberg. Stochastic hillclimbing as a baseline method for evaluating genetic algorithms. Technical Report Technical Report CSD-94-834. Computers Science Department, University of California at Berkeley, USA, 1995.
17. E.D. de Jong and J.B. Pollack. Multi-objective methods for tree size control. *Genetic Programming and Evolvable Machines*, 4(3):211–233, September 2003.
18. A. Ekárt and S. Gustafson. A data structure for improved GP analysis via efficient computation and visualisation of population measures. In M. Keijzer et al., editors, *Genetic Programming, Proceedings of the 6th European Conference*, volume 3003 of *LNCS*, pages 35–46, Coimbra, Portugal, April 2004. Springer-Verlag.

Using Genetic Programming for Multiclass Classification by Simultaneously Solving Component Binary Classification Problems

William Smart and Mengjie Zhang

School of Mathematics, Statistics and Computer Sciences,
Victoria University of Wellington, P. O. Box 600, Wellington, New Zealand
{smartwill, mengjie}@mcs.vuw.ac.nz

Abstract. In this paper a new method is presented to solve a series of multiclass object classification problems using Genetic Programming (GP). All component two-class subproblems of the multiclass problem are solved in a single run, using a multi-objective fitness function. Probabilistic methods are used, with each evolved program required to solve only one subproblem. Programs gain a fitness related to their rank at the subproblem that they solve best. The new method is compared with two other GP based methods on four multiclass object classification problems of varying difficulty. The new method outperforms the other methods significantly in terms of both test classification accuracy and training time at the best validation performance in almost all experiments.

1 Introduction

Object classification problems abound in daily life, and a computer system that can solve object classification problems is very desirable. The advantages of using a computer to solve such problems, over a human expert, include lower cost, higher speed and higher reliability in the face of large throughput. However, building automatic computer systems for object and image classification tasks that are reliable and can achieve desirable performance is very difficult.

GP research has considered a variety of kinds of evolved program representations for classification tasks, including decision tree classifiers and classification rule sets [1, 2, 3]. Recently, a new form of classifier representation – numeric expression classifiers – has been developed using GP [4, 5, 6, 7]. This form has been successfully applied to real world classification problems such as detecting and recognising particular classes of objects in images [5, 6, 8], demonstrating the potential of GP as a general method for classification problems.

The output of a numeric expression classifier is a single numeric value (the program output), and problems arise when attempting to convert this value into a class label. For binary problems, one reasonable solution is to assign one class if the program output is negative, and the other otherwise [4, 5, 6, 9]. However, the problem is much more complicated when three or more classes exist (multiclass

M. Keijzer et al. (Eds.): EuroGP 2005, LNCS 3447, pp. 227–239, 2005.

problems), as multiple boundaries need to be found to divide the numeric space into three or more class regions.

Statically-assigned class boundary methods have been used widely, but are seen to unnecessarily constrain programs, leading to long search time and low final accuracy [4, 7]. In previous research a *probabilistic* method has been used to avoid the setting of class boundaries [10], however this method still constrains each program to solve the entire problem, even when the problem has many classes.

To avoid the above problems, the multiclass classification task may be decomposed into many binary tasks [4]. However, in the past each binary task requires a separate GP evolution, leading to a long total time for evolution even though each evolution is quite short [4].

The goal of this paper is to construct a method to decompose a multiclass classification problem into a number of two-class (binary) subproblems, solve all these subproblems in a single GP run, then combine the binary subproblems to solve the whole multiclass problem. A secondary goal is to evaluate the new method on a variety of problems of varying difficulty, comparing it with two other GP based methods.

This paper is organized as follows. In section 2 two existing fitness functions for comparison with the new method are presented. In section 3 the new method is described. In section 4 the data sets and settings used for experiments are given. In section 5 the results of experiments are presented and discussed. In section 6 conclusions are drawn, and some directions are given for future research.

2   Two Existing Fitness Functions

The new approach described in this paper will be compared with two existing fitness functions: Program Classification Map (PCM) [7, 11] and Probabilistic Multiclass (PM) [10].

2.1   Program Classification Map

In PCM, the floating-point output of a program is converted directly into a class label, depending on the numeric region it falls into. Thresholds are set at even spacing (one unit) on the number line from some negative number to the same positive number. For an N class problem, there will be $N - 1$ thresholds. The classes are assigned to the regions before, between and after the thresholds, in order of magnitude. For example, figure 1 shows the numeric regions of a five class problem. The fitness of a program is found by subtracting the training set accuracy from 100%.

2.2   Probabilistic Model of Program Output

Based on the feature values of the training examples for a particular class, the mean and standard deviation of the program output values for that class can

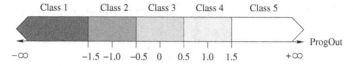

Fig. 1. Program Classification Map (PCM) for a five-class problem

be calculated. In this way, we can attain the mean and standard deviation for each class. These program statistics are compared in order to get a fitness value indicating how well the program separates the classes.

Probabilistic Binary. Figure 2 shows three examples of normal curves that may be gained by modeling three program's results on the training examples in the binary classification problems. In the figure, the leftmost program's normal curves are substantially "overlapped", so there is a high probability of misclassification. The rightmost program's normal curves are well "separated", so there is a low probability of misclassification and this program represent a good classifier.

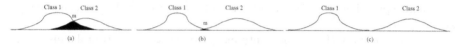

Fig. 2. Example normal distributions for a binary problem. (a) a bad discerner between the two classes, (b) an acceptable discerner, (c) a good discerner

In the binary problem case, equation 1 is used to determine the distribution distance (d) between the classes in standard deviations.

$$d = 2 \times \frac{|\mu_1 - \mu_2|}{\sigma_1 + \sigma_2} \tag{1}$$

where μ_i and σ_i are the mean and standard deviation of the program outputs for class i in the training set. For programs that distinguish between the two classes well, the distance d will be large. In such a case, the probability of misclassification would be small as the distribution overlap occurs at a high standard deviation.

To be consistent with the PCM method, we convert the distribution distance measure d to a standardised fitness measure ds, as shown in equation 2, which indicates the misclassification of a genetic program in the binary classification problem.

$$ds = \frac{1}{1 + d} \tag{2}$$

Probabilistic Multiclass (PM). In the PM method, the fitness of a program for the multiclass (N-class) classification problem is defined as the sum of the standardised distribution distance of all binary classification problems, as shown below:

$$Fitness(PM) = \sum_{i=1}^{C_N^2} ds_i \qquad (3)$$

For instance, for a four class problem there are $C_4^2 = 6$ binary classification subproblems. For each subproblem, we can calculate its ds value. The fitness of the whole four class problem is the sum of all the six ds values.

3   Communal Binary Decomposition

For presentation convenience, the new approach developed in this paper is called Communal Binary Decomposition (CBD).

CBD also uses a *probabilistic* method to model the outputs of programs, as PM does. However, while a solution program in PM must separate the distributions of all classes in the multiclass problem, in CBD each program only needs to separate two classes. In CBD the program's fitness depends on its performance at separating just one pair of classes for a particular binary classification subproblem.

The separation of the problem into many two-class (binary) problems is similar to Binary Decomposition [4]. However in CBD all the problems are solved in one evolution using a multi-objective fitness function, which is why it is called "communal".

3.1   Getting Fitness

In each generation, the following steps are taken to calculate the fitness of a program:

1. For each pair of classes in a binary classification problem, calculate and store the separation distance d (equation 1).
2. For each pair of classes in a binary classification problem, sort the programs in the population based on the separation distance values in a descending order.
3. Calculate the fitness of each program based on its position in the list of programs where the program achieves the best performance (position) in all the binary classification problems.

Figure 3 shows an example of using this method to evaluate the fitness of four programs in a three class problem.

The main table in the figure lists the separation distances of all programs on all binary problems. Below the main table, the entries for each binary problem are sorted from best program to worst. For example, for the binary problem of separating class one from class two, program D was best at a distance of 2.01, so it is first in the list. Then each program is assigned a fitness to the position in the sorted list where it achieves the best performance (occurs earliest, as indicated by arrows).

Program Seperations for Class Pairs

	1 vs. 2	1 vs. 3	2 vs. 3	Best results	Fitnesses
Prog. A	1.22	1.83	1.02	A was first in (1 vs 3)	1
Prog. B	0.81	0.23	1.67	B was second in (2 vs 3)	2
Prog. C	0.93	0.69	0.56	C was third in (1 vs 2)	3
Prog. D	2.01	1.65	2.63	D was first in (1 vs 2)	1
Sorted Lists	DACB	ADCB	DBAC		

Arrows indicate best positions

Fig. 3. Evaluating the fitness of four example programs in a three class system

With this fitness function, each program is encouraged to excel in separating one pair of classes, although the program is free to do so to any pair it chooses. Any program is free to separate more than one pair of classes, but only the pair where the program performs the best is used to calculate the fitness.

3.2   Solving the Multiclass Problem

In order that the multiclass problem is solved, a group of *expert* programs is assembled during evolution. An expert program is stored for each of the binary problems.

In each generation, for each binary problem, the program in the population that has the best separation is compared to the expert for the binary problem. If it is found to be better than the expert, it replaces the expert.

If the standardised distribution distance value ds (equation 2) for the (possibly new) expert program falls below (is better than) a parameter *solveAt*, then the binary problem is marked as solved. A solved binary problem differs from an unsolved problem in that programs are no longer encouraged to solve it. Solved binary problems are not included in the calculation of the best position for each program (discussed in section 3.1). When all binary problems have been solved, the problem is considered solved.

Note that the best program at separating classes of a solved binary problem is still found, for comparison with the current expert at the problem. As such, experts of solved binary problems can still be replaced and improved upon.

3.3   Combining CBD Expert Programs to Predict Class

To find the accuracy of the system on the test set or validation set, the experts gained for all the binary problems are combined to predict the class of each object image in the test set or validation set.

Equation 4 is used to find the probability density at points in the normal curve.

$$P_{e,c,o} = \frac{e^{\left(\frac{-(res_{e,o} - \mu_{e,c})^2}{2\sigma_{e,c}^2}\right)}}{\sigma_{e,c}\sqrt{2\pi}} \tag{4}$$

where $P_{e,c,o}$ is the probability density calculated for the expert program e using the normal distribution for class c and features for object o. $res_{e,o}$ is the output result of evaluating expert program e on object o. $\mu_{e,c}$ and $\sigma_{e,c}$ are the mean and standard deviation, respectively, of the output results of the expert program e for all training object examples for class c.

Using the function in equation 4, equation 5 shows a probability value of object o belonging to class a in the class pair (a,b) (binary classification problem), and equation 6 shows the same value expressed in a different form.

$$\frac{P_{e(a,b),a,o}}{P_{e(a,b),a,o} + P_{e(a,b),b,o}} = Pr(cls = a | cls \in \{a,b\}) \tag{5}$$

$$= \frac{Pr(cls = a \cap (cls \in \{a,b\}))}{Pr(cls \in \{a,b\})}$$

$$= \frac{Pr(cls = a)}{Pr(cls \in \{a,b\})} = \frac{Pr(cls = a)}{Pr(cls = a) + Pr(cls = b)} \tag{6}$$

where the expert for discerning class a from class b is called $e(a,b)$ and $Pr(x)$ is the probability that the condition x is true.

To obtain the probability of any object in the test set belonging to class c with the expert programs in a multiclass (N-class) problem, we consider the inverted sum of all binary classification subproblems associated with class c that the multiclass problem can be decomposed into:

$$\sum_{i=1}^{c-1} \frac{1}{Pr(cls = c | cls \in \{i,c\})} + \sum_{j=c+1}^{N} \frac{1}{Pr(cls = c | cls \in \{c,j\})}$$

$$= \sum_{i=1}^{c-1} \frac{Pr(cls = i) + Pr(cls = c)}{Pr(cls = c)} + \sum_{j=c+1}^{N} \frac{Pr(cls = c) + Pr(cls = j)}{Pr(cls = c)}$$

$$= \frac{1 + (N-2) \cdot Pr(cls = c)}{Pr(cls = c)} = N - 2 + \frac{1}{Pr(cls = c)}$$

Note that the sum of the probability of a particular object belonging to each of all the possible classes is equal to one, i.e. $\sum_{i=1}^{N} Pr(cls = i) = 1$.

Accordingly, the probability of an object that the GP system classifies to class c is $Pr(cls = c)$ or p_c:

$$p_c = \frac{1}{\sum_{i=1}^{c-1} \frac{Pr(cls=i)+Pr(cls=c)}{Pr(cls=c)} + \sum_{j=c+1}^{N} \frac{Pr(cls=c)+Pr(cls=j)}{Pr(cls=c)} - (N-2)}$$

$$= \frac{1}{\sum_{i=1}^{c-1} \frac{1}{Pr(cls=c|cls\in\{i,c\})} + \sum_{j=c+1}^{N} \frac{1}{Pr(cls=c|cls\in\{c,j\})} - (N-2)} \tag{7}$$

Based on equations 7 and 5, we can calculate the probability of an object being of any class. The class with the highest probability for the object is used as the class the GP system classified into.

4   Data Sets and Evolutionary Settings

4.1   Image Data Sets

We used four data sets providing object classification problems of increasing difficulty in the experiments. Example images are shown in figure 4.

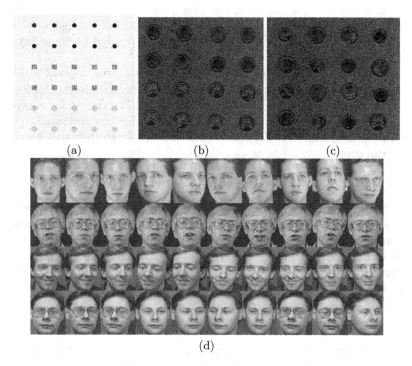

(a)   (b)   (c)

(d)

Fig. 4. Dataset examples: (a) Shapes, (b) 3-class coins, (c) 5-class coins, and (d) Faces

The first set of images (figure 4a) was generated to give well defined objects against a relatively clean background. The pixels of the objects were produced using a Gaussian generator with different means and variances for each class. Three types of shape were drawn against the light grey background: black circles, dark grey squares and light circles. The *three-class shape* data set was created by cutting out 720 objects (also called cutouts), from these images.

The second and the third sets of images (figure 4b and 4c) contain scanned New Zealand coins. The coins were located in different places with different orientations and appeared in different sides (head and tail). In addition, the background was quite cluttered. We need to distinguish different coins, with different sides, from the background. Two data sets were created from these images, one with three classes and one with five classes. The *three-class coin* data set has 576 cutouts that are either 10c heads, 10c tails or background. The *five-class coin* data set has 480 cutouts that are either 5c heads, 5c tails, 10c heads, 10c tails or background. The classification problem in the coin data sets is much harder than that of the shape data set. The problem is very hard due to the cluttered background and similarity of the classes as well as the low resolution of images.

The fourth data set consists of 40 images of four human faces (figure 4d) taken at different times, varying lighting slightly, with different expressions (open/closed eyes, smiling/non-smiling) and facial details (glasses/no-glasses). These images were collected from the first four directories of the ORL face database [12]. All the images were taken against a dark homogeneous background with limited orientations. The task here is to distinguish those faces into the four different people.

For the shape and the coin data sets, the objects were equally split into three separate data sets: one third for the training set used directly for learning the genetic program classifiers, one third for the validation set for controlling overfitting, and one third for the test set for measuring the performance of the learned program classifiers. For the faces data set, due to the small number of images, the standard ten-fold cross validation technique was applied.

4.2   GP Environment and Settings

In all experiments, programs used the tree-structured representation with all nodes returning single, floating-point numbers. Reproduction, mutation and crossover were used as the genetic operators. Initial programs and the subtrees created in the mutation operator were generated by the half-and-half method. A tournament selection mechanism was used with a tournament size of three.

The GP parameters used in experiments are shown in table 1.

Table 1. Parameters used for GP training for the three data sets

Parameter Names	Shapes	coins	faces	Parameter Names	Shapes	coins	faces
population-size	500	500	500	reproduction-rate	10%	10%	10%
initial-max-depth	5	5	5	cross-rate	60%	60%	60%
max-depth	7	7	7	mutation-rate	30%	30%	30%
max-generations	40	40	40	cross-term	30%	30%	30%
object-size	16×16	70×70	92×112	cross-func	70%	70%	70%

In this approach, the GP search was terminated when one of the following events occurred:

– The number of generations exceeds *max-generations*.
– The training problem was considered solved.

4.3   Terminals and Functions

Terminals. Two forms of terminals were used: numeric and feature terminals. Each numeric terminal returned a single constant value, set initially by randomly sampling a standard normal distribution.

Each feature terminal returned the value of a particular input feature. It is the feature terminals that form the inputs of the GP system from the environment. The features used are simple pixel-statistics derived from the object image. Two statistics (mean and standard deviation) for each of two regions (whole square cutout and centre square of half side-length) were used, making four features altogether.

Functions. Five functions were used, including addition, subtraction, multiplication, protected division and a continuous conditional (soft if) function. Each of the first four functions took two arguments, and applied simple mathematical operations. The soft if function returned the value r in equation 8. This allowed either the second or third argument to control the output, depending on the value of the first argument.

$$r = \frac{a_2}{1 + e^{2a_1}} + \frac{a_3}{1 + e^{-2a_1}} \tag{8}$$

where r is the output of the soft if, and a_i is the value of the i'th argument.

5   Results and Discussion

This section presents a series of results of the new method on the four object classification data sets. These results are compared with those for the PCM and PM methods.

Throughout evolution, the accuracy of the system on the validation set was monitored. The accuracy shown in results (except those for the Face data set) is the accuracy of the system on the test set, at the point of best validation set accuracy. This method is employed to avoid overfitting. For the Face data set, ten-fold cross-validation (TFCV) was used, and the accuracy reported is the maximum test set accuracy, using TFCV, found during evolution. The "run time" reported is the total time required for evolution.

For all parameter settings, the GP process was run 50 times with different random seeds. The mean results were presented, with standard deviation included for accuracy results.

5.1   Overall Results

Table 2 shows a comparison of the best classification results obtained by both the new method (CBD) and the other approaches (PCM and PM) using the same sets of features, functions and parameters. The *solveAt* parameter to the new method was set to 0.01.

On the shape and the first coin data set, all the three GP methods achieved good results, reflecting the fact that the classification problems are relatively easy in these two data sets. However, the new method achieved perfect test set accuracy for all 50 runs of evolution for the shape data set where the other two existing GP methods did not. On both classification tasks, the new CBD method achieved the best classification accuracy. On the very hard five-class coin problem, the new method gained almost ideal performance (99.19% accuracy), while the PM method achieved 95.92% and PCM only 73.26%. For the face data set, new method achieved comparable results with the PM method, but greatly outperformed the PCM method.

The number of generations required to gain maximum validation set accuracy was very small with the new method. For the shape and three-class coin data sets, this number was well below one for the new method, indicating that the problems could be solved in the initial generation at most of the time due to the good fitness measure in the new CBD method.

The new method normally took a longer time to solve the training problem and terminate the run than the other methods. However, it often found a peak in the validation set accuracy in a shorter time than the PM method. It is expected that the new method would outperform the other methods if only a very limited time or number of generations was allowed for the run.

Table 2. Comparison between PCM, PM and CBD

Dataset	Classes	Method	Gens. at best Val. Acc. (s)	Time to Best Val. Acc. (s)	Run Time (s)	Test Acc. at Best Val. Acc. (%)
Shapes	3	PCM	3.02	0.41	0.45	99.82 ± 0.43
		PM	0.92	0.52	0.53	99.97 ± 0.14
		CBD	0.04	0.53	22.70	100.00 ± 0.00
Coins	3	PCM	8.74	0.94	1.29	99.44 ± 0.88
		PM	1.18	0.50	0.53	99.91 ± 0.27
		CBD	0.06	0.44	25.00	99.97 ± 0.12
	5	PCM	31.32	3.60	4.72	73.26 ± 7.45
		PM	28.82	9.85	14.23	95.92 ± 2.40
		CBD	5.06	2.82	21.69	99.19 ± 0.82
Faces	4	PCM	5.91	0.22	1.75	81.65 ± 13.49
		PM	5.63	0.44	2.78	97.75 ± 7.15
		CBD	2.12	0.36	5.93	96.45 ± 8.73

5.2 Different Problem Solving Criteria in the New Approach

Table 3 shows a comparison of the results on the four data sets using different values for the *solveAt* parameter in the new method. The *solveAt* parameter indicates the separation measure value at which to consider a binary problem solved. Five values were examined for *solveAt* over the four data sets.

Values smaller than 0.01 showed no considerable improvement in accuracy, but increased run time. Values larger than 0.01 degraded performance slightly, but did decrease the time per run. From these experiments, a value of 0.01 for the *solveAt* parameter seems a good starting point.

Table 3. Comparison between values of the *solveAt* threshold, in CBD

Dataset	Classes	solveAt	Gens. at best Val. Acc. (s)	Time to Best Val. Acc. (s)	Run Time (s)	Test Acc. at Best Val. Acc. (%)
Shapes	3	0.003	0.04	0.53	28.24	100.00 ± 0.00
		0.010	0.04	0.53	22.70	100.00 ± 0.00
		0.030	0.02	0.52	4.46	100.00 ± 0.00
		0.100	0.00	0.51	0.51	99.39 ± 4.26
		0.300	0.00	0.50	0.51	99.39 ± 4.26
Coins	3	0.003	0.06	0.44	24.82	99.97 ± 0.12
		0.010	0.06	0.44	25.00	99.97 ± 0.12
		0.030	0.06	0.44	23.90	99.97 ± 0.12
		0.100	0.04	0.43	4.23	99.97 ± 0.12
		0.300	0.00	0.41	0.41	99.96 ± 0.14
	5	0.003	5.72	3.16	21.99	99.16 ± 0.87
		0.010	5.06	2.82	21.69	99.19 ± 0.82
		0.030	5.22	2.88	21.32	99.17 ± 0.98
		0.100	3.78	2.08	21.69	99.29 ± 0.91
		0.300	0.70	0.65	1.71	99.04 ± 1.02
Faces	4	0.003	2.12	0.36	5.93	96.45 ± 8.73
		0.010	2.12	0.36	5.93	96.45 ± 8.73
		0.030	2.07	0.35	5.91	96.40 ± 8.78
		0.100	1.96	0.33	5.93	96.15 ± 9.02
		0.300	1.55	0.28	4.86	95.30 ± 9.77

6   Conclusions

The goal of this paper was to construct a method to decompose a multiclass classification problem into multiple two-class subproblems, solve all these subproblems in a single GP run, then combine the subproblems to solve the multiclass problem. This goal was achieved in the CBD method, which allows many component binary classification problems to be solved in one run of the GP process.

A secondary goal was to evaluate the new method on a variety of problems of varying difficulty, comparing it with two other methods. This goal was achieved by a series of experiments comparing the new method with a basic GP approach (PCM) and a previous probabilistic GP method (PM). The new method was seen to outperform both methods when applied to most data sets.

Like many existing GP approaches for multiclass classification problems, the new method can solve a problem of any number of classes in a single GP run. However, unlike most of these approaches, each evolved program produced in the new method is not required to solve the entire multiclass problem. Instead, each program only needs to solve a single component binary subproblem to gain good fitness. The fitness function gives each program a fitness related to its best performance for a binary subproblem.

A mathematically derived method was found to determine the most probable class of a test example, based on the results of a number of expert programs for solving those binary classification subproblems.

The new method requires a parameter, which specifies when a component problem is considered solved. While it does not appear to have a reliable way to obtain a good value for this parameter for different problems which usually needs empirical search, our experiments suggest that 0.01 is a good starting point.

Although developed for multiclass object classification problems, this approach is expected to be able to be applied to general classification problems.

In future work, we will investigate the new method on other general classification problems and compare the performance with other long established methods such as decision trees and neural networks.

References

1. John R. Koza. *Genetic programming : on the programming of computers by means of natural selection.* Cambridge, Mass. : MIT Press, London, England, 1992.
2. John R. Koza. *Genetic Programming II: Automatic Discovery of Reusable Programs.* Cambridge, Mass. : MIT Press, London, England, 1994.
3. Wolfgang Banzhaf, Peter Nordin, Robert E. Keller, and Frank D. Francone. *Genetic Programming: An Introduction on the Automatic Evolution of computer programs and its Applications.* Morgan Kaufmann Publishers, 1998.
4. Thomas Loveard and Victor Ciesielski. Representing classification problems in genetic programming. In *Proceedings of the Congress on Evolutionary Computation*, volume 2, pages 1070–1077. 2001. IEEE Press.
5. Andy Song, Vic Ciesielski, and Hugh Williams. Texture classifiers generated by genetic programming. In David B. Fogel, et al. editors, *Proceedings of the 2002 Congress on Evolutionary Computation*, pages 243–248. IEEE Press, 2002.
6. Walter Alden Tackett. Genetic programming for feature discovery and image discrimination. In Stephanie Forrest, editor, *Proceedings of the 5th International Conference on Genetic Algorithms*, pages 303–309. 1993. Morgan Kaufmann.
7. Mengjie Zhang and Victor Ciesielski. Genetic programming for multiple class object detection. In Norman Foo, editor, *Proceedings of the 12th Australian Joint Conference on Artificial Intelligence*, pages 180–192, Sydney, Australia, December 1999. Springer-Verlag. LNAI 1747.
8. Mengjie Zhang and Will Smart. Multiclass object classification using genetic programming. In Guenther R. Raidl, et al. editors, *Applications of Evolutionary Computing*, Volume 3005, *LNCS*, pages 367–376, 2004. Springer Verlag.
9. Daniel Howard, S. C. Roberts, and R. Brankin. Target detection in SAR imagery by genetic programming. *Advances in Engineering Software*, 30:303–311, 1999.
10. Will Smart and Mengjie Zhang. Probability based genetic programming for multiclass object classification. In *Proceedings PRICAI 2004, LNAI Vol. 3157*, pages 251–261, Springer-Verlag, 2004.
11. Mengjie Zhang, Victor Ciesielski, and Peter Andreae. A domain independent window-approach to multiclass object detection using genetic programming. *EURASIP Journal on Signal Processing, Special Issue on Genetic and Evolutionary Computation for Signal Processing and Image Analysis*, 2003(8):841–859, 2003.

12. F. Samaria and A. Harter. Parameterisation of a stochastic model for human face identification. In *2nd IEEE Workshop on Applications of Computer Vision*, Sarasota (Florida), July 1994. ORL database is available at: www.cam-orl.co.uk/facedatabase.html.

13. Will Smart and Mengjie Zhang. Classification strategies for image classification in genetic programming. In Donald Bailey, editor, *Proceeding of Image and Vision Computing Conference*, pages 402–407, New Zealand, 2003.

Context-Based Repeated Sequences in Linear Genetic Programming

Garnett Carl Wilson and Malcolm Iain Heywood

Faculty of Computer Science,
Dalhousie University, Halifax, NS B3H 1W5
{gwilson, mheywood}@cs.dal.ca

Abstract. Repeating code sequences are found in both artificial and natural genomes as an emergent phenomenon. These patterns are of interest in research-
ing both how evolution reuses code segments to create superior individuals and
whether building blocks are used in the formation of repeated sequences. We
describe a GP representation using a special type of crossover that is more con-
ducive to the formation of repeated sequences than traditional GP. We then establish that the repeated sequence phenomenon in the implementation displays
traits of building blocks by establishing associated regularity of genotype and
phenotype elements.

1 Introduction

The presence of code bloat has been a widely studied phenomenon in the genetic programming (GP) field [1]. Another potentially important phenomenon that is not as
thoroughly investigated is the emergence of repeating sequences of instructions found
to be present in individuals evolved using linear genetic programming (L-GP). In a
recent paper [2], Langdon and Banzhaf discovered that hierarchical repeating sequences are evolved by L-GP in linear time series prediction programs. Langdon and
Banzhaf also suggest that the length of the programs (bloat) appears to be more important than crossover in establishing repetitive sequences for their analysis of the
Mackey-Glass benchmark using two types of 2 point crossover and headless chicken
crossover.

Langdon and Banzhaf [2] raise a number of interesting questions in their paper, including whether or not (1) there are new representations of GP that might be better
able to generate repeated sequences and (2) building blocks are involved in the formation of repeating sequences. We present an answer to (1) by using a particular variation of the crossover operator. We also construct an argument that the repeating sequences in this implementation constitute building blocks, thus providing a possible
response to (2). We thus present an implementation that is particularly good at forming repeating sequences using building blocks. To investigate the nature of repeating
sequences, we apply L-GP to the San Mateo version of the Artificial Ant Trail [3,
chapter 12]. This provides a difficult benchmark problem that is also sufficiently
straightforward to establish relationships between cause (instruction execution) and

M. Keijzer et al. (Eds.): EuroGP 2005, LNCS 3447, pp. 240–249, 2005.

effect (efficient food eating strategy), for fitness increases as the instruction pointer moves from beginning to end of an individual's genome.

Two L-GP implementations, each with a different crossover operator, are then applied to this problem. The repeated sequences resulting from a fixed-size (FS) crossover operator are compared to the sequences resulting from a variable size (VS) crossover operator, showing the FS crossover operator to be superior to the VS in repeated sequence creation. Code regularity in solutions is thus established, and an analysis of fitness accumulation resulting from the execution of particular sections of the genome is then completed. The result is that both a code regularity (repeating code patterns) and an associated regularity in fitness accumulation (result of code regularity) are established for the FS implementation. In the end, the combination of code regularity and performance of genome sections points to the existence of building blocks within a GP implementation that uses the crossover operator as the means of creating the repeating sequences within individuals of a fixed size. However, this finding does not mean that bloat is unable to contribute to sequence formation more so than crossover in implementations where fixed length individuals are not used. Therefore, the sequence formation due to bloat over crossover, as expected by Landon and Banzhaf, may still be realized in other implementations.

In Section 2 of the paper, we outline the Linear GP implementation of the San Mateo Trail. Section 3 describes the specifics of the operators used in the two forms of linear GP that are applied to the problem. Section 4 examines the distribution of instruction sequences across the two forms of linear GP, establishing the presence of repeating patterns (code regularity) in the genotype. Section 5 analyses the nature of the solutions found in terms of fitness gain per instruction section of the genome and establishes regularity in the form of the solutions (phenotype) for the FS implementation. Section 6 concludes the paper.

2 San Mateo Trail with Linear GP

Koza described several versions of the Ant Trail problem of varying degrees of difficulty [4]. Moreover, it has been shown that specific instances of the problem represent deceptive problems for fixed length (tree) GP [5], [6, Chapter 9]. The instance used for our research is the San Mateo Trail, proposed by Koza in 1994, and represents the most difficult version of the problem on account of the additional discontinuities, extended length and non-toroidal nature of the grid on which the trail is laid [3, chapter 12]. Specifically, the trail consists of a series of nine 13×13 grids on which 9 to 12 food items are distributed in discontinuous sequences of various patterns. Should an ant attempt to wander off the edge of a grid, it re-appears on the next grid at the default start position (center of the first row pointing south). Likewise, should an ant successfully eat all the food items on a grid, it is then repositioned on the next grid at the default start position. In total there are 96 items of food. Table 1 summarizes the key features of the problem [3]. The ant is capable of moving right, left, or ahead in the direction it is facing, with a limit of 120 RIGHT or LEFT turns and 80 MOVE_AHEADs.

Table 1. Table of San Mateo Trail Problem Specifics using Linear Genetic Programming

Experiment Parameter	Definition or Acceptable Values
Objective	Find program to control artificial ant such that all 96 food items are located.
Terminal set	(right), (left), (move)
Functional set	IF-FOOD-AHEAD
Fitness cases	9 fitness cases, each with 13 × 13 grid with 9-12 food items
Raw fitness	Sum of food eaten over the nine fitness cases within 120 RIGHT/LEFT or 80 MOVE_AHEADs.
Standardized fitness	Total food (96) less the raw fitness
Hits	As per raw fitness
Wrapper	None
Success predicate	Program with the maximum number of hits

The linearly-structured GP employed in this case uses the instruction format illustrated in Figure 1 for a four register ant. Note that the first bit in the instructions (the mode bit) switches between a "load register" instruction, in which the content of a register is over-written with the two least significant bits of the instruction, or an "execute register" instruction. In the latter case the instruction of a register is executed in conjunction with the 'IF-FOOD-AHEAD' test, the contents of different registers being executed depending on the results of the test. The 'IF-FOOD-AHEAD' test returns true if the ant is facing food, and false otherwise. Note that 12 bits were used to allow for future specification of further registers, where 2 bits would be needed to differentiate 4 registers, 3 bits to differentiate 8 registers, and so on. The pseudocode in Figure 1 corresponds to 4 register ants, and thus uses two bits.

```
if (modebit == 1) then
   if (food_ahead == true) then
      execute instruction in register # bits 6 to 7
   else // no food ahead
      execute instruction in register # bits 11 to 12
else // mode bit == 0
      load instruction in bits 11 to 12 into register #
      bits 6 to 7
```

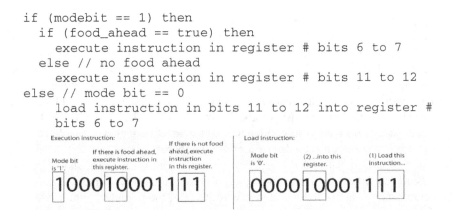

Fig. 1. Instruction format and pseudocode for interpretation of the twelve bit instructions comprising a 4 register ant's program

One implication of this instruction set is that an individual may evolve a program that does not consist of any 'IF-FOOD-AHEAD' tests. This means that the ant never finishes the trail, because it never moves anywhere. In order to penalize this property, an additional constraint is introduced in which the number of register references made per grid is limited to the maximum number of instructions. Table 2 summarizes the parameters used to define the linear GP runs, where the two different crossover types are explained in Section 3. Note that both implementations will contain individuals of the same lengths in the initial population. The implementations use the same seed to generate the lengths with a uniform random distribution, with an individual's minimum length being 4 instructions (1 page) and maximum length being either 64 instructions (16 pages) or 128 instructions (32 pages).

Table 2. Parameterization of Linear GP for the San Mateo Trail

Generic Linear GP Parameters			
Population size	125	P(crossover)	0.9
Number of trials	50	P(mutate)	0.5
Selection method	Steady state	P(swap)	0.9
Number of registers	4 or 8	Maximum number of	50 000
Tournament size	4	Generations	
Fixed Size Linear GP (FS L- GP)			
Maximum number of pages	16 or 32	Instructions / Page	4
Variable Size Linear GP (VS L-GP)			
Instruction limits at initialization	[1...16] x 4 Or [1...32] x 4	Maximum number of instructions	64 or 128

3 Crossover Operators

Comparison is made between two linear GP crossover types. In the first case, the exchange of arbitrary instruction sequences takes place under the control of crossover as outlined by Heulsbergen [7]: Instead of exchanging n instructions from one individual with m from another, where it was not necessarily the case that $n = m$, we ensured that l instructions were exchanged between each individual where $1 \leq l \leq min(length\_indiv1, length\_indiv2)$. This results in individuals of fixed length, i.e., the only way that the population distribution of lengths changes is through children from parents of one length over-writing less fit individuals of a different length. The length l was chosen from a uniform random distribution, along with a start point for each parent, sp_1, sp_2, also selected from a uniform random distribution. Then, beginning at the respective start points in each parent, l instructions were exchanged. If l exceeded the lengths of one or both of the individuals based on start point, l and the start points were re-chosen. Hereafter this is referred to as variable size (VS) crossover.

In the second case, called "fixed size" (FS) crossover, individuals are defined in terms of a number of pages where each page consists of a fixed number of instructions (always 4 in these experiments). Crossover may now only take place between single pages from each parent, resulting in fixed length individuals [8]. In both cases, the mutation operator selects an instruction with uniform probability and performs an EX-OR operation with a randomly created bit sequence. Field specific mutation was not used since it was not found to result in an improvement for two register addressing formats, as is used in this research [8]. Finally, an additional 'swap' mutation operator interchanges two randomly chosen instructions within the same individual. This operator was found to provide a useful method for code re-ordering in linear GP [8]. In summary, the only difference between the two forms of linear GP is the two point crossover operator.

4 Code Regularity (Repeating Sequence) Analysis

In this section we are interested in establishing whether *code* regularity is being established to a greater extent in fixed size (FS) linear GP or variable-size (VS) linear GP, where Langdon and Banzhaf [2] already recognize the presence of repeated sequences in linear GP genomes in general. If regular code sequences could be established, in conjunction with regularity found in the nature of the solutions, we will establish that fixing the crossover points enables the evolution of code with respect to page-boundaries to provide building blocks that form repeating sequences.

In order to measure code regularity, the frequency of seeing different instruction sequences (of a given length) within final solutions is recorded. This is then normalized with respect to the total number of sequences of that length, to provide results independent of instruction count. To do so, the concept of instruction type needs to be established. Here, an instruction with a mode bit of '0' (load instruction, Figure 1) could be type 1, 2 or 3, depending on the last two bits (which dictates what is loaded into the register). Any instruction with a mode bit of '1' (act on information in the registers, Figure 1) is always a type 4 instruction. Register information cannot be used in this instruction type because which register is used with mode bit '1' depends on a factor not determined by an individual's instructions—namely the 'IF_FOOD_AHEAD' test (see Section 2). This parameterization is designed to reveal how instructions relevant to ant behavior may be repeated in order to use those behaviors in the formation of solution strategies. The relevant fields are selected such that all components of ant behavior are accounted for: executing and loading, including the placing of particular commands in memory registers so they will be executed. In other words, this parameterization allows an analysis of instruction sequence as opposed to simple bit sequences. In this respect, all possible information on instruction sequences is gathered. The instruction types are outlined in Table 3.

Table 3. Instruction Type Definitions

Mode Bit	Register Field	Instruction Type
0 (load)	00 (right)	1
	01 (left)	2
	10, 11 (move)	3
1 (act on register)	N/A	4

Instruction sequences identified take the form of any combination of instruction type over a pre-specified length of two, three, four and pairs with a wildcard (three instructions with one unidentified). This results in 16, 64, 256 and 48 different combinations of instructions respectively. All the instruction sequences present for each of the four lengths are ordered and the number of times each sequence occurs for each of FS and VS is recorded. Finally, in order to summarize this large amount of information, a count is made of the number of times that either page or variable size crossover results in a larger count than the other for each possible instruction type sequence. This provides a concise way of seeing which implementation possessed higher levels of code regularity. The number of times the larger count occurs for FS and VS for each sequence of some length is then normalized by dividing it by the total number of possible sequences of that length to yield a percentage. Figures 2 and 3 summarize this information for the case of a maximum instruction limit of 128 instructions over the 50 trials for the 4 register and 8 register ants, respectively. Note that if the total percentage of largest counts accounted for by both VS and FS crossover cases do not total 100% for a particular sequence length, the remaining difference is the percentage of the sequences where the counts were tied between the FS and VS implementations.

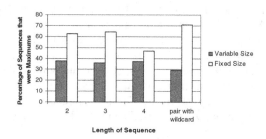

Fig. 2. Number of maximum occurrences/highest code regularity for either VS or FS for each instruction sequence possible for a given sequence length in 4 register ants with a maximum instruction limit of 128

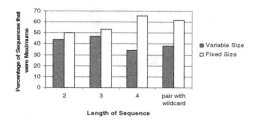

Fig. 3. Number of maximum occurrences/highest code regularity for either VS or FS for each instruction sequence possible for a given sequence length in 8 register ants with a maximum instruction limit of 128

Figure 2 and 3 indicate that the fixed size case records a high number of maximum counts. Moreover, the distinction becomes increasingly more apparent as sequence length increases, where specificity (of code) increases with increasing sequence length. Thus, fixed size crossover does indeed solve problems with higher levels of code regularity because its solutions contain more repeated sequences of each length.

However, does this higher code regularity come at a cost of disruption in fitness? To test this, we compared parent and child fitness and command difference with each tournament round. To compute the command difference between parent and child the number of different instructions with respect to each location is simply counted. Since both FS and VS crossover enforce a fixed length chromosome for parent and child, a simple count will always suffice. Each command pair must have the same bit string and position in each individual for the two commands to be considered the same. This is reasonable given that the action of the ant given the bit string relies so heavily on the previous instructions to dictate its current place on the grid and its register contents. Fitness difference and command difference are then plotted with the percentage of total evaluations that account for that fitness-command difference pair. The ensuing data structure details the fitness change that results from a difference in genome sequence, as well as how likely that change is. Finally, the difference in the percentage of cases (fixed size – variable size) is plotted to indicate whether FS crossover is responsible for more / less cases of fitness decline, as well as the magnitude of those declines.

The results for individuals differing by 1 to 8 commands are shown in Figure 4 for the 4 register, 128 (maximum) instruction experiment of 50 trials. The command difference is restricted to the range 1 to 8 because a maximum of 8 commands can be different between two ants in the fixed size implementation in the situation where two pages (4 instructions each) have been crossed over and none of the corresponding instructions for those pages match in either individual. The possibility of 0 commands being different is not of interest, for the fitness difference will necessarily be 0. Fitness differences greater than zero favor FS and those less than zero favor VS.

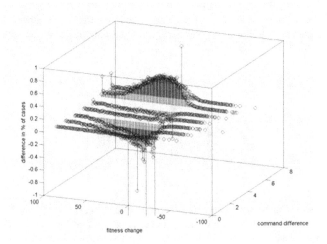

Fig. 4. Fitness change difference (Fixed size – Variable size) given command difference and the percentage of evaluations in which it occurs over 50 trials for the 4 register, 128 maximum instruction experiment

From Figure 4 it is noted that there is only a fitness difference in +/- 1 % of the cases (except for only 3 outliers) between the fixed size and variable size implementations. Furthermore, higher (lower) command difference changes consistently favor the fixed size (variable size) implementation. When considering all data projected onto only the fitness change and difference axes, a t-test indicated no significant difference between fixed size and variable size (0.803). The percentage of cases that were declines, gains, or had no change in fitness over 50 trials for the fixed size and variable size implementations is shown below in Table 4. The numbers are almost identical. The overall analysis indicates that the higher code regularity of the fixed size implementation is certainly not at a cost of fitness disruption.

Table 4. Percentage of Cases that were Declines, Gains, or No Change for Fixed size and Variable size Implementations over 50 Trials for 4 Register, 32 Page Maximum Ants

Fitness case	Fixed size	Variable size
Declines	86.80%	86.96%
Positives	6.32%	6.41%
No Change	6.88%	6.63%

5 Solution Regularity Analysis

The purpose of this section is to identify whether fixed size crossover has any implications for the strategy learned. To do so, we examine the fitness measure, which is simply the number of pieces of food 'eaten' by the ant, at each instruction throughout a typical tournament in relation to the genome. Firstly, a point is plotted in 3 dimensions for each piece of food eaten per instruction per tournament round. Since this data is hard to visualize in 2 dimensions, we illustrate these properties more formally by projecting all the data points on to the fitness and instruction axes, normalize to a range of 0 – 100% and construct a series of ten histogram bins at intervals of 10%. The process is repeated for all converging 50 trials and a student t-test performed for the independence of the distribution associated with each bin across the trials. This is summarized in Figure 5, whereas Figure 6 plots the mean accumulation of fitness in each bin (note the log axis implies that the *shorter* the bar, the *greater* the accumulation of points).

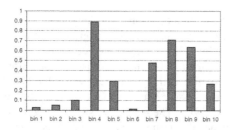

Fig. 5. T-test for hypothesis of same fitness means per histogram bin

It is now apparent that the fixed size approach uses more points in the bins 1, 2, 4-7, Figure 6, where this is statistically significant at the 90% confidence interval in bins 1, 2, and 6 (Figure 5). In effect, the sharper, more immediate fitness accumulation by the genome for FS crossover has a greater accumulation over the first 2 bins, whereas the VS crossover results in a more gradual fitness accumulation as instructions are executed, indicated by the higher accumulation for VS in bin 3. The earlier achievement of fitness accumulation for FS is then apparent in the higher accumulation for bins 4 to 7. While the FS instruction set has mostly finished its fitness accumulation toward the end of the genome, the VS scheme dominates the remaining three bins, Figure 6.

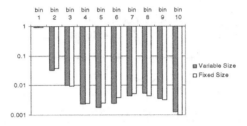

Fig. 6. Histogram for fitness (log scale)

The accumulation of fitness points is thus markedly different in VS as opposed to FS. The FS individuals with repeating instruction sets provide the same fast initial fitness accumulation strategy behavior in each case, whereas the VS individuals simply accumulate fitness gradually across the genome. Thus, the FS individuals match code (genotype) regularity with solution (phenotype) regularity. Section 4 and 5 together thus show that the FS implementation yields more repeating sequences than VS and that FS also repeatedly exhibits a strategy of more efficient fitness gathering within the solution. Recall from Section 2 that the minimum and maximum lengths of individuals are identical for VS and FS, the initial populations consist of individuals of identical length, and all operators were identical except for crossover. Thus, the fitness accumulation strategies must be attributable to the function of the different crossover operators.

6   Conclusion and Future Work

Two linear GP (L-GP) implementations were examined using a benchmark problem chosen for the characteristic that it involved gradual fitness accumulation with increasing instruction count. Thus, performance could be attributed to particular areas of the genome. One implementation used a crossover operator that allowed exchange of a variable number of instructions between two individuals (VS), while the other implementation used a crossover operator that only permitted the exchange of a fixed number of instructions (FS). The results are summarized below in Table 5. Examining Table 5, it establishes the argument for the presence of repeated sections of code that result in problem solving strategy formation without fitness disruption (building blocks).

Table 5. Summary of results of all experiments conducted

Experimental Factor	Fixed Size	Variable Size
Genotype regularity	Always higher	Always lower
Phenotype regularity	Efficient fitness gathering	Sparse fitness gathering
Fitness decline	Approximately equal	Approximately equal

Section 4 of the paper established rows 1 and 3: An analysis of the genotype of the FS and VS individuals found that the FS implementation generated individuals with a higher number of repeating instruction sequences than the VS implementation. The comparison of genotypes also established that the higher code regularity of the FS implementation did not come with an associated cost of fitness disruption. Section 5, where the actual solutions produced by the genotypes, i.e., the phenotypes, were analyzed establishes row 2. Here, the FS program was found to form a particularly efficient fitness gathering strategy as opposed to the VS implementation where fitness gathering simply occurred gradually across the genome.

The appearance of code repeatability points to the presence of re-usable modules. These repeated sections that can be modularized were evolved in our implementation rather than having the modules specified *a priori*. Future work will involve ways of effectively identifying the modularity of the repeated sections and their use as evolved functions within the individual's program.

Acknowledgements. The authors gratefully acknowledge the support of a NSERC PGS-B and Honorary Izaak Walton Killam Scholarship (Garnett Wilson), and the CFI New Opportunities and NSERC research grants (Dr. M. Heywood).

References

1. Soule, T., Heckendorn R.B.: An Analysis of the Causes of Code Growth in Genetic Programming, Genetic Programming and Evolvable Machines, 3(3) (2002) 283-309
2. Langdon, W. B., Banzhaf W.: Repeated Sequences in Linear GP Genomes, GECCO Late-breaking Papers CD-ROM (2004)
3. Koza J.R.: Genetic Programming II: Automatic Discovery of Reusable Programs. MIT Press (1994)
4. Koza J.R.: Genetic Programming: On the Programming of Computers by Natural Selection. MIT Press (1992)
5. Langdon W.B., Poli R.: Why Ants are Hard. Gentic Programming 1998: Proceedings of the Third Annual Conference (1998) 193-201
6. Langdon W.B., Poli R.: Foundations of Genetic Programming. Springer-Verlag (2002)
7. Huelsbergen L.: Toward Simulated Evolution of Machine-Language Iteration, Proceedings of the 1st Conference on Genetic Programming (1996) 315-320
8. Heywood M.I., Zincir-Heywood A.N.: "Dynamic Page-Based Crossover in Linear Genetic Programming," IEEE Transactions on Systems, Man and Cybernetics—Part B: Cybernetics, 32(3) (2002) 380-388

Evolution of a Strategy for Ship Guidance Using Two Implementations of Genetic Programming

Eva Alfaro-Cid[1], Euan William McGookin[2], and David James Murray-Smith[2]

[1] Instituto Tecnológico de Informática,
Universidad Politécnica de Valencia, 46022,Valencia, Spain
evalfaro@iti.upv.es
[2] Centre of Systems and Control,
Dept. of Electronics and Electrical Engineering,
University of Glasgow, G12 8LT, Glasgow, UK

Abstract. In this paper the implementation of *Genetic Programming* (GP) to optimise a controller structure for a supply ship is assessed. GP is used to evolve control strategies for manoeuvring the ship. The optimised controllers are evaluated through computer simulations and real manoeuvrability tests in a water basin laboratory. In order to deal with the issue of generation of numerical constants, two kinds of GP algorithms are implemented. The first one chooses the constants necessary to create the control structure by random generation . The second algorithm includes a *Genetic Algorithm* (GA) for the optimisation of such constants. The results obtained illustrate the benefits of using GP to optimise propulsion and navigation controllers for ships.

1 Introduction

In order to ensure the safe navigation of surface vessels their motion (i.e. navigation and propulsion capabilities) has to be controlled accurately. This can be achieved through the design and implementation of automatic control systems. The performance of the control techniques depends not only on the control structure but also on the values of the controller's parameters. Conventionally, these parameters are manually tuned by the designer. This relies on an ad hoc approach to tuning, which depends on the experience of the designer. A solution to this problem (widely used in the field of control engineering [1]) is to use evolutionary optimisation techniques such as *Genetic Algorithms* (GAs) that tune such parameters automatically.

However, GAs are parameter optimisers and in the majority of cases do not vary the structure of the optimising subject. In the context of controller optimisation they are presented with the structure of a particular control methodology and vary the associated parameters to obtain the desired performance for the system [1-3].

Genetic Programming (GP) [4] evolves candidate solutions without specifying *a priori* their size, shape or structure. By using GP, the optimisation problem of finding a near-optimal controller is taken a step forward in that the structure of the whole controller is optimised and not only the parameters that define such a structure.

M. Keijzer et al. (Eds.): EuroGP 2005, LNCS 3447, pp. 250–260, 2005.

The particular application used in this research is a scale model of an oil platform supply ship called *CyberShip II* (CS2) [5]. The optimisation problem for the GP is to provide a control strategy that governs the heading and propulsion dynamics of CS2.

The GP optimisation of the control structures has been conducted through computer simulations in Matlab using a mathematical model of CS2. The optimised controllers have been implemented and tested on the physical model of CS2.

In order to deal with the issue of the generation of numerical constants, two kinds of GP algorithms have been implemented. The first one chooses the constants necessary to create the controller structure by random generation (GP+RG) [4]. The second GP algorithm includes a GA technique for the optimisation of such constants (GP+GA) [6]. The results obtained from both methods are presented and compared.

2 CyberShip II

The control subject used in this work is CS2, which is a scale model (scale 1/70th approx.) of an oil platform supply ship. This test vessel has been developed at the Marine Cybernetics Lab (MCLab) at the Norwegian University of Science and Technology (NTNU) in Trondheim, Norway. The MCLab is a purpose built experimental laboratory for testing of ships and underwater vehicles. For more information about the MCLab and CS2 refer to [5], [7] and [8].

Prior to the real testing the non-linear hydrodynamic model of the vessel has been used for the simulations of the design stage. The model corresponds to the non-linear state space equation [10]:

$$\dot{\mathbf{x}} = \mathbf{A}(\mathbf{x}) \cdot \mathbf{x} + \mathbf{B} \cdot \boldsymbol{\tau} \qquad (1)$$

Here $\mathbf{x} = [\boldsymbol{v}, \boldsymbol{\eta}]^{\mathrm{T}}$, where $\boldsymbol{v} = [u, v, r]^{\mathrm{T}}$ is the body-fixed linear and angular velocity vector and $\boldsymbol{\eta} = [x, y, \psi]^{\mathrm{T}}$ is the Earth-fixed position and orientation vector. $\boldsymbol{\tau} = [\tau_1, \tau_2, \tau_3]^{\mathrm{T}}$ is the input vector (τ_1, τ_2 and τ_3 are the forces and torque along the X, Y and Z-axis). These are the inputs to CS2 that are used to control its motion.

In order to create a more realistic environment, wind generated waves are simulated during the manoeuvres used to evaluate each tree during the optimisation. These are the most relevant disturbances experience by surface vessels and they can be realistically reproduced in the MCLab during tests.

The model that has been used to simulate the waves' action on the vessel derives from the forces and moments induced by a regular sea on a block-shaped ship [9]. It forms a vector called $\boldsymbol{\tau}_{\mathrm{waves}}$ that is directly added to the input vector, $\boldsymbol{\tau}$, in (1).

3 Genetic Programming

The optimisation criterion used in this study is defined by the cost function in (2). Since the objective of the controllers is to make the vessel track desired heading and propulsion trajectories with the minimum actuator effort, the components of the cost function must reflect these design objectives [11].

$$C = \sum_{i=0}^{tot} \left[\left[\left(\Delta \psi_i \right)^2 + \lambda_1 \left(\tau_{3i} \right)^2 + \mu_1 \left(\frac{\Delta \tau_{3i}}{\Delta t} \right)^2 \right] + \left[\left(\Delta u_i \right)^2 + \lambda_2 \left(\tau_{1i} \right)^2 + \mu_2 \left(\frac{\Delta \tau_{1i}}{\Delta t} \right)^2 \right] \right] \qquad (2)$$

Here, $\Delta \psi_i$ is the ith heading angle error between the desired and obtained heading, τ_{3i} is the ith yaw thrust force, Δu_i is the ith surge velocity error between the desired and obtained surge velocity and τ_{1i} is the ith surge thrust force. Therefore, the quantities $\Delta \psi$ and Δu give an indication of how good the tracking is between the actual and the desired heading and surge velocity, and the input components τ_3 and τ_1 are used to keep the actuators to a minimum so that they are within the operating limits.

The third and sixth terms of (2) introduce a measurement of the inputs rates [11]. The minimisation of these terms helps reduce the oscillations in the inputs, avoiding unnecessary wear and tear of the actuators that shortens their operational lifespan.

Also, in (2), *tot* is the total number of simulation time steps and λ_1, λ_2, μ_1 and μ_2 are scaling factors. As the input force and torque are always larger than the output errors near the optimum, they dominate the cost values in this critical area of the search space. It leads to solutions that provide very small thruster effort, but very poor tracking of the desired responses. In order to avoid this, these four weighting coefficients are introduced, so that an equally balanced trade-off between the six terms of the cost function is obtained. It is a single objective, multi-aspect criterion.

3.1 GP Operators

Selection. In this research *tournament selection* has been used [12].

Crossover. In this work *subtree crossover* [4] has been used and the probability of crossover has been chosen to be 80% due to the satisfactory results obtained with this probability in a study comparing the performance of various crossover probabilities and mutation probabilities presented in [3].

Mutation. The tree structure of GP solutions allows a variety of mutation operators. In this study a combination of two methods is employed i.e. *subtree mutation* [4] and *point mutation* [19]. Mutation occurs with a probability of 0.1 [3]. Once a tree is chosen for mutation, the probability of undergoing subtree or point mutation is 0.5.

3.2 GP Coding in Matlab

The GP algorithm used in this paper is coded in Matlab. The difficulty of coding GP in Matlab lies in the lack of pointers. This requires a different coding approach. The whole population is stored in a cell array, every cell storing one individual. The tree structure is represented by a matrix (see Fig. 2) in which the number of rows is the number of internal nodes and it is evolved along the GP generations. Every internal node is encoded as a 1x5 vector (see Fig. 1).

Node number	Node type	1st argument	Function	2nd argument

Fig. 1. Internal node representation. The first element in the node vector is the node number. The second element distinguishes if the arguments of the function are terminal nodes or internal nodes (for example, a value of 0 indicates that both arguments are terminal nodes and a value of 1/2 indicates that the $1^{st}/2^{nd}$ argument is an internal node). All the internal nodes have arity 1 or 2. The three last elements provide the arguments of the functions in columns 3 and 5 and the function itself in column 4. If an argument is a terminal, it is included in the correspondent position; otherwise the numeric value refers to the number of the internal node that is rooted there [3]

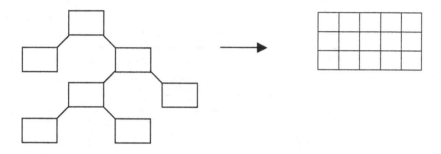

Fig. 2. Matrix representation of a tree. The tree structure needs to be flattened so every internal node is assigned a number. The nodes of the tree are counted from left to right, upwards. A node is not counted until all the nodes of the subtrees rooted in it are counted

3.3 GP Application to the Control Problem

Every solution to the stated control problem consists of two independent trees: one for heading control and other for propulsion control (i.e. decoupled controllers).

For this application the terminal set consists of 4 common terms: error, state, reference and one numerical constant as shown in Table 1.

Table 1. Terminal sets for propulsion and heading. The propulsion terminal set consists of the surge error ($\Delta u = u_d - u$), surge (u) and desired surge (u_d) plus one numerical constant. The heading terminal set consists of the heading error ($\Delta \psi = \psi_d - \psi$), heading ($\psi$) and desired heading ($\psi_d$) plus one numerical constant (R)

Propulsion	Heading
surge error (ε_p)	*heading error (ε_h)*
u	ψ
u_d	ψ_d
R	R

The probability of generating a numerical constant is 0.5 since the number of numerical constants required to create a control structure is larger than the number of variables.

The function set is formed by eleven functions. The four basic arithmetic operations {+, -, *, /} are routinely included in most GP algorithms. The *integral* and *derivative* functions are included to account for a PID type of structure. The *hyperbolic tangent* and *sign* functions allow the construction of *switching terms* equivalents, which are similar to *Sliding Mode* (SM) control [2]. The *place command* is included as a *Pole Placement* (PP) technique [14]. In addition, the *sine* and *exponential* functions give more versatility to the algorithm.

Table 2. Function set. The function set includes functions that are related to the following control techniques: *PID, Sliding Mode* and *Pole Placement*

2-argument functions	
arg1 arg2	arg1 + arg2
arg1 - arg2	arg1 / arg2
arg1 $tanh(h'(x - x_d) / $ arg2))	
1-argument functions	
$\int$ arg dt	d(arg) / dt
sin(arg)	exp(arg)
arg $sign(h'(x - x_d))$	
1/2-argument functions	
place(-arg)	
place(0, -arg1, -arg2)	

In the hyperbolic tangent and sign formulas from Table 2, $x = [u]$ and $x_d = [u_d]$ for the propulsion control tree, while $x = [v, r, \psi]^T$ and $x_d = [v_d, r_d, \psi_d]^T$ for heading. The **h** matrix is the right eigenvector associated with a zero pole for the desired closed-loop system matrix calculated based on the best solution found in a GA optimization of a decoupled SM controller [2, 3].

The place command returns the value -**k·x**, where **x** is as defined before and **k** is the feedback vector obtained by executing *place (0, -arg1, -arg2)* in the heading control or *place (-arg)* in the propulsion control.

In order to ensure that the closure property is met, the poles to be assigned by the *place* command are always real numbers and some of the functions have a protection mechanism. Thus, the hyperbolic tangent returns *arg1* when *arg2* is 0 and the *place* command returns 0 if there is any error flag (e.g. if the poles are too close).

3.4 Random Generation of the Numerical Constants

In the first implementation every time the random constant R in the set of terminals is chosen a random number is generated and associated with that terminal node [4]. The GP should be able to generate other constants needed by using arithmetic operations.

As opposed to Koza's GP that does not use mutation, in this work point mutation has been included as an operator. This enables the GP to modify the terminal values.

Thus, a numerical constant can change its value and a terminal occupied by a variable can be mutated into a numerical constant.

3.5 GAs Optimisation of the Numerical Constants

Various authors have pointed out that the random generation of numerical constants is not a very efficient way of creating new constants [6, 15, 16]. The main drawback of this approach is that the number of constants depends totally on the initialisation of the trees.

Various approaches can be found in the literature that address this issue. In [6] the authors combined a GP with a GA for the tuning of the numeric parameters in the GP tree. They associate a GA-like fixed-length chromosome that represents the numerical values of the solution, although they may or may not be present in the tree. The chromosome is evolved together with the tree and is submitted to crossover and mutation. The main problem of this approach is that the fixed-length of the chromosome determines the maximum number of numerical constants that can be found in the tree. This requires *a priori* knowledge of the solution. Also, if the chromosome is made to be very long just to account for any additional constant, the length of the chromosome hampers the correct evolution of solutions and increases the computational cost.

In this work, the second GP algorithm tested uses a GA as a parametric optimisation technique. The aim is that the GP+GA algorithm provides a better parameter tuning and better results. The GP+GA method used is basically different from the mechanism presented by [6]. Instead of associating a GA chromosome with a GP tree and evolving them together, GP+GA combines a GP evolution process with a GA learning process, i.e. every time a tree is evaluated a mini-GA is run to optimise the values of the numerical constants present in that tree. With this approach the maximum number of constants in the tree does not need to be fixed and only those constants that are in the tree are encoded, reducing the size of the chromosomes.

The GP+GA has been coded so that the total number of tree evaluations is the same as in the GP+RG case, providing a good basis for comparison. The number of trees in the population used in GP+RG is 120 and the number of generations is 31. In order to get the same number of evaluations for the GP+GA optimisation, the GP has a population of 31 individuals and it runs for 8 generations. Each GA has a population of 5 individuals and it runs for 3 generations.

4 Results

The manoeuvre used for the GP optimisation in the evaluation of the candidate solutions has been a double step manoeuvre of 45° for heading while increasing the speed from rest to 0.2 m/s and back to rest [3].

The best results found in each optimisation are validated after the optimisation. This validation test is used to verify that the resulting tree is actually performing a control task, not merely generating a signal shaped in the right way for this manoeuvre but totally wrong for any other.

The manoeuvre used in the validation test consists of two turning circles linked together. This manoeuvre has been chosen following the recommendations of the Maritime Safety Committee (Resolution MSC.137(76)) for ship performance testing.

The resulting best controllers have been tested on the real vessel.

4.1 GP+RG Optimisation: Simulated and Real Results

The cost functions obtained after the GP optimisation are presented in Table 3. It can be observed that the best overall result is obtained from run 1. Some of the good results from the GP optimisation failed the validation test (e.g. results from runs 8, 19 and 20). This corroborates the importance of the validation test.

Table 3. GP results obtained in the optimisation using the double step manoeuvre and the validation test using a double turning point

Run	Double Step	Turning	Run	Double Step	Turning
1	12.90	79.78	11	285.32	841.22
2	79.15	213.40	12	53.37	534.41
3	123.78	360.49	13	132.25	445.48
4	18.09	194.95	14	114.01	705.51
5	7.63	289.88	15	275.93	954.56
6	51.55	646.29	16	95.92	150968.05
7	120.11	423.66	17	327.72	53316.41
8	12.34	593943.52	18	351.56	53180.82
9	370.40	949.33	19	30.52	735485.71
10	328.79	843.70	20	47.73	1501055.7

The structures of the controllers from run 1 are based on an hyperbolic tangent:

$$\tau_{1com} = -203.5368 \cdot tanh\left(\frac{h_p^{'} \cdot (u - u_d)}{11.8223}\right) + 2 \cdot sin(u_d) \tag{3}$$

$$\tau_{3com} = 14.752 \cdot tanh\left(\frac{\mathbf{h}_h^{'} \cdot (\mathbf{x}_h - \mathbf{x}_{hd})}{-0.7931}\right) \tag{4}$$

By analysing the range of values that perform as argument of the hyperbolic tangents it is easy to see that the functions are not reaching the saturations limits, therefore the hyperbolic tangents are acting as proportional controllers [2]:

$$\tau_{1com} \approx -17.5 \cdot h_p^{'} \cdot (u - u_d) + 2 \cdot sin(u_d) \tag{5}$$

$$\tau_{3com} \approx -19 \cdot \mathbf{h}_h^{'} \cdot (\mathbf{x}_h - \mathbf{x}_{hd}) \tag{6}$$

Since at low speed $sin(u_d) \approx u_d$, in this case, the sine term is acting as a feedforward control (i.e. provides the same control effort regardless of the current surge value).

Fig. 3 illustrates the simulated performance of the controllers from Eq. (3) and (4).

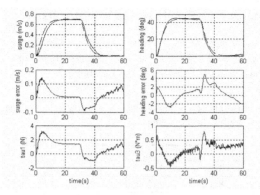

Fig. 3. Simulated results of the GP+RG optimisation. The upper subplots represent the desired (*dashed line*) and measured (*solid line*) outputs, u (*left*) and ψ (*right*). The subplots in the middle represent the output errors, i.e. the surge error, $u_d - u$ (*left*), and the heading error, $\psi_d - \psi$ (*right*). Finally, the subplots at the bottom depict the control signals τ_1 (*left*) and τ_3 (*right*)

It can be seen that the performance of the controller for the evaluation manoeuvre is quite good. In the propulsion control there is a slight steady-state error caused by the lack of integral term. Also, the gain of the controller is quite low, which explains the slow transient response. The heading control performs better that the propulsion one. The tracking is very good apart from a slight overshooting. Both controllers keep the actuator signals within limits and are not significantly affected by the disturbances, apart from the ripple in the actuators forces.

Fig. 4 shows the results obtained when the controllers are tested in the water tank. The real responses obtained are satisfactory. The tracking is quite good, especially for the heading response, although the same overshooting that has been observed in the simulated response can also be observed. The surge tracking is satisfactory. Both controllers keep the actuators signals within the limits. A delay is introduced by the system that induces very high initial control signals until the manoeuvre starts.

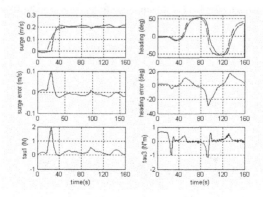

Fig. 4. Real results of the GP+RG optimisation. The format of the figure is as in Fig. 3

4.2 GP+GA Optimisation: Simulated and Real Results

The best cost values obtained in the runs of the GP+GA optimisation are shown in Table 4. In general, the GP+GA scheme has converged to worse cost functions. The best overall cost has been achieved from run number 1.

Table 4. GP+GA results obtained in the optimisation using the double step manoeuvre and the validation test using a double turning point

Run	Double Step	Turning	Run	Double Step	Turning
1	50.13	239.91	11	228.93	47737.78
2	143.78	332.39	12	953.27	53225.26
3	267.23	750.16	13	629.91	1411779.6
4	262.70	38420482.8	14	229.21	640.17
5	115.50	323.28	15	319.63	871.85
6	353.10	834.67	16	330.65	66907.90
7	394.20	837.06	17	530.52	1487.08
8	118.84	347.29	18	561.61	54399.76
9	128.37	687.91	19	908.00	54393.05
10	127.65	721.92	20	747.22	55726.58

The structure of the best result consists of a hyperbolic tangent for the heading control and a proportional controller for the propulsion. As before, the hyperbolic tangent in Eq. (8) acts as a proportional controller and can be expressed as in Eq. (9):

$$\tau_{1com} = 93.6 \cdot \varepsilon_p \tag{7}$$

$$\tau_{3com} = -57.4927 \cdot \tanh\left(\frac{h_h' \cdot (x_h - x_{hd})}{4.0842}\right) \tag{8}$$

$$\tau_{3com} \approx -14 \cdot h_h' \cdot (x_h - x_{hd}) \tag{9}$$

The heading controller in Eq. (8) is equivalent to that of Eq. (4) but with a smaller gain. The propulsion controller substitutes the feedfoward term by a higher gain.

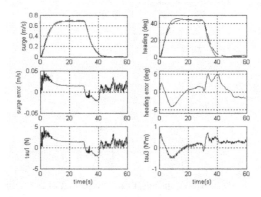

Fig. 5. Simulated Results of the GP+GA Optimisation. The format of the figure is as in Fig. 3

While comparing the simulated performance of the controllers from Eq. (7) and (8) shown in Fig. 5 with the performance of the controllers obtained in the GP+RG optimisation (see Fig. 4) it can be observed that the heading response is very similar. Regarding the propulsion control, although this controller achieves a faster transient response and better tracking of the reference (due to a higher gain), the control effort and the rippling in the signals is worse, which leads to a higher cost function.

Fig. 6 shows the results obtained when the controllers are tested in the real plant. The heading response is similar to that of the GP+RG optimisation (see Fig. 4). The propulsion control effort reaches the actuator limits causing peaks in the error signal.

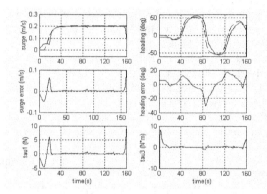

Fig. 6. Real Results of the GP+GA Optimisation. The format of the figure is as in Fig. 3

5 Conclusions

The results obtained in the GP optimisations are very satisfactory. The manoeuvring performance of the controllers illustrated in the figures also proves their adequacy.

Although the numerical cost values obtained with the GP+RG optimisation are better, both GP implementations have converged to trees that provide very similar control strategies. The best results obtained in both sets of runs are based on a hyperbolic tangent function providing the heading control and a proportional term or a hyperbolic function acting as a proportional term providing the propulsion control.

The terminal values chosen by the search method as arguments for the hyperbolic functions for these best results make this function operate in its proportional range instead of in the switching area. Thus, in the case of the propulsion control, since the subsystem is of 1st order, the hyperbolic tangent provides an outcome proportional to the surge speed error (i.e. a proportional term). In the case of the heading control, the resulting commanded force is effectively of the form: $\tau_{3com} \approx -\mathbf{k} \cdot \mathbf{h}' \cdot (\mathbf{x} - \mathbf{x_d})$, i.e. a full state feedback control with a feedback matrix and a conditioning matrix equal to $\mathbf{k} \cdot \mathbf{h}'$.

References

1. Chipperfield, A., Fleming P.: Genetic Algorithms in Control Systems Engineering. Control and Computers, Vol. 23 (1995) 88-94
2. McGookin, E.W.: Optimization of Sliding Mode Controllers for Marine Applications: A Study of Methods and Implementation Issues. PhD Thesis. University of Glasgow, UK (1997)
3. Alfaro-Cid, E.: Optimisation of Time Domain Controllers for Supply Ships Using Genetic Algorithms and Genetic Programming. PhD Thesis. University of Glasgow, UK (2003)
4. Koza, J.R.: Genetic programming: On the Programming of Computers by Means of Natural Selection. MIT Press, Cambridge, MA (1992)
5. Lindegaard, K.P., Fossen, T.I.: Fuel Efficient Rudder and Propeller Control Allocation for Marine Craft: Experiments with a Model Ship. IEEE Transactions on Control Systems Technology, Vol. 11 (2003) 850-862
6. Howard, L.M. and D'Angelo, D.J.: The GA-P: A Genetic Algorithm and Genetic Programming Hybrid. IEEE Expert, Vol. 10 (1995) 11-15
7. Corneliussen, J.: Implementation of a Guidance System for CyberShip II. Master Thesis. NTNU, Trondheim, Norway (2003)
8. Sveen, D.A.: Robust and Adaptive Tracking Control of Surface vessel for Synchronization with an ROV: Practical Implementation on CyberShip II. Master Thesis. NTNU, Norway (2003)
9. Zuidweg, J.K.: Automatic Guidance of Ships as a Control Problem. PhD Thesis. Delft University of Technology, The Netherlands (1970)
10. Fossen, T.I.: Guidance and Control of Ocean Vehicles. John Wiley& Sons Ltd, Chichester (1994)
11. Alfaro-Cid, E., McGookin, E.W., Murray-Smith, D.J.: Genetic Algorithm Optimisation of a Supply Ship Propulsion and Navigation Systems. Proceedings of the MTS/IEEE Oceans Conference. Honolulu, USA (2001) 2645-2652
12. Luke, S.: Issues in Scaling Genetic Programming: Breeding Strategies, Tree Generation, and Code Bloat. PhD Thesis. University of Maryland, USA (2000)
13. O'Reilly, U.: An Analysis of Genetic Programming. PhD Thesis. Carleton University, Canada (1995)
14. Kaustky, J., Nichols, N.K., Van Dooren, P.: Robust Pole Assignment in Linear State Feedback. International Journal of Control, Vol. 41 (1985) 1129-1155
15. Esparcia-Alcazar, A.I.: Genetic Programming for Adaptive Digital Signal Processing, PhD Thesis. University of Glasgow, UK (1998)
16. Fernandez, T., Evett, M.: Numeric Mutation as an Improvement to Symbolic Regression in Genetic Programming. EP98 Lecture Notes in Computer Science, Vol. 1447 (1998) 251-260

Evolution of Vertex and Pixel Shaders

Marc Ebner, Markus Reinhardt, and Jürgen Albert

Universität Würzburg, Lehrstuhl für Informatik II,
Am Hubland, 97074 Würzburg, Germany
ebner@informatik.uni-wuerzburg.de
http://www2.informatik.uni-wuerzburg.de/staff/ebner/welcome.html

Abstract. In real-time rendering, objects are represented using polygons or triangles. Triangles are easy to render and graphics hardware is highly optimized for rendering of triangles. Initially, the shading computations were carried out by dedicated hardwired algorithms for each vertex and then interpolated by the rasterizer. Todays graphics hardware contains vertex and pixel shaders which can be reprogrammed by the user. Vertex and pixel shaders allow almost arbitrary computations per vertex respectively per pixel. We have developed a system to evolve such programs. The system runs on a variety of graphics hardware due to the use of NVIDIA's high level Cg shader language. Fitness of the shaders is determined by user interaction. Both fixed length and variable length genomes are supported. The system is highly customizable. Each individual consists of a series of meta commands. The resulting Cg program is translated into the low level commands which are required for the particular graphics hardware.

1 Motivation

In computer graphics three dimensional objects are usually represented using polygons. Polygons in turn can be broken down to triangles. A triangle is to computer graphics what the atom is to chemistry. Even curved objects such as spheres or cylinders are approximated with triangles. The surface nevertheless appears round due to special shading techniques. The advantage of using triangles is that the graphics pipeline can be highly optimized. A triangle consists of three vertices. Each vertex is assigned a number of attributes such as color, reflectance properties or a normal vector. Initially, graphics libraries used fixed algorithms to compute the color of a vertex using the assigned reflectance properties of the material it is supposed to represent [1, 6, 24]. After the color of the vertex is calculated, the polygon or triangle is filled by interpolating the colors computed for the vertices. This method is called Gouraud shading. In todays graphic hardware these shading algorithms are no longer fixed, they can be reprogrammed by the user. This is done using pixel and vertex shaders [3, 10]. A vertex shader is a small program which computes or modifies attributes such as position, normal vector, or reflectance properties. These attributes are interpolated to obtain the data for each pixel. A pixel shader is used to compute

M. Keijzer et al. (Eds.): EuroGP 2005, LNCS 3447, pp. 261–270, 2005.

the color of each pixel from these attributes. Both, vertex and pixel shaders are programs which can be evolved. Use of genetic programming [2, 7, 8] to evolve shaders was originally suggested by Kenton Musgrave [12].

We have developed a system which allows us to evolve pixel and vertex shaders by user interaction [17]. The system starts off with a number of randomly created pixel shaders and a fixed vertex shader or vice versa. The pixel and vertex shaders are applied to an object which is shown to the user. The user can then judge how good the pixel respectively vertex shader is and set its fitness value. Genetic operators are applied and new shaders are created. Again the shaders are presented to the user which has to rate the quality of the shaders. This process can continue for as long as the user wants. In the following we will first summarize some background material on vertex and pixel shaders. Then we describe our system and the experiments we have made.

2   Vertex and Pixel Shaders

Vertex and pixel shaders can be reprogrammed using a custom assembly language. A vertex shader receives its input, the attributes of a vertex, through a fixed number of registers. This input is read-only. A vertex shader processes four dimensional data. Each register contains four floating point numbers which map naturally to the three color bands red, green, and blue. The fourth component describes how transparent the object is. A set of output registers is used to store the modified attributes. Another set of registers can be used during the computation. A small amount of memory can also be accessed read-only.

A vertex shader program consists of a sequence of commands. Originally, a vertex shader could contain a maximum of 128 commands. The set of commands included standard arithmetic operators such as addition, subtraction, multiplication and computation of the scalar product between two vectors. Apart from the standard operators, some commands also addressed the special needs in computer graphics such as the computation of coefficients for ambient, diffuse and specular lighting or the computation of coefficients for light attenuation. Originally, there were no explicit flow control statements such as if, for, while or goto. However, it was possible to implement if-then-else operations within the simple instruction set given. Subsequently vertex shader instructions now also contain flow-control instructions to jump forward, loop a fixed number of times, and call subroutines [1].

A pixel shader is used to compute the color of every pixel of a fragment. It receives the interpolated components such as diffuse and specular light which was computed by the vertex shader as input. The vertex shader also has access to multiple textures and can combine the diffuse and specular components with this texture data. The registers of a pixel shader contain four values where the red, green, blue and alpha components of a color are stored. There are also a number of registers where temporary data may be stored and some address registers through which texture data can be accessed. Like the vertex shader, a pixel shader is a small program. The difference is that the commands of the pixel shader are tailored for texture access. It consists of two set of commands, arithmetic operations and operations for texture addressing. No explicit flow control statements are included. Using a pixel shader it is also possible to change the depth of a pixel or even end further processing of a patch.

3   Evolution of Vertex and Pixel Shaders

Both vertex and pixel shaders are basically short sequences of commands which can be evolved. Evolution of shaders was originally proposed by Kenton Musgrave [12]. Loviscach and Meyer-Spradow [9, 11] evolved vertex and pixel shaders using a one-to-one fixed length representation between the genotype and the assembly language of the hardware. In contrast to the system by Loviscach and Meyer-Spradow, we evolve shaders using NVIDIA's high level Cg shader language. This allows us to define arbitrary computer architectures for which shaders may be evolved.

We have used linear genetic programming [13, 14] to evolve vertex and pixel shaders. Each individual consists of a sequence of numbers from the range [0, 255]. We work with both fixed and variable length individuals. The information stored in an individual is mapped into a program as shown in Figure 1. A reading head moves along the individual and parses byte after byte. The first number is treated as an opcode of a command. Depending on the type of command we either need none, one, two or more arguments. If no arguments are needed then we proceed with the next byte and map this value into another command. Otherwise we fetch the required number of arguments from the individual and map these bytes to the corresponding variables. We then proceed with the next byte. In order to perform the mapping from bytes to commands respectively variables, we have defined two tables. One table lists the set of commands, the other lists the set of variables. A modulo operation is used in both cases to map any value from the range [0, 255] to a valid command respectively to a valid variable.

We do not use a particular graphics hardware as our target. Instead, we have chosen to use NVIDIA's Cg Toolkit [15] to perform the final mapping to the graphics hardware. This allows us to produce vertex and pixel shaders for all current and hopefully all future hardware. When mapping individuals to vertex or pixel shaders we create an output in the Cg language. The Cg language is a high level language similar to C. A vertex or pixel shader is constructed by first defining a wrapper. This wrapper is the same for all individuals. The wrapper consists of a header and a footer. A shader is created by taking the header, appending the commands as specified by the genotype, and finally appending the footer.

Our system is extremely versatile in that we can define arbitrary computer architectures using the table of commands and the table of variables. These tables are not stored internally but can be modified by the user. By modifying the tables we can vary the basic architecture of the programs to be evolved. We can define architectures with either a one-address instruction, a two-address

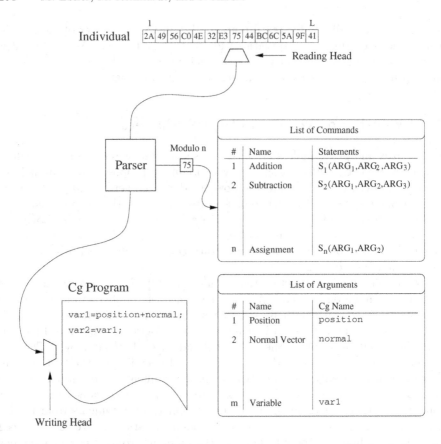

Fig. 1. An individual is mapped to a Cg program by moving a reading head along the bytes stored in the genotype. After reading the first byte we look up the corresponding set of statements and the number of required arguments in the list of arguments. The byte is fed through a modulo operation to access the table. Then the arguments are looked up in the second table. For each argument we parse another byte from the individual. After all arguments are available we append the statements to the Cg program. This process continues until all bytes of the individual have been parsed or we run out of bytes

instruction or even stack based architectures. It is also possible to create new meta commands which are not part of the original language. Each entry of the command table can contain an arbitrary sequence of statements of the Cg language. The required arguments are accessed through escape sequences. When parsing the individual these escape sequences are mapped to the corresponding variables.

Currently, we use perceptual selection [20, 21] to evaluate the individuals. We work with small population sizes. All individuals are presented to the user for evaluation. The vertex respectively pixel shaders are applied to a three dimensional object for viewing. If the user likes what he sees, he sets the fitness of these individuals to a high value. Figure 2 shows one step of the evolutionary

Parent Population   Offspring

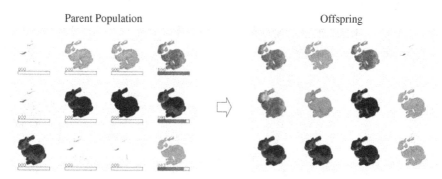

Fig. 2. A population of 12 individuals is shown on the left. The user manually sets the fitness of any individuals he likes. On the right we see the offspring population created from the three parent individuals

algorithm. On the left, we see a 4×3 matrix of individuals. The color bar below each individual is used to specify the fitness of an individual. Similar methods of interactive evaluation were used by Dawkins [4] when evolving his classic Biomorphs. Perceptual selection is also frequently used in evolutionary art systems [5, 18, 19, 22, 25]. Rowbottom gives a review of many of these systems [19].

When the user has finished evaluating the individuals (not all individuals have to be evaluated), the next generation of individuals is created. Both crossover and mutation operators are used. First we decide if a crossover is applied at all using a single crossover probability. Then we chose the actual crossover operator (one point or two point crossover) with uniform probability. After the two offspring are created, the individuals are mutated. The mutation probability is specified per byte. The type of mutation actually used is selected with uniform probability. Two types of mutation operations were defined for fixed length individuals: flip mutation and swap mutation. Flip mutation replaces one byte with a new value. Swap mutation swaps one byte with another byte of the same string. For variable length individuals, two additional mutation operators were used: insert and delete. The insert mutation operator inserts a new byte. The delete mutation operator deletes a byte from the individual.

When implementing evolutionary algorithms we usually want that parent and offspring are closely related. However our opcodes or meta commands require a variable number of arguments. If a single mutation were to change an opcode which requires a single argument to an opcode with a different number of arguments then this would have a large effect on the resulting Cg code. The entire code following the locus where the mutation occurs would be changed. Arguments would be interpreted as opcodes and vice versa. What we probably want is that a single mutation is able to either change the opcode or the argument into some other opcode or argument. Therefore we implemented a second parsing mode. Let k be the maximum number of arguments over all commands.

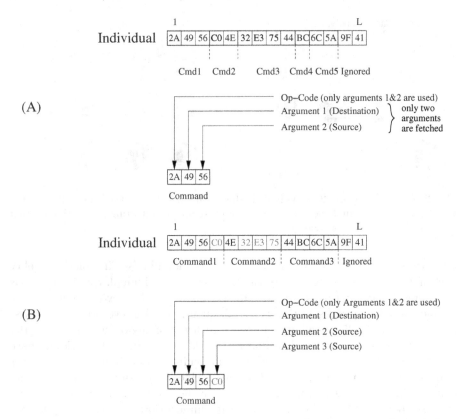

Fig. 3. Each individual consists of a sequence of commands. The commands themselves can either be considered to have variable length (A) or a fixed length (B). Using a fixed length for commands has the advantage that opcodes and arguments are always registered. Therefore, it is possible to change a single argument of an individual using a point mutation. A crossover exchanges opcodes with opcodes and arguments with arguments. This is not the case for (A). If variable length commands are used a single point mutation may have a large effect on the resulting Cg code

In this case, we fetch $k + 1$ bytes (one byte for the opcode and k bytes for the arguments) from the individual. The two parsing modes are shown in Figure 3. Since we do not know which method of parsing the individuals leads to better results, we have implemented both methods. Again, the choice on how the individuals are evaluated, is left to the user.

4   Experiments

Figure 4 shows a collection of evolved vertex shaders. Each row shows the results for a single shader applied to four different shapes: a plane, a sphere, a torus and the Stanford bunny. The shaders were evolved during two runs lasting 130 and

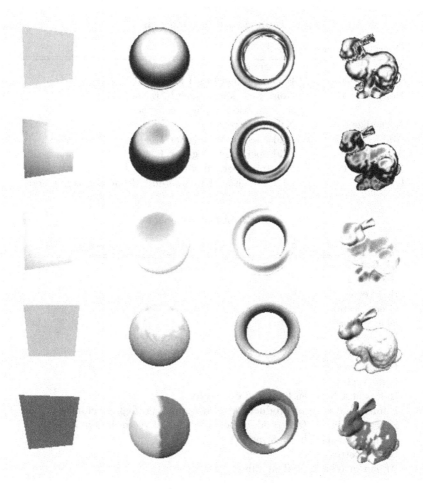

Fig. 4. A collection of evolved vertex shaders. Each row shows the results for a single vertex shader. A vertex shader is applied to four different objects: a plane, a sphere, a torus and the Stanford bunny

185 generations. Both runs used a two register machine model. The contents of the two registers can be exchanged using a swap operation. Instructions include addition, subtraction, multiplication, protected division, and a sine function. A normalize function and the popular noise function [16] was also included. The set of meta commands is shown in Table 1. The list of arguments is shown in Table 2. Arguments include the two registers, three color vectors red, green and blue, the current position of the vertex, normal vector, eye vector, light vector as well as some pre-calculated values such as the half vector, diffuse and specular lighting. An animator (a float value which changes periodically) is also included. The animator can be used to create animated shaders. Output of register 1 is used as the color of the vertex. We have used a population size of 12. All shaders shown in Figure 4 were evolved using fixed length individuals of length 20. Mutation

Table 1. List of commands

Operator	Operands	Result
Nop	0	No operation
Swap	0	Exchange registers 1 and 2
Noise	0	Noise function applied to register 1
Sin	0	Sine function applied to register 1
Normalize	0	Normalize register 1
Add	1	Add operand to register 1
Subtract	1	Add operand to register 1
Multiply	1	Multiply operand with register 1
Divide	1	Divide register 1 by operand, if operand is non-zero

Table 2. List of arguments

Argument
Register 1
Register 2
Red vector
Green vector
Blue vector
Position of vertex
Normal vector
Eye vector
Light vector
Half vector between light and eye vector
Diffuse lighting (dot product between normal and light vectors)
Specular lighting (dot product between normal vector and half vector)
Diffuse and specular lighting
Eye to vertex vector
Animator (changing float value)

probability was set to $\frac{1}{20}$ which resulted in one mutation per offspring. Crossover probability was set to 0.9. Roulette wheel selection was used to select offspring for breeding.

Although we were able to evolve some nice shaders we also noticed some limitations with the current approach. It is hard to evolve towards a particular target. For instance, it would be nice to be able to select two individuals and then obtain offspring which contain traits from both parents. If two individuals are selected, offspring may have interesting traits but may not have the intended look to them. I.e., if one selects a textured individual and another individual with a different color then the next generation will contain all types of individuals but not necessarily an individual with both the texture of one parent and the color of the second parent. This may be caused by a number of factors. First of all we are working with very small population sizes because fitness has to be determined by the user. Another cause may be the use of linear individuals. It may be that a tree based genetic representation is more amenable to evolution in this case. Rowbottom [19] noted that most evolutionary art systems have a certain signature to them. This also seems to be the case here. The evolved individuals seem to be largely a function of the type of commands and arguments used.

5   Conclusion

Vertex and pixel shaders are an exciting concept of computer graphics. We have developed a system to evolve vertex and pixel shaders via user interaction. Individuals are interpreted as linear sequences of commands which are translated into a high level computer graphics language. Individuals are applied to four different objects and presented to the user who then decides which individuals get to produce offspring.

Our system is highly customizable. With this system it is possible to define virtual intermediate architectures. At present, it is not known which architecture is best suited to evolve vertex and pixel shaders. Our initial experiments focused on the evolution of linear programs. It would be interesting to see how tree based genetic programming compares to linear genetic programming for the evolution of vertex and pixel shaders.

Another possible extension would be the automatic evolution of shaders for animated effects. One could take a short sequence of a movie taken with a digital camera and then evolve shaders which mimic the effect seen in the video. Other than evolving vertex and pixel shaders for computer graphics the concept may also be of interest to other researches who want to speed up their genetic programming experiments. It may be possible to use vertex or pixel shaders in other areas such as evolution of classifiers.

Acknowledgments

Our system uses the GAlib genetic package version 2.4, written by Matthew Wall at the Massachusetts Institute of Technology [23].

References

1. T. Akenine-Möller and E. Haines. *Real-Time Rendering*. A K Peters, Natick, MA, 2nd ed., 2002.
2. W. Banzhaf, P. Nordin, R. E. Keller, and F. D. Francone. *Genetic Programming - An Introduction: On The Automatic Evolution of Computer Programs and Its Applications*. Morgan Kaufmann Publishers, San Francisco, CA, 1998.
3. NVIDIA Corporation. Nvidia nfinitefx engine: Programmable vertex shaders.
4. R. Dawkins. *The Blind Watchmaker*. W. W. Norton & Company, New York, 1996.
5. A. E. Eiben, R. Nabuurs, and I. Booij. The Escher evolver: Evolution to the people. In P. J. Bentley and D. W. Corne, eds., *Creative Evolutionary Systems*, pp. 425–439. Morgan Kaufmann Publishers, 2001.

6. J. D. Foley, A. van Dam, S. K. Feiner, and J. F. Hughes. *Computer Graphics: Principles and Practice. 2nd Ed. in C.* Addison-Wesley Publishing Company, Reading, MA, 1996.

7. J. R. Koza. *Genetic Programming. On the Programming of Computers by Means of Natural Selection.* The MIT Press, Cambridge, MA, 1992.

8. J. R. Koza. *Genetic Programming II. Automatic Discovery of Reusable Programs.* The MIT Press, Cambridge, MA, 1994.

9. J. Loviscach and J. Meyer-Spradow. Genetic programming of vertex shaders. In *Proc. of EuroMedia 2003*, pp. 29–31, 2003.

10. C. Maughan and M. Wloka. Vertex shader introduction. Technical report, NVIDIA Corporation, 2001.

11. J. Meyer-Spradow and J. Loviscach. Evolutionary design of BRDFs. In M. Chover, H. Hagen, and D. Tost, eds., *Eurographics 2003 Short Paper Proceedings*, pp. 301–306, 2003.

12. F. Kenton Musgrave. Genetic textures. In D. S. Ebert, F. Kenton Musgrave, D. Peachey, K. Perlin, and S. Worley, editors, *Texturing and Modeling: A Procedural Approach. 2nd Ed.*, pp. 373–385, Cambridge, 1998. AP Professional.

13. P. Nordin. A compiling genetic programming system that directly manipulates the machine code. In K. E. Kinnear, Jr., ed., *Advances in Genetic Programming*, pp. 311–331, Cambridge, MA, 1994. The MIT Press.

14. P. Nordin and W. Banzhaf. An on-line method to evolve behavior and to control a miniature robot in real time with genetic programming. *Adaptive Behaviour*, 5(2):107–140, 1997.

15. NVIDIA. Cg toolkit. user's manual. a developer's guide to programmable graphics. Technical report, NVIDIA Corporation, Santa Clara, CA, 2002.

16. K. Perlin and E. M. Hoffert. Hypertexture. *SIGGRAPH '89 Conference Proceedings, Computer Graphics, Boston, MA*, 23(3):253–262, 1989.

17. M. Reinhardt. *Evolution von Pixel- und Vertex-Shader Programmen.* Projektpraktikum, Universität Würzburg, Institut für Informatik, Lehrstuhl für Informatik II, July 2004.

18. S. Rooke. Eons of genetically evolved algorithmic images. In P. J. Bentley and D. W. Corne, eds., *Creative Evolutionary Systems*, pp. 339–365. Morgan Kaufmann Publishers, 2001.

19. A. Rowbottom. Evolutionary art and form. In P. J. Bentley, ed., *Evolutionary Design by Computers*, pp. 261–277, San Francisco, 1999. Morgan Kaufmann.

20. K. Sims. Artificial evolution for computer graphics. *Computer Graphics*, 25(4):319–328, 1991.

21. K. Sims. Interactive evolution of dynamical systems. In F. J. Varela and P. Bourgine, eds., *Toward a practice of autonomous systems: Proc. of the 1st Europ. Conf. on Artificial Life*, pp. 171–178, Cambridge, MA, 1992. The MIT Press.

22. S. Todd and W. Latham. The mutation and growth of art by computers. In P. J. Bentley, ed., *Evolutionary Design by Computers*, pp. 221–250, San Francisco, 1999. Morgan Kaufmann.

23. M. Wall. *GAlib: A C++ Library of Genetic Algorithm Components, Version 2.4.* Mechanical Engineering Department, Massachusetts Institute of Technology, 1996.

24. A. Watt. *3D Computer Graphics.* Addison-Wesley, Harlow, England, 2000.

25. M. Witbrock and S. Neil-Reilly. Evolving genetic art. In P. J. Bentley, ed., *Evolutionary Design by Computers*, pp. 251–259, San Francisco, 1999. Morgan Kaufmann.

Evolve Schema Directly Using Instruction Matrix Based Genetic Programming

Li Gang, Lee Kin Hong, and Leung Kwong Sak

Department of Computer Science and Engineering,
The Chinese University of Hong Kong, Shatin, N.T. Hong Kong
{gli, khlee, ksleung}@cse.cuhk.edu.hk

Abstract. This paper proposes a new architecture for tree-based genetic programming to evolve schema directly. It uses fixed length hs-expressions to represent program trees, keeps schema information in an instruction matrix, and extracts individuals from it. In order to manipulate the instruction matrix and the hs-expression, new genetic operators and new matrix functions are developed. The experimental results verify that its results are better than those of the canonical genetic programming on the problems tested in this paper.

Key words: IMGP, hs-expression, instruction matrix, schema evolution, Genetic Programming.

1 Introduction

As a branch of Evolutionary Computation(EC), Genetic Programming(GP)[1][2] automatically constructs computer programs. A popular theory behind EC is the schema theory[3] in Genetic Algorithms (GA). It has been extended to GP by [4][5]. Although the schema theory explains why GP works and helps to design GP systems, not much GP work has been done to make use of it to evolve schema directly.

To evolve the schema directly, this paper proposes a new GP architecture, the instruction matrix based GP (IMGP). It keeps the information of schema in an instruction matrix, and hence it eliminates the explicit population. IMGP extracts the individuals from the instruction matrix, executes the genetic operations on them, and updates the instruction matrix accordingly. The experiments verify that IMGP performs better than the canonical GP on 4 testing problems.

The rest of the paper is organized as follows. Section 2 reviews related work. Section 3 gives the outline of IMGP. Section 4 describes some of its operators and functions in detail. Section 5 presents the experiments and the results. Section 6 discusses the relationship between IMGP and the schema theory. Section 7 concludes the paper.

2 Related Background

In GP, the evolution subjects are usually the individual programs and the population. A novel alternative is to evolve the instructions directly and use them to construct programs.

M. Keijzer et al. (Eds.): EuroGP 2005, LNCS 3447, pp. 271–280, 2005.
© Springer-Verlag Berlin Heidelberg 2005

Probabilistic Incremental Program Evolution (PIPE)[6] applies Estimation of Distribution Algorithms [7] to GP. It has a probability tree, each of whose nodes is a probability vector, which keeps the probabilities of the functions and terminals of the node. In each generation, PIPE creates a population by constructing trees based on the probability tree, and updates the probability tree with the information from the best individual in the population. However, updating the probability tree only with one individual may be unable to express the information of the rest population. Besides, no linkage information between nodes is kept in the probability tree.

Competent Genetic Programming[8] combines Compact Genetic Algorithm[9] and PIPE as a multivariate probabilistic model building of GP. Its significance is that it partitions the tree into subtrees, and builds the probabilistic model for each subtree. Therefore, it is not only able to calculate the frequency of the nodes, but the frequency of the subtrees as well. The computation is high as it uses greedy search heuristic which calculates the complexity of each possible subtree to identify good building blocks.

Grammar Model-based Program Evolution(GMPE)[10] evolves GP programs with Probabilistic Context-free Grammar, in which each grammar has a production probability. It updates the grammars with the superior individuals in the population, and use the grammars to generate new individuals. A grammar can generate a single node or a whole subtree, so GMPE also keeps the information of some building blocks. The grammar has no position information as regard to the whole tree, so the position of its derivative are not fixed.

3   Architecture

To calculate the fitness of the instructions and the subtrees so as to evolve schema directly, we propose the Instruction Matrix based Genetic Programming (IMGP). IMGP uses a fixed length expression to represent a program tree, employs a matrix instead of a population, and extracts programs from the matrix. The other features include,

1. IMGP has multiple copies of the same instruction for each tree node.
2. It updates with all the individuals besides the best individual.
3. It keeps crossover and mutation, which are modified though.
4. The instructions have fixed positions in the tree.

3.1   hs-Expression

Rather than using the *s-expression* in the canonical GP[1], we propose a new tree representation, *hs-expression*. It is similar to the array in Heap Sort, but the larger-than relation changes to the parent-of relation. We use a $2^{D+1} - 1$ long array to store a binary tree of depth D at most, and every possible node has a corresponding element in the array even if the node does not exist. The tree root is element 0 in the hs-expression. For the given element at locus k in the hs-expression, its left and right children are the 2k+1th and 2k+2th elements.

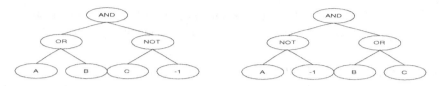

Fig. 1. hs-expression (and,or,not,a,b,c,-1)   **Fig. 2.** hs-expression (and,not,or,a,-1,b,c)

a	b	c	d	e	AND	or	not	and	or	not		AND
a	b	c	d	e	and	OR	not	and	or	not		OR
a	b	c	d	e	and	or	not	and	or	NOT		NOT
A	b	c	d	e	and	or	not	and	or	not		A
a	B	c	d	e	and	or	not	and	or	not		B
a	b	C	d	e	and	or	not	and	or	not		C
a	b	c	d	e	and	or	not	and	or	not		-1

Fig. 3. Instruction Matrix and Extracted hs-expression. The IM keeps multiple instructions for each element of hs-expression. Each element of hs-expression is extracted from the corresponding row in the IM

If it has no children, those elements are set to -1 instead. Figs. 1 and 2 are two examples. Unlike the trees represented by s-expression, the trees of hs-expressions of the same length have exactly the same shape if -1 is viewed as a virtual node, and the elements at the same locus in the hs-expressions always correspond to the nodes at the same position in the trees.

3.2   Instruction Matrix

In IMGP, the population is encoded in the instruction matrix (IM), whose cells are instructions. Each row of the matrix corresponds to an element of hs-expression, and each row contains multiple copies of functions and terminals. The matrix height is $2^{D+1} - 1$, equal to the length of hs-expression, if the tree's maximum depth is D, and its width is the number of functions and terminals in a row. Fig. 3 shows a sample IM and an hs-expression extracted from it. Basically, the element at locus k in the hs-expression is extracted from row k in the IM. The details are described in Section 4.1.

Besides the instruction code, the matrix cell keeps some auxiliary data. A pseudo code of its initialized data structure is shown below. Please note, the smaller the fitness is, the better it is. The specific usages of these fields are explained in detail in Section 4.

```
struct instruction {
    opcode_type incode; //the instruction code
    double best_fit = MAX_FITNESS; //the best fitness
    double fitness = 0; //the average fitness
    int left = -1; //the left child of its best subtree
```

```
int right = -1; //the right child of its best subtree
int eval_num = 0; //the times it has been selected};
```

3.3   Algorithm Outline

Algorithm 1 is the main process of evolution. In each generation, IMGP does the following steps repeatedly while counting the number of individuals evaluated. Firstly IMGP extracts two individuals from the IM and evaluates them. Then it tries crossover and mutation on them to produce offspring, which it evaluates as well. For any individual, its fitness is fed back to its instructions in the IM. At this point, all these individuals are destroyed. After evaluating P individuals, IMGP finishes a generation by shuffling every row in the IM. When it completes N generations, the best program is reported as the result.

Algorithm 1: The Main Program

Output: the best individual
initialize the instruction matrix;
for *gen from 0 to N* **do**
    while *num < P* **do**
        the number of individuals evaluated ← 0;
        extract two individuals i and j from the matrix;
        calculate their fitness respectively;
        update their instructions with the fitness;
        if *crossover i with j successfully* **then**
            evaluate the offspring and update its instructions;
        else if *mutate i successfully* **then**
            evaluate the offspring and update its instructions;
        end
        if *crossover j with i successfully* **then**
            evaluate the offspring and update its instructions;
        else if *mutate j successfully* **then**
            evaluate the offspring and update its instructions;
        end
        num ← num + the number of individuals evaluated;
    end
    shuffle the instruction matrix;
end

---

4   Operators

In IMGP, crossover and mutation are similar to those of the canonical tree-based GP. However, as IMGP keeps the fitness of the instructions in the IM, when an individual crossovers or mutates, it replaces a subtree only with a better one so that the offspring might be better, and the two subtrees of the parents must reside at the same position to reduce the macro-mutation effect of the standard crossover[2]. Due to the new representations of the individual and the population, we have also designed three new operators to handle them as follows.

4.1 Individual Extraction

IMGP extracts the individuals from the IM probabilistically. Firstly IMGP constructs an empty hs-expression filled with -1, and aligns it vertically with the IM. Then it starts to extract the root from row 0, and puts it at locus 0 in the hs-expression. Every tree node is extracted from the corresponding row using binary tournament selection, i.e. comparing the average and best fitness of two randomly selected instructions, and placed at the corresponding locus in the hs-expression. If the selected instruction is a function, IMGP proceeds to select its left child from the 2k+1th row, and its right child, if any, from the 2k+2th row. It does so recursively until all the branches are completed. For instance, in Fig. 3, the words of capital letters are the selected instructions, and the extracted hs-expression is on the right. The corresponding tree is depicted in Fig.1

After a node is selected, IMGP occasionally checks whether its best subtree should be selected as a whole so that the nodes in the best subtree are extracted directly without binary tournament selection. How often it does so depends on its best and average fitness. The bigger the difference between them is, the more likely its subtree is selected. The assumption is that if the best fitness is much better than the average fitness, the tree constructed with the best subtree is likely to be much better than the tree constructed without it. With this method, IMGP makes use of the good building blocks, i.e. the subtrees in the case of GP, in the new individuals.

Extraction makes IMGP avoid focusing on a small search space. In the canonical GP, when an individual changes by crossover or mutation, it replaces one of its subtrees with a new one. The offspring is different from its parent, but it is still in the neighborhood of the parent. Therefore, the search space of the canonical GP is largely determined by the initial population. Conversely, IMGP does not generate an individual from an existing parent. It extracts a completely new individual from the IM, and the individual has slim similarity with the previous individuals. Therefore, IMGP searches a relatively large space.

4.2 Instruction Evaluation

In IMGP, the individual is evaluated using the post-order recursive routine. Since the individual is destroyed later, it cannot carry its fitness as in the canonical GP. Instead, the fitness is fed back to its instructions as their new fitness, so the fitness of the instructions can be used as selection criterion in extraction later. The feedback comes in two ways:

1. The new fitness is averaged out with the old average fitness of the instruction, so we know how good the instruction is at the fixed position on average.

2. If the new fitness is better than the best fitness of the instruction, its left and right pointers are changed to those of the current individual accordingly. This actually keeps good subtrees in the IM, which can be selected in the new individuals later.

4.3   Matrix Shuffle

GP converges by spreading good instructions over the population to produce fit individuals, as some instructions appear in many individuals. IMGP uses matrix shuffle to propagate good instructions, and consequently to increase the probability to select them together in an individual. It shuffles the whole matrix row by row. In each row, it replaces the worse instructions with the better ones in terms of their average fitness. Therefore, while the IM evolves, some good instructions emerge to dominate the rows.

GP converges at the cost of diversity. As the population converges, the individuals become alike by and by, which means that the majority of the population consist of the same instructions, while the other instructions seldom exist. In IMGP, however, matrix shuffle prohibits the fitter instructions from reproducing too many instances, and reserves a minimum number of the less fit instructions. This thus maintains diversity effectively and easily, so it is unlikely for the individuals extracted later to have the same instructions.

5   Experiments

This section presents the performance of IMGP on 4 benchmark problems to verify its superiority to the canonical GP.

The first problem is the symbol regression problem which searches for a mathematical expression $y = x^4 + x^3 + x^2 + x$, where x is an integer uniformly and randomly generated from the range of [0, 20). The fitness used is the hit count which is incremented by one if the difference between the program output and the correct result is larger than a predefined threshold. The second problem is to discover the even-5-parity expression, $\neg(a \oplus b \oplus c \oplus d \oplus e)$. The training cases are all the 2^5 combinations of the 5 binary variables. The fitness is calculated as the sum of the wrong results produced by the individual program. The third problem is the artificial ant on the Santa Fe Trail. Executing the problem solution repeatedly enables the ant to eat all the 89 food pellets on the trail within 400 steps. The number of the food not eaten by the ant is used as the fitness. The fourth problem is boolean 11-multiplexer. Among the 11 variables, three are used as the address to select the output from one of the other 8 variables. However, GP has no idea of which ones are the address. Similar to even-5-parity, the training cases are all the 2^{11} combinations of the 11 binary variables, and the fitness is the number of incorrect output.

Table 1 lists the parameter setting in the experiments. IMGP has no population, but for convenient comparison with the canonical GP, we refer to the number of the individuals evaluated between matrix shuffles (generations) as the population size, i.e. P. Please note that in Artificial Ant and 11-Multiplexer,

Table 1. The Experiment Settings

Parameters	Symbol Regression	Even-5-Parity	Artificial Ant	11-multiplexer
Terminals	{x}	{a,b,c,d,e}	{move,left,right}	{a,b,...,k}
Functions	{+,-,*,/}	{and,or,nand,nor}	{if,progn2,progn3}	{if,and,or,not}
Population	500	1000	2000	4000
Matrix Width	40	405	90	150
Matrix Height	63	1023	1093	3280
Generations	100	100	100	100

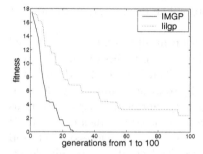

Fig. 4. The result for $x^4 + x^3 + x^2 + x$

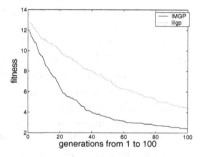

Fig. 5. The result for $\neg a \oplus b \oplus c \oplus d \oplus e$

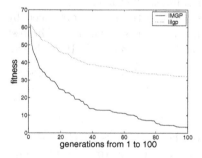

Fig. 6. The result for Artificial Ant

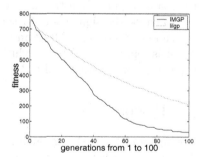

Fig. 7. The result for 11-multiplexer

some functions require 3 arguments, which means the maximum arity of the program tree is 3 in stead of 2. Therefore, their matrix heights and hs-expression lengths increase to $\frac{3^{D+1}-1}{2}$.

We use lilgp[11] as the canonical tree-based GP system. For fair comparison, the Ephemeral Random Constant(ERC) is removed from lilgp. The maximum number of generations is 100. The tournament size is 2 as in IMGP. The other parameters of lilgp are set according to [1]. Both lilgp and IMGP use the same random seeds, which themselves are randomly generated. Figs. 4, 5, 6 and 7

Table 2. t-test compares the results of canonical GP and IMGP on the 4 problems. t is the t-statistic of the difference between the means of the results of canonical GP and IMGP. P is the probability that the actual means of their results in a number of runs are the same

	Symbol Regression	Even-5-Parity	Artificial Ant	11-multiplexer
t	1.8283	4.7019	13.7893	12.1930
P	0.0415	2.23×10^{-5}	2.57×10^{-16}	5.27×10^{-15}

are the experiment results for the 4 problems, which plot the fitness of the best individuals from generation 1 to generation 100. The results are averaged over 20 independent runs.

In symbol regression, IMGP finds the solution in all the 20 runs compared with 17 successful runs in lilgp. In terms of the average fitness of the best individuals, IMGP also converges faster than lilgp. In Even-5-parity, IMGP finds the solution in 3 runs, however, lilgp fails to find the solution in any of the 20 runs. Regarding the convergence speed, IMGP also outperforms lilgp significantly as its average best fitness is 2.4 while lilgp's is nearly twice. In artificial ant, IMGP finds 12 ants eating all the food pellets. lilgp cannot find any successful ant, and its average fitness of its best individuals is 31.8, far from 0. In 11-multiplexer, IMGP finds the perfect multiplexer for 13 times, while lilgp fails all the time. And once again, IMGP's average best fitness is much better than lilgp's, although their starting fitness in the first generation are almost the same.

To check the significance of the difference between the results produced by the canonical GP and IMGP on the 4 problems, a t-test is performed. The test result is reported in Table 2, which verifies that the results of IMGP are better than those of the canonical GP.

6   Discussions

Rosca[4] and Poli & Langdon[5] introduce two schema theories for GP independently. Their schema is a contiguous tree fragment starting from the tree root. A tree can have only one instance of a certain schema and the position of the schema is fixed. However, the "don't care" symbol # in Rosca's schema theory represents a set of subtrees, while the "don't care" symbol = is exactly one tree node in Poli & Langdon's schema theory. Figs. 8 and 9 illustrate these two position schema theories.

In IMGP, an instruction's fitness is averaged over the fitness of all the trees containing it at the fixed position. Considering Rosca's schema theory, we think the fitness of *and* at the third row of the matrix is actually the fitness of the schema (#, #, *and*, #, #), which has *and* as the root of the right subtree, whose left and right subtrees can be anything except -1, as depicted in Fig. 10. Generally, the fitness of an instruction in row k of the matrix is the fitness of the order 1 schema with the instruction at the position k. This way, IMGP maintains the fitness of all the order 1 schema.

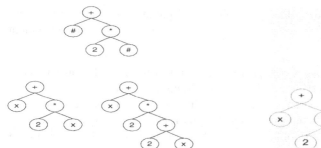

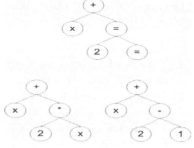

Fig. 8. Rosca Schema

Fig. 9. Poli & Langdon Schema

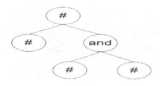

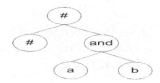

Fig. 10. A Node Schema

Fig. 11. A Subtree Schema

Additionally, an instruction has its best fitness together with its best subtree. Suppose the function *and* at the third row of the matrix has its best left child pointing to a, and the best right child pointing to b, then the best fitness of *and* is actually the best fitness of the schema $(\#, \#, and, a, b)$. Its root can be any function, its left subtree can be anything except -1, and its right subtree is (and,a,b), as depicted in Fig. 11. This way, IMGP is able to remember the best fitness of the schema of order larger than 1.

According to the extraction criterion in section 4.1, if an instruction's fitness is better than the others', which means its 1-order schema is better than the other 1-order schema with different instructions at the same position, it will be selected more often than the other instructions, i.e. more programs will sample its schema. Similarly, if an instruction's best fitness is much better than its average fitness, this will not only increase the chance of selecting this instruction, but if it is indeed selected, more trees will sample the schema containing its best subtree. Therefore, although IMGP is unable to keep the information of all the schema, it does evolve some of them, and use them to generate new individuals.

7 Conclusion

This paper has proposed a new tree-based genetic programming architecture to evolve schema directly. Rather than gathering individuals to form the population, IMGP extracts individuals from the instruction matrix, and stores the program in a new hs-expression. The experiments have shown that IMGP is superior to the canonical GP on the problems tested in this papar. This makes us believe IMGP is worthy of further research. We do have several ideas of improvement in mind.

1. We intend to adaptively change its parameters during evolution so that a general parameter setting can be used in different problems.
2. We plan to record the times that a specific instruction has been selected, and select more often the rarely selected instruction.
3. We can keep in each row only one copy of each instruction, which remembers multiple subtrees, so that the matrix size decreases greatly.

Acknowledgements

The work described in this paper is supported by a grant from the Research Grants Council of the Hong Kong Special Administrative Region, China (Project No. CUHK4192/03E).

References

1. Koza, J.R.: Genetic Programming: On the Programming of Computers by Means of Natural Selection. MIT Press, Cambridge, MA, USA (1992)
2. Banzhaf, W., Nordin, P., Keller, R.E., Francone, F.D.: Genetic Programming – An Introduction; On the Automatic Evolution of Computer Programs and its Applications. Morgan Kaufmann, dpunkt.verlag (1998)
3. Holland, J.H.: Adaptation in Natural and Artificial Systems. The University of Michigan Press, Ann Arbor, Michigan (1975)
4. Rosca, J.P.: Analysis of complexity drift in genetic programming. In Koza, J.R., Deb, K., Dorigo, M., Fogel, D.B., Garzon, M., Iba, H., Riolo, R.L., eds.: Genetic Programming 1997: Proceedings of the Second Annual Conference, Stanford University, CA, USA, Morgan Kaufmann (1997) 286–294
5. Langdon, W.B., Poli, R.: Foundations of Genetic Programming. Springer-Verlag (2002)
6. Salustowicz, R.P., Schmidhuber, J.: Probabilistic incremental program evolution. Evolutionary Computation 5 (1997) 123–141
7. Larranga, P., Lozano, J.A., eds.: Estimation of Distribution Algorithms: A New Tool for Evolutionary Computation. Kluwer Academic Publishers (2002)
8. Sastry, K., Goldberg, D.E.: Probabilistic model building and competent genetic programming. (2003) 205–220
9. Harik, G.R.: Linkage learning via probabilistic modeling in the ECGA. Technical report, University of Illinois at Urbana-Champaign, Urbana IL, US (1999)
10. Shan, Y., McKay, R.I., Baxter, R., Abbass, H., Essam, D., Nguyen, H.: Grammar model-based program evolution. In: Proceedings of the 2004 IEEE Congress on Evolutionary Computation, Portland, Oregon, IEEE Press (2004) 478–485
11. Zongker, D., Punch, B.: lilgp 1.01 user's manual. Technical report, Michigan State University, USA (1996)

Evolving Defence Strategies by Genetic Programming

David Jackson

Dept. of Computer Science, University of Liverpool,
Liverpool L69 3BX, United Kingdom
d.jackson@csc.liv.ac.uk

Abstract. Computer games and simulations are commonly used as a basis for analysing and developing battlefield strategies. Such strategies are usually programmed explicitly, but it is also possible to generate them automatically via the use of evolutionary programming techniques. We focus in particular on the use of genetic programming to evolve strategies for a single defender facing multiple simultaneous attacks. By expressing the problem domain in the form of a 'Space Invaders' game, we show that it is possible to evolve winning strategies for an increasingly complex sequence of scenarios.

1 Introduction

In military contexts, implementing a defence strategy for an autonomous entity often involves highly complex computer programming. The 'intelligence' that must be built into such systems is often derived from detailed warfare simulation and the application of game theory. Indeed, it has long been recognised that there is an extensive overlap of true battlefield strategy with various forms of game playing.

In many modern computer games, artificial intelligence techniques are extensively used to increase the sophistication of the behaviour of computer opponents, and to heighten the sense of reality of the game-playing experience [1]. Such techniques are usually programmed in by the game's developers, but it is becoming increasingly apparent that evolutionary computation techniques may also have a role to play in this regard. Evolutionary programming has been used to evolve strategy for a large number of games, including chess [2], checkers [3], poker [4], Othello [5], and backgammon [6]. Many of these are board or card games involving the evolution of 'mini-max' strategies [7], but there has also been work done on problem domains with more obvious militaristic connections, including air strike planning [8], pursuer-evader scenarios [9], minesweeping [10], and missile firing [11,12].

In general, however, much less research work has been done on evolving defence strategy for responding to multiple simultaneous enemies launching unpredictable attacks in real time. In our paper, we wish to take a closer look at some of the issues involved in evolving defence strategy via genetic programming. To achieve this we couch the problem in the form of the well-known arcade game Space Invaders. This game, one of the earliest and most successful computer games ever written, is deceptively simple in concept and yet frustratingly difficult to master.

M. Keijzer et al. (Eds.): EuroGP 2005, LNCS 3447, pp. 281–290, 2005.

For those not familiar with the game, the idea is that there is a lone defender at the bottom of the screen, and a large number of aliens who descend from the skies, dropping bombs as they move. The defender's task is to shoot down all the aliens before they land, whilst avoiding being destroyed by the bombs. To achieve this, the defender has only three actions available: move left, move right, and fire a missile. However, protection may be sought beneath a number of fortifications until they too are destroyed.

The simplicity of the game makes it an ideal subject for studying the potential for devising defence strategy via evolutionary programming. More specifically, the research question we wish to address is whether, through the use of genetic programming, we can evolve programs for the defender that will enable it to win sets of increasingly complex games. In the next section we will describe the experimental approach we have used in more detail. This is followed by a description of the experiments themselves and their outcomes, and then some concluding remarks.

2 Experimental Approach

The programs that we shall evolve in the following experiments are viewed as directing the behaviour of a single defender operating within a square 'arena,' as depicted in Fig. 1. The arena takes the form of a grid of cells; for simplicity, Fig. 1 depicts this to be of size 10x10, although the experiments described here actually used an arena size of 20x20.

In evaluating the fitness of an individual, the program code is executed over a set of random tests (games). The number of such games is usually set at 50. For each test of a candidate program, the defender is placed randomly somewhere on the bottom

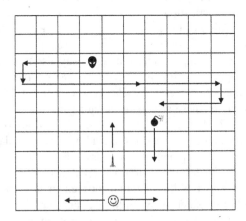

Fig. 1. Movement of entities within the battle arena

initially travelling in a randomly-chosen direction. Each alien descends in a boustrophedon pattern, i.e. across the arena, down a row, across in the opposite direction, and so on. As the aliens travel, they may release bombs. Again for simplicity, Fig. 1 shows a single alien and a single bomb, and also a missile that the defender has launched to attack the alien. Although there may be many bombs in the air simultaneously, there can be at most only one missile; the defender must wait until an existing missile hits an enemy or leaves the arena before it launches the next.

The program code corresponding to a member of the population being evolved is built from the three primitives representing a move to the cell on the immediate left, a move right, and the firing of a missile, together with other primitives for all the decision-making associated with the strategy. A game advances in 'steps.' Execution of the left, right and fire primitives cause the game to move on by one step. The other nodes of a program tree – those corresponding to the decision making – are viewed as consuming negligible 'thinking' time, and do not cause a game step to be taken.

In each game step, all aliens, bombs and missiles move ahead one cell in their respective directions of travel. As there are no looping constructs available in the programming primitives, each program is evaluated repeatedly until a game ends. To ensure progression of 'think-only' programs, i.e. those which do not employ any of the left, right or fire primitives, a game is also advanced by a step at the end of each program iteration. In general, a game ends when all aliens are destroyed, when an alien succeeds in landing, or when a bomb hits the defender (unlike the original arcade game, this defender gets no second chances !)

In performing the following experiments, we have elected to use steady-state evolution. Candidates for both reproduction and deletion within the population are selected via tournament, acting on a sample size of 5. The population size is 500; this is initialised using the 'ramped half-and-half' method advocated by Koza [13], in which an equal number of program trees is randomly generated for each tree depth up to an initial maximum (6 in our case). For each set of trees of a given depth, half are generated 'fully,' i.e. all branches are of the same complete depth, and half are 'grown,' i.e. branches may end in a terminal node before the set depth is reached. There are no duplicates in this initial population, although no attempt is made to identify or prevent them in subsequent generations.

Evolution proceeds over 50 generations using the standard operators of crossover and fitness proportionate reproduction, selected probabilistically so that 90% of individuals are created via crossover. Crossover points are selected randomly, but with a distribution of 90% applied to function nodes and 10% applied to terminal nodes. Mutation was not used in these experiments.

3 The Experiments

3.1 Experiment 1

The first experiment begins things simply, with the defender opposed by just a single alien. Moreover, the alien has no weapons, so that a game ends when it either manages to land or is hit by a missile. The terminal set for this experiment is as follows:

{LEFT, RIGHT, FIRE, Y_DIST, X_DIST}

The first two terminals in this set, LEFT and RIGHT, simply move the defender one cell in the appropriate direction, unless the defender is against the edge of the arena that makes such movement impossible. FIRE launches a missile from the defender's current coordinates, unless a missile is already in the air. Execution of LEFT, RIGHT and FIRE nodes during fitness evaluation all cause the game to be advanced one step, and all three nodes return a zero result to the tree evaluation function. Y_DIST returns the current vertical distance of the alien above 'ground level' and X_DIST returns its horizontal distance from the defender. The value of X_DIST is positive if the alien is currently approaching the defender, negative if it is receding.

The function set for the experiment looks like this:

{IF, EQ, PROGN2, PROGN3}

The IF function takes three arguments and is defined as:

if <arg1> then <arg2> else <arg3>

The EQ function evaluates its 2 arguments and tests them for equality, returning 1 if the results are the same and zero otherwise. PROGN2 and PROGN3 are connectives, as used in problems such as the Santa Fe artificial ant trail [13]. They simply cause each of their sub-tree arguments to be evaluated in turn, PROGN2 having two arguments and PROGN3 having three. PROGN2, PROGN3 and the IF function all return a zero result when executed.

The next problem to consider is how to define the fitness metric. Since the primary aim of the defender is to shoot down the alien, we can define fitness in terms of how close the defender's missiles come to hitting the alien. When a missile achieves the same Y-axis value as the alien, the distance considered is the absolute difference in their X-coordinates. The best of these measurements forms the value recorded for that game. This means that, if the defender manages to score a direct hit with any one of its missiles during a game, the score for that game is zero. A non-zero value at the end of the game indicates that none of the missiles hit their target, and the alien managed to land. These values are then summed over the 50 games to give a final fitness score. It follows that this score will be zero only if all games are won by the defender.

In executing the GP system for this initial problem, it was found that it was not difficult to evolve solutions. Almost every run led to a solution, usually within a handful of generations. It was also found that the winning strategies that generally appeared involved firing missiles as often as possible, whether on-target or not. Part of the reason for this is that, as long as one of the missiles hits the alien, it does not matter where the others go. If a kill is achieved, the shots that go wide will not affect the perfect game score of zero.

However, it is not sufficient merely to stand still and fire repeatedly. Programs which did this let too many aliens slip through to ground level. Rather, a successful strategy involves firing coupled with movement. Most of the zero-fitness programs achieved this by gradually migrating to the left or right edge of the arena, and then firing continually from there. The reason for this seems to be that the leftmost and rightmost columns are the only ones in which the alien performs any vertical motion, and the extra time spent lingering in those columns while the descent is achieved seems to improve the defender's chances of a direct hit. The following 41-node solution is one that works in such a way. It moves to the right edge, and then fires repeatedly from there, occasionally jumping one cell to the left and back again:

PROGN3 (IF (IF (LEFT RIGHT X_DIST) PROGN3 (FIRE RIGHT RIGHT)
EQ (RIGHT RIGHT)) PROGN2 (IF (FIRE PROGN3 (EQ (X_DIST EQ
(RIGHT RIGHT)) IF (FIRE RIGHT LEFT) PROGN3 (LEFT LEFT LEFT))
RIGHT) IF (RIGHT LEFT Y_DIST)) EQ (Y_DIST IF (RIGHT LEFT FIRE)))

In an attempt to encourage solutions in which accurate aiming would become more
of a priority, we made a small alteration to the fitness metric. Instead of basing the
fitness score on the *best* missile proximity of a game, we tried basing it on the
proximities of *all* the shots fired. In this modified version of the experiment, any
misses whatsoever lead to a non-zero fitness, so that zero (optimal) fitness is achieved
only if each and every missile hits its target. We also alter our termination criterion so
that success is defined in terms of the number of kills, rather than zero fitness. This
means that a run is judged successful if it evolves a program that wins all 50 games.
This 'best' program may or may not have zero fitness. The situation is similar to that
used in, for example, symbolic regression problems, where the best program is often
regarded as the one which scores the most matches of inputs to outputs, even though
the overall error term representing its fitness value may be worse than that of other
population members. Finally in this modified version of the experiment, we also
introduce an additional penalty (200 points) to be added to the score each time an
alien manages to land.

As before, little computational effort was required to evolve solutions, and
approximately 90% of these were of the 'rapid-fire' type seen previously. The other
10%, however, were much more considered in their shooting strategies, preferring to
wait until the enemy was in range before firing a missile. The following 20-node
solution is one such 'aimer':

IF (EQ (Y_DIST X_DIST) PROGN3 (FIRE LEFT X_DIST) IF (PROGN3
(PROGN3 (Y_DIST Y_DIST RIGHT) LEFT PROGN2 (LEFT LEFT)) Y_DIST
Y_DIST))

Execution of this program causes the defender to migrate to the left edge of the
arena, from where it fires only when it calculates that a direct hit is possible. The
decision is based on the X and Y distances of the alien from the defender.

3.2 Experiment 2

In this experiment, the alien fights back! It is now given the ability to drop a bomb
when it is directly above the defender. However, only one bomb can be in the air at a
time. This bomb moves down one cell each time the game advances, moving in
lockstep with the alien, any missile, and possibly the defender. The defender cannot
shoot down a bomb; it can only move out of its way. To assist it in identifying when it
needs to do this, the terminal set is expanded slightly:

{LEFT, RIGHT, FIRE, Y_DIST, X_DIST, ATTACKED}

The ATTACKED terminal returns 1 if the defender is directly below a bomb, zero
otherwise. We also impose a penalty of 200 points if the defender is hit by a bomb.

This experiment was tried with each of the two approaches to fitness evaluation
described for the previous experiment. Using the 'best proximity' approach, in which
only the closest shot in a game counts towards the final score, it proved very difficult
to evolve solutions. In one set of 100 runs, only 5 of those runs resulted in solutions.

The lack of accuracy inherent in the 'rapid-fire' approach coupled with the newly-introduced need to spend time dodging the enemy bombs usually resulted in the defender being killed first or the alien landing.

The 'all-missile' approach, in which the proximity of each and every missile counts towards the fitness score, resulted in a much better performance, with a success rate of 33% in one set of 100 runs. In one run, the following 46-node solution was produced in generation 26:

IF (EQ (IF (IF (EQ (IF (ATTACKED ATTACKED Y_DIST) EQ (X_DIST Y_DIST)) LEFT EQ (EQ (X_DIST Y_DIST) ATTACKED)) RIGHT LEFT) EQ (X_DIST Y_DIST)) PROGN2 (PROGN2 (RIGHT Y_DIST) EQ (IF (ATTACKED IF (LEFT RIGHT LEFT) EQ (X_DIST Y_DIST)) EQ (X_DIST Y_DIST))) EQ (ATTACKED PROGN3 (FIRE ATTACKED RIGHT)))

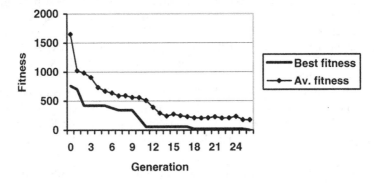

Fig. 2. Fitness graph for Experiment 2

Monitoring of the execution of this program reveals that it avoids bombs by repeatedly performing little 'dances' that involve taking two steps to the left and then one step to the right. The graph of Figure 2 shows how the best fitness and average fitness of the population changed during the evolution of this particular program.

3.3 Experiment 3

This experiment is similar to the previous one, in that there is still just a single alien and defender, but the alien is now capable of dropping a lot more bombs. As before, the alien will tend to drop a bomb when directly above the defender, but it will also drop other bombs at random on each pass across the arena. Within the GP system, the maximum number of bombs in the air at any point is controlled by a parameter MAX_BOMBS. Another restriction is that bombs cannot share the same X coordinate, i.e. only one bomb can be present in each column of the arena.

In performing this experiment we abandoned the 'best proximity' approach to measuring fitness, which performed so poorly in the previous experiment. Using the 'all missiles' approach instead, we were able to evolve programs that could cope with large numbers of bombs. In one set of runs, for example, we found that it was possible to evolve programs that could cope with MAX_BOMBS set to 10, i.e. with up to half the sky containing bombs. In one such run, the following 19-node program evolved in generation 35:

IF (ATTACKED PROGN2 (ATTACKED LEFT) IF (EQ (X_DIST Y_DIST)
IF (EQ (X_DIST Y_DIST) IF (Y_DIST FIRE X_DIST) Y_DIST) X_DIST))

In executing this strategy, the defender moves as little as possible – enough to evade the bombs. It does this as soon as a bomb appears above it, while the bomb is still high in the air. This gives the defender time to check if it is still in danger after moving, so that it can take further evasive action if required. As soon as the enemy comes into range, the defender launches a missile to bring it down.

3.4 Experiment 4

In this final experiment, we take a step closer to the original game by introducing multiple aliens. A consequence of this is that the function set (F) and terminal set (T) are altered as follows:

T = {LEFT, RIGHT, FIRE, ATTACKED, TARGET_LEFT, TARGET_RIGHT}
F = {IF, PROGN2, PROGN3}

It will be seen that the tokens X_DIST, Y_DIST and EQ have all been removed from these sets. The existence of the first two of these in particular no longer makes any sense, since they refer to a single alien, and we now have many. Moreover, their primary purpose was to enable the evolution of exact targeting, and it has already been demonstrated in the previous experiments that such an ability can be readily evolved.

In their place we now have TARGET_LEFT and TARGET_RIGHT. These terminals return 1 if a missile launched from one place to the defender's left/right at the present time would lead to a better shot (i.e. closer proximity to any alien) than one launched from the current position; otherwise they return zero. These primitives act only as 'hints' as to how to move in order to improve the chances of a missile launch being on target, since they take no account of the time-consuming steps that may be associated with the other nodes of the program tree. By the time the defender has physically moved to a new position that it calculates to be better, the game may have advanced several steps and the aliens will all have changed positions. This makes life more difficult for the defender, but it may also encourage the evolution of strategies that involve more proactive 'pursuit' of the invaders.

The effort involved in evolving solutions to this problem is obviously affected by the value of MAX_BOMBS, and by a new parameter MAX_ALIENS. A useful metric here is provided in Koza's definition of computational effort [13], which calculates the minimum number of individuals that need to be processed in order to attain a probability of 0.99 that a solution will be evolved for a given population size. If MAX_BOMBS and MAX_ALIENS are both set to 3, for example, the success rate over 100 runs is 47%, and the computational effort required turns out to be 58,500 individuals, representing 13 runs to generation 8. If MAX_BOMBS and MAX_ALIENS are each increased to 5, the success rate degrades to 12%, while the computational effort jumps to 675,000 individuals (75 runs to generation 17). Figure 3 depicts the changes in best and average fitness during the evolution of one of these latter programs over 26 generations.

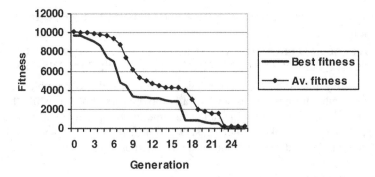

Fig. 3. Fitness graph for Experiment 4

For most of the solutions produced, movement of the defender is usually associated with dodging the bombs that rain down. However, when the bombs are very few in number, these evasion tactics become less important, while the need to destroy the aliens before they land becomes more so. In such scenarios, the pursuit strategy mentioned earlier may become apparent, with the defender attempting to place itself in optimum positions for ensuring the accuracy of its missiles.

4   Conclusions

It has been shown that by using genetic programming it is possible to evolve defence strategies to cope with battle games of significant difficulty. Such strategies incorporate many of the aspects of evasion, pursuit and targeting behaviour found in human game players. It has also been interesting to see the evolution of individuals exhibiting varying forms of game strategy, such as the 'rapid-firers' and the 'aimers.'

Although the individuals that satisfy the termination criteria of a run are often referred to in this paper as 'solutions,' this is in fact a slight exaggeration of the truth. To be accurate, what these programs represent are defence strategies that have demonstrated their ability to win a sequence of 50 consecutive random games. It is sometimes found that such programs can and do lose games when tested further; however, experimentation suggests that the 50 game test sequence is sufficient to promote fairly good generality. For example, the 'solutions' produced in Experiment 4, with MAX_BOMBS and MAX_INVADERS both set to 3, were further tested in another sequence of 50 games that had not been used as a 'training set' during the evolutionary process. It was found that, on average, 45 of these 50 games could still be won, and that 10% of the programs won all of these 50 additional games. It would of course be possible to increase confidence in the robustness of strategies by simply raising the number of tests performed during fitness evaluation; this has to be balanced against the additional computational expense incurred.

As we have moved through the series of experiments described above, we have gradually increased the level of sophistication of the games. Future plans are to see how much further this can be taken. An obvious future addition is the inclusion of buildings behind which the defender can seek refuge. We would also like to investigate the effects of heightening the sensory and decision-making powers of the defender. One way of achieving this would be to endow it with the ability to sense the distance of bombs, and not just their mere presence, so that it might be able to make more informed decisions about exactly when it should dodge an approaching bomb. Another way might be to allow it somehow to analyse the positions of all the aliens simultaneously and assign priorities to them, so that perhaps the most threatening enemies are eliminated first. Yet another suggestion is to endow the aliens with more sophisticated strategies, perhaps to be co-evolved with those of the defender.

Finally, it should be noted that, since fitness evaluation of the strategies being evolved essentially involves execution of a set of games from start to finish, it is possible to bolt a simple visual interface onto the fitness function to view these games. This was in fact done in the experiments described above, providing a fascinating insight into defence strategies created entirely without explicit human programming. A screen-shot of one of the games being played out for a solution evolved in Experiment 4 is shown in Figure 4.

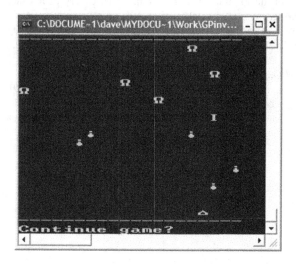

Fig. 4. Visual interface to the GP system during fitness evaluation

References

1. Laird, J.E.: Research in Human-Level AI Using Computer Games. Communications of the ACM 45(1) (2002) 32-35
2. Gross, R., Albrecht, K., Kantschik, W., Banzhaf, W.: Evolving Chess Playing Programs. In: Langdon, W.B. et al (eds.): GECCO 2002. Morgan Kaufmann, San Francisco, CA (2002) 740-747
3. Chellapilla, K., Fogel, D.B.: Evolving an Expert Checkers Playing Program without Using Human Expertise. IEEE Trans. on Evolutionary Computation 5(4) (2001) 422-428
4. Kendall, G., Willdig, M.: An Investigation of an Adaptive Poker Player. In: 14th Australian Joint Conf. on Artificial Intelligence, Lecture Notes in Artificial Intelligence, vol. 2256. Springer-Verlag, Berlin Heidelberg (2001) 189-200

5. Eskin, E., Siegel, E.V.: Genetic Programming Applied to Othello: Introducing Students to Machine Learning Research. In: Proc. 30th Technical Symposium of the ACM Special Interest Group in Computer Science Education (SIGCSE), New Orleans, LA, USA (1999) 242-246

6. Pollack, J., Blair, A.D., Land, M.: Coevolution of a Backgammon Player. In: Langton, C.G. and Shimohara, K. (eds.): Artificial Life V: Proc. 5th Int. Workshop on the Synthesis and Simulation of Living Systems, MIT Press, Cambridge, MA, USA (1996) 92-98

7. Koza, J.R.: Genetic Evolution and Co-Evolution of Game Strategies. In: International Conf. on Game Theory and its Applications, Stony Brook, New York (1992)

8. Miles, C., Louis, S.J., Cole, N.: Learning to Play Like a Human: Case Injected Genetic Algorithms Applied to Strategic Computer Game Playing. In: Congress on Evolutionary Computation (CEC 2004), Portland, Oregon, USA (2004)

9. Moore, F.W., Garcia, O.N.: A Genetic Programming Approach to Strategy Optimization in the Extended Two-Dimensional Pursuer/Evader Problem. In: Koza, J.R. et al (eds.): Genetic Programming 1997: Proceedings of the Second Annual Conference. Morgan Kaufmann, San Francisco, CA, USA (1997) 249-254

10. Koza, J.R.: Genetic Programming II: Automatic Discovery of Reusable Programs. MIT Press, Cambridge, MA, USA (1994)

11. Moore, F.W.: A Methodology for Missile Countermeasures Optimization under Uncertainty. Evolutionary Computation 10(2). MIT Press, Cambridge, MA, USA (2002) 129-149

12. Nyongesa, H.O.: Generation of Time-Delay Algorithms for Anti-air Missiles using Genetic Programming. In: Boers, E.J.W. et al (eds.): EvoWorkshop 2001, Lecture Notes in Computer Science, vol. 2037. Springer-Verlag, Berlin Heidelberg (2001) 243-247

13. Koza, J.R.: Genetic Programming: On the Programming of Computers by Means of Natural Selection. MIT Press, Cambridge, MA (1992)

Extending Particle Swarm Optimisation via Genetic Programming

Riccardo Poli, William B. Langdon, and Owen Holland

Department of Computer Science, University of Essex, UK

Abstract. Particle Swarm Optimisers (PSOs) search using a set of interacting particles flying over the fitness landscape. These are typically controlled by forces that encourage each particle to fly back both towards the best point sampled by it and towards the swarm's best. Here we explore the possibility of evolving optimal force generating equations to control the particles in a PSO using genetic programming.

1   Introduction

The class of complex systems sometimes referred to as *swarm systems* is a rich source of novel computational methods that can solve difficult problems efficiently and reliably. When swarms solve problems in nature, their abilities are usually attributed to swarm intelligence; perhaps the best-known examples are colonies of social insects such as termites, bees and ants. In recent years it has proved possible to identify, abstract and exploit the computational principles underlying some forms of swarm intelligence, and to deploy them for scientific and industrial purposes. One of the best-developed techniques of this type is Particle Swarm Optimisation (PSO) [9].

In PSOs, which are inspired by flocks of birds and shoals of fish, a number of simple entities — the particles — are placed in the parameter space of some problem or function, and each evaluates the fitness at its current location. Each particle then determines its movement through the parameter space by combining some aspect of the history of its own fitness values with those of one or more members of the swarm, and then moving through the parameter space with a velocity determined by the locations and processed fitness values of those other members, along with some random perturbations. The next iteration takes place after all particles have been moved. Eventually the swarm as a whole, like a flock of birds collectively foraging for food, is likely to move close to the best location.

This simple model can deal with difficult problems efficiently. Naturally, different variations of the basic recipe have been tried and compared to existing techniques and different application areas have been investigated. However, the situation overall is still as it was in 2001 when Kennedy and Eberhart wrote: "...we are looking at a paradigm in its youth, full of potential and fertile with new ideas and new perspectives...Researchers in many countries are experimenting with particle swarms...Many of the questions that have been asked have not yet been satisfactorily answered." [9]

M. Keijzer et al. (Eds.): EuroGP 2005, LNCS 3447, pp. 291–300, 2005.

This research is part of a project that aims to systematically explore the extension of particle swarms by including strategies from biology, by extending the physics of the particles, and by providing a solid theoretical and mathematical basis for the understanding and problem-specific design of new particle swarm algorithms. In this paper, in an effort to extend PSO models beyond real biology and physics and push the limits of swarm intelligence into the exploration of swarms as they could be, we study the possibility of evolving, through the use of genetic programming (GP) [10, 12], the force generating equations to control the particles in a PSO. This gives us a methodology for routinely "inventing" specialised PSOs which are near-optimum for specific domains. Our aim is to verify the feasibility of this approach and to understand, through the analysis of the evolved components, what types of PSOs are best for different landscapes.

Section 2 provides a review of the work to date on particle swarms, while Section 3 describes how to use GP to automatically generate PSO tailored to particular tasks. The results section (4) is followed, in Section 5, by a brief restatement of our findings and future direction.

2   Particle Swarm Optimisation

The initial ideas of James Kennedy (a social psychologist) and Russ Eberhart (an engineer and computer scientist) were essentially aimed at producing computational intelligence by exploiting simple analogues of social interaction, rather than purely individual cognitive abilities. Their 1995 simulations, influenced by Heppner's work [7], involved analogues of bird flocks searching for corn. Their continuing emphasis on the social nature of intelligence, even in humans, can be seen in their book, Swarm Intelligence [9].

In the simplest (and original) version of PSO, each particle is moved by two elastic forces, one attracting it with random magnitude to the fittest location so far encountered by the particle, and one attracting it with random magnitude to the best location encountered by any member of the swarm.[1] Suppose we are dealing with an N dimensional problem. Each particle's position, velocity and acceleration, can each be represented as a vector with N components (one for each dimension). Starting with the acceleration vector, $a = (a_1, \cdots, a_N)$, each component, a_i, is given by

$$a_i = \psi_1 R_1 (x_{s_i} - x_i) + \psi_2 R_2 (x_{p_i} - x_i)$$

where x_{s_i} is the ith component of the best point visited by the swarm, x_i is the ith component of the particle's current location, x_{p_i} is the ith component of its personal best, R_1 and R_2 are two independent random variables uniformly distributed in [0,1] and ψ_1 and ψ_2 are two learning rates. ψ_1 and ψ_2 control the relative proportion of cognition and social interaction in the swarm. The same formula is used independently for each dimension of the problem.

---

[1] Here we limit ourselves to the case where particles can socially interact freely. More general models constrain inter-particle interactions via a neighbourhood structure.

The velocity of a particle $v = (v_1, \cdots, v_n)$ and its position are updated every time step using the equations:

$$v_i(t) = v_i(t-1) + a_i \qquad x_i(t) = x_i(t-1) + v_i(t)$$

This system can lead the particles to become unstable, with their speed increasing without control. This is harmful to the search and need to be controlled. The standard technique is to bound velocities so that $v_i \in [-V_{max}, +V_{max}]$.

Early variations in PSO techniques involved the addition of analogues of physical characteristics to the members of the swarm, such as the "inertia weight" ω [17], where the velocity update equation is modified as follows:

$$v_i(t) = \omega v_i(t-1) + a_i$$

In vector notation, the velocity change can be written as $\Delta v = a - (1 - \omega)v$. That is, the constant $1 - \omega$ acts effectively a friction coefficient.

Following Kennedy's graphical examinations of the trajectories of individual particles and their responses to variations in key parameters [8] the first real attempt at providing a theoretical understanding of PSO was the "surfing the waves" model presented by Ozcan [14]. Shortly afterwards, Clerc developed a comprehensive 5-dimensional mathematical analysis of the basic PSO system [5]. A particularly important contribution of that work was the use and analysis of a modified update rule:

$$v_i(t) = \kappa(v_i(t-1) + a_i)$$

where the constant κ is called a constriction coefficient. If κ is correctly chosen, it guarantees the stability of the PSO without the need to bound velocities.

More recently, there have been further explorations of physics-based effects in the swarm. For example, Blackwell has investigated quantum swarms [3] and charged particles [2], and Poli and Stephens have proposed a scheme in which the particles do not "fly above" the fitness landscape, but actually slide over it [15]. Krink and collaborators have looked at a range of modifications of PSO, including ideas from physics (spatially extended particles [11], self-organised criticality [13]) and biology (e.g. division of labour [16]). Finally, there has also been cross-fertilisation between PSOs and evolutionary computing. For example, Angeline [1] introduced selection and Brits *et al.* [4] have explored niching.

3   Evolution of PSOs via Genetic Programming

In the original PSO and well as in most improvements proposed in the literature (see previous section), the equation controlling the particles is of the form:

$$a_i = F(x_i, x_{s_i}, x_{p_i}, v_i)$$

for some function F. Our approach to explore the space of possible PSOs is to use GP to evolve the function F so as to maximise some performance measure.

Clearly, we cannot expect to be able to evolve a PSO that can beat all other PSOs on all possible problems [18]. We should, however, be able to evolve PSOs that can outperform known PSOs on specific classes of problems of interest.

Evolving specialised search algorithms is typically a heavy computational task. So, a very efficient implementation of both PSO and GP were required for this work. Since nothing particularly fancy is required of the GP environment (except efficiency), we used a small and highly efficient C implementation of GP. For efficiency, we implemented our own minimalist PSO engine in C. This has to be compact and efficient in that it is invoked multiple times and for many time steps during the fitness evaluation of each GP program.

The function set for GP included the functions $+, -, \times$ and the protected division DIV. I.e. if $|y| <= 0.001$ DIV$(x, y) = x$ else DIV$(x, y) = x/y$. The terminal set included the coordinate of a particle x_i, the corresponding component of the velocity v_i, the coordinate of the best point visited by the particle x_{p_i}, the coordinate of the best point visited by the swarm x_{s_i}, the numerical constants 1.0, -1.0, 0.5, -0.5, and finally a zero-arity function R which returns random numbers uniformly distributed within the range $[-1, 1]$.

In order to evolve PSOs that are able to solve a class of problems as opposed to just one problem, we need to build a fitness function which uses the program being evaluated as the F function in a PSO, and evaluates the performance of the resulting PSO on a training set of problems taken from the given class.

In our study we considered two classes of benchmark problems, the city-block sphere problem class and the Rastrigin's problem class, of two different dimensions, $N = 2$ and $N = 10$. Problem instances from the city-block sphere class have the following form:

$$f(\mathbf{x}) = \sum_{i=1}^{N} |x_i - c_i|.$$

Every city-block problem has a single global optimum (at $\mathbf{x} = (c_1, \ldots, c_N)$, where $f(\mathbf{x}) = 0$) and no local optima. Problem instances from the Rastrigin's class have the following form:

$$f(\mathbf{x}) = 10N + \sum_{i=1}^{N} \left((x_i - c_i)^2 - 10\cos(2\pi(x_i - c_i)) \right)$$

This has many local optima and one global optimum $\mathbf{x} = (c_1, \ldots, c_N)$ with $f(\mathbf{x}) = 0$.

At the beginning of each GP run, 10 random problems were generated from the chosen class of functions (either the city-block sphere or the Rastrigin function class) by choosing the values c_i uniformly at random from the range $[-1, 1]$. To limit the computational load of the simulations, during fitness evaluation we used PSOs with 10 particles and run them for only 30 iterations on each of the 10 problems. The initial coordinates for the particles were drawn uniformly at random from the interval $[-5, 5]$. Since performance can vary substantially with the initial random position of the particles, for each problem the PSO was run 5 times with different initial random positions for the 10 particles. Initial particle velocity is 0. To ensure stability we updated velocities using Clerc's update rule (see previous section), using a constriction factor $\kappa = 0.7$. We also clipped the components v_i of particle velocities within the range $[-2.0, +2.0]$.

In each of the $10 \times 5 = 50$ training cases, the performance of the PSO needs to be evaluated. More than one performance measure could be used. For example, if one wanted to encourage the convergence of the swarm at the global optimum, performance could be evaluated as the sum of the distances between the global optimum and each particle at the end of the 30 PSO iterations. If one is only interested in the ability of the PSO to find the global optimum, then the distance between the swarm best and the global optimum at the end of the PSO iterations should be used as a performance measure. If one does not care about global optima, but is only interested in achieving good values of the objective function, then the difference between the best objective function value observed in a PSO run and the objective function value at the global optimum could be used as a performance measure. Fitness functions that encourage the swarm not to collapse onto the swarm best could be defined for dynamic problems which require the ability to track moving optima. And so on.

We experimented with two fitness functions: a) we measured and accumulated (over 50 fitness cases) the (city-block) distance between the swarm best and the global optimum at the end of each PSO run

$$\sum_i |x_{s_i} - c_i|$$

and b) we measured and accumulated (over 50 fitness cases and 10 particles) the (city-block) distance between each particle's position and the global optimum at the end of each PSO run

$$\sum_x \sum_i |x_i - c_i|$$

The negation of either (a) or (b) minus a parsimony pressure term (see below) was returned as the fitness of the GP program controlling the PSO.

In the GP system we used steady state binary tournaments for parent selection and binary negative tournaments to decide who to remove from the population. The initial population was created using the "grow" method with max depth of 6 levels (the root node being at level 0). We used population sizes of 1000 and 5000 individuals. We used 90% standard sub-tree crossover (with uniform random selection of crossover points) and 10% point mutation with a 2% chance of mutation per tree node. Runs were terminated either manually when fitness appeared to be sufficiently good or automatically at generation 100. To favour readability and understandability of the evolved solutions, a mild parsimony pressure (parsimony coefficient=0.01) was applied to the fitness function to encourage the evolution of a simpler F.

4   Results

In order to be able to evaluate the PSOs produced by GP, we compared them with a number of human-designed update rules, most of which have previously appeared in the literature. The update rules included:

PSO a version of the standard PSO where $\psi_1 = \psi_2 = 1.0$, that is

$$F = R_1(x_{p_i} - x_i) + R_2(x_{s_i} - x_i)$$

PSOD1 a deterministic (no random coefficients) 100% social version ($\psi_1 = 0$, $\psi_2 = 1$) of the standard PSO:

$$F = (x_{s_i} - x_i)$$

PSOR0 a PSO controlled by random forces

$$F = 2.0\dot{R} - 1.0$$

PSOR1 a 100% social ($\psi_1 = 0, \psi_2 = 1$) version of the standard PSO

$$F = R(x_{s_i} - x_i)$$

In our GP runs we evolved several high performance PSOs. Three of the most interesting ones are:

PSOG1 was evolved when the training set was the shifted city-block sphere functions of dimension $N = 2$, and so it was expected to perform well on unimodal objective functions:

$$F = (x_{s_i} - x_i) - (v_i \dot{R})$$

This is particularly interesting since it includes a deterministic, 100% social component as well as a random friction component.

PSOG2 was also evolved when the training set included shifted city-block sphere functions of dimension $N = 2$. Its equation is equivalent to

$$F = 0.5\left((x_{s_i} - x_i) + (x_{p_i} - x_i) - v_i\right)$$

This is interesting because it is completely deterministic (particles are attracted towards the middle between swarm best and particle best) and because it includes standard friction.

PSOG3 was evolved when the training set included shifted Rastrigin functions of dimension $N = 2$, and so it was expected to perform well on highly multimodal objective functions. Its equation is equivalent to:

$$F = R_1(x_{s_i} - x_i) - 0.75R_2R_1x_ix_{s_i}^2 - 0.25R_3R_2R_1x_ix_{s_i}$$

Table 1. City Block Sphere. Normalised mean (standard deviation) of the distance between best location found by each PSO and global optima. Best results in bold

N	C	PSO	PSOD1	PSOR0	PSOR1	PSOG1	PSOG2	PSOG3
2	1	.046 (.089)	.002 (.0002)	.26 (.029)	.003 (.0003)	**.001** (.0016)	.048 (.014)	.01 (.007)
2	2	.054 (.093)	.002 (.0002)	.27 (.036)	.003 (.0004)	**.001** (.0019)	.066 (.029)	.04 (.028)
10	1	.52 (.47)	.31 (.025)	1.6 (.021)	.27 (.022)	.45 (.025)	.65 (.023)	**.17** (.037)
10	2	.62 (.45)	.38 (.028)	1.6 (.036)	**.31** (.027)	.55 (.045)	.8 (.056)	.45 (.096)

Table 2. Rastrigin. Normalised mean (standard deviation) of the distance between best location by each PSO and global optima. Best results in bold

N	C	PSO	PSOD1	PSOR0	PSOR1	PSOG1	PSOG2	PSOG3
2	1	.66 (.22)	.81 (.081)	.77 (.072)	.64 (.072)	.94 (.084)	.72 (.07)	**.28** (.066)
2	2	.71 (.2)	.85 (.098)	.79 (.066)	.66 (.091)	.94 (.12)	.75 (.12)	**.47** (.16)
10	1	1.2 (.35)	1.3 (.05)	1.9 (.042)	1.3 (.07)	1.3 (.055)	1.1 (.054)	**.59** (.057)
10	2	1.4 (.28)	1.4 (.073)	1.9 (.049)	1.4 (.064)	1.4 (.079)	1.3 (.063)	**.94** (.11)

This is interesting for a number of reasons. Firstly, it does not use information about each particle's best. Probably this is because, in a highly multimodal landscape, particles should not trust their own observations too much. Their personal best is likely to belong to the basin of attraction of a deceptive local optimum. The swarm best, instead, has a higher chance of being in the basin of attraction of the global optimum, and so particles should aim at exploring its surroundings. **PSOG3**'s second term is also interesting. Unless the swarm best is near the origin, this component will tend to push the particles towards the origin. The push is very mild if the swarm best is not too far from the origin, but it becomes quite strong otherwise. Clearly, the reason why this component is useful is that GP has found a regularity in the training set: global optima tend to be near the origin, and so that is an area of the search space that should be explored preferentially.

In order to compare the behaviour of the hand-designed and evolved PSOs, we tested them on 30 random problems taken from the city-block sphere and Rastrigin function classes for two and ten dimensions. The problems were generated by selecting the components of the global optimum c_i uniformly at random from the interval $[-C, +C]$, with $C = 1.0$ and $C = 2.0$. Note that **PSOG1** and **PSOG2** were evolved using the city-block sphere problem class, $N = 2$ and $C = 1.0$, and that **PSOG3** was evolved using the Rastrigin function problem class, $N = 2$ and $C = 1.0$. Thus, all other conditions represent off-sample test problems and are useful to assess the generalisation capabilities of these PSOs.

For each problem instance we performed 30 independent runs of each PSO. The results are reported in Tables 1 and 2. The tables show the average over the 30 problems and the standard deviation (in brackets) of the mean (over 30 independent runs) absolute error between the coordinates of the swarm best and the coordinates of the global optimum (i.e., $\sum_i |x_s - c_i|/N$) at the end of

30 PSO iterations. This gives an idea of how far the swarm best was from the global optimum in each dimension. Data in boldface represent the PSO with the best average performance in each condition.

On the two-dimensional city block sphere problem class, all PSOs do quite well, with the exception of **PSOR0** which cannot really be expected to do very well on unimodal functions. **PSOG1** is better than the standard PSO. However, to our surprise also **PSOG3** (which wasn't evolved on sphere functions) does better than the standard PSO (and in fact is even better than **PSOG1** for ten dimensions), suggesting **PSOG3** may be a good all-rounder. On the Rastrigin function problem class, **PSOG3** outperforms all other PSOs by a considerable margin. In ten dimensions **PSOG2** is second best, while in two dimensions **PSOR1** is second best. In other tests (not reported) where each PSO was run for 300 iterations instead of 30, we obtained essentially the same results.

It is interesting to compare the behaviour of **PSOG3** and **PSO**. Figure 1 illustrates the behaviour of **PSO** on a city block sphere function with $N = 2$. The

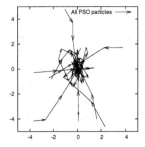

Fig. 1. Trajectories of the particles in one prototypical run of **PSO** on the 2–D city-block sphere problem

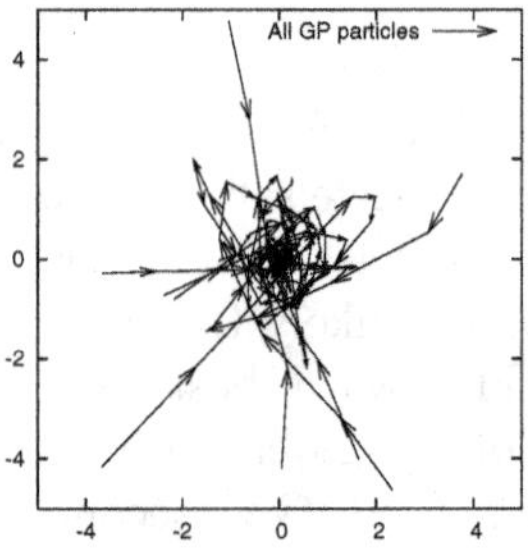

Fig. 2. A run of **PSOG3** on the 2–D city-block sphere problem (same initial conditions as in Fig.1)

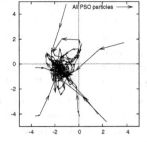

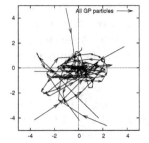

Fig. 3. A run of **PSO** on the 2–D Rastrigin function problem (global optimum at the intersection of cross hairs)

Fig. 4. A run of **PSOG3** on the 2–D Rastrigin function problem

particles tend to rapidly focus towards the global optimum at the origin. Figure 2 shows the behaviour of **PSOG3** in exactly the same conditions (including same starting positions for the particles). Here the particles tend to focus less rapidly.

Figures 3 and 4 illustrate the behaviour of **PSO** on a Rastrigin function with $N = 2$. Here the swarm controlled by **PSO** is quickly attracted towards a deceptive local optimum, while the particles in **PSOG3** perform bigger orbits and eventually discover and start converging towards the global optimum at the origin.

5   Conclusions

GP has been able to evolve a variety of particle swarm optimisers that work as well or considerably better than standard human-designed PSOs. Analysis of the evolved programs has led to new insights in the design of PSOs tailored for specific classes of landscapes.

To evolve our PSOs we have used the state of the art in GP, but we have not proposed a great deal in terms of new GP technology. However, in the more general context of machine intelligence, this work represents an important step within a new research trend: using search algorithms to discover new search algorithms. This approach has become possible thanks to the growth in computing power. We can already foresee that the results of this may be spectacular (see, for example, the case of Fukunaga's award winning work on evolving human-competitive SAT problem solvers [6]). The main contribution of this paper is to show that genetic programming can evolve better than human-designed PSOs in a few hours on a standard PC. A second contribution is to give us new ideas on what types of particle behaviours are most appropriate for which type of landscape.

In future research we intend to apply the approach to a variety of problem domains (including real-world problems) and to extend it by allowing GP to use more information on the past history of the swarm to control the particles and by allowing the evolution of coupled force-generating equations. We also intend to explore the effects and benefits of using different performance measures for PSO evolution.

Acknowledgments

The authors would like to thank EPSRC (grant GR/T11234/01) for financial support.

References

1. P. J. Angeline. Using selection to improve particle swarm optimization. In *IEEE World Congress on computational intelligence, ICEC-98*, pages 84–89, Anchorange, Alaska, 1998.
2. T. M. Blackwell and P. J. Bentley. Dynamic search with charged swarms. In *GECCO 2002: Proceedings of the Genetic and Evolutionary Computation Conference*, pages 19–26, New York, 9-13 July 2002. Morgan Kaufmann Publishers.

3. T. M. Blackwell and J. Branke. Multi-swarm optimization in dynamic environments. In *Applications of Evolutionary Computing*. Springer, 2004.

4. R. Brits, A. P. Engelbrecht, and B. Bergh. A Niching Particle Swarm Optimizer. In *Proceedings of the 4th Asia-Pacific Conference on Simulated Evolution and Learning (SEAL'02)*, volume 2, pages 692–696, Orchid Country Club, Singapore, Nov. 2002. Nanyang Technical University.

5. M. Clerc and J. Kennedy. The particle swarm-explosion, stability, and convergence in a multidimensional complex space. *IEEE Transactions on Evolutionary Computation*, 6(1):58–73, 2002.

6. A. S. Fukunaga. Evolving local search heuristics for SAT using genetic programming. In *Genetic and Evolutionary Computation – GECCO-2004, Part II*, volume 3103 of *Lecture Notes in Computer Science*, pages 483–494, Seattle, WA, USA, 26-30 June 2004. Springer-Verlag.

7. F. Heppner and U. Grenander. A stochastic nonlinear model for coordinated bird flocks. In *The ubiquity of Chaos*. AAAS publications, Washington DC, 1990.

8. J. Kennedy. The behavior of particles. In *Evolutionary Programming VII: Proceedings of the Seventh Annual Conference on evolutionary programming*, pages 581–589, San Diego, CA, 1998.

9. J. Kennedy and R. C. Eberhart. *Swarm Intelligence*. Morgan Kaufmann Publishers, San Francisco, California, 2001.

10. J. R. Koza. *Genetic Programming: On the Programming of Computers by Means of Natural Selection*. MIT Press, Cambridge, MA, USA, 1992.

11. T. Krink, J. S. Vesterstrøm, and R. Riget. Particle swarm optimisation with spatial particle extension. In *Proceedings of the 2002 Congress on Evolutionary Computation CEC2002*, pages 1474–1479. IEEE Press, 2002.

12. W. B. Langdon and R. Poli. *Foundations of Genetic Programming*. Springer-Verlag, 2002.

13. M. Lovbjerg and T. Krink. Extending particle swarm opimisers with self-organized criticality, July 11 2002.

14. E. Ozcan and C. K. Mohan. Particle swarm optimization: surfing the waves. In *Proceedings of the IEEE Congress on evolutionary computation (CEC 1999)*, Washington DC, 1999.

15. R. Poli and C. R. Stephens. Constrained molecular dynamics as a search and optimization tool. In *Genetic Programming 7th European Conference, EuroGP 2004, Proceedings*, volume 3003 of *LNCS*, pages 150–161, Coimbra, Portugal, 5-7 Apr. 2004. Springer-Verlag.

16. J. Riget, J. S. Vesterstrm, and K. Krink. Division of labor in particle swarm opimisation, July 11 2002.

17. Y. Shi and R. C. Eberhart. A modified particle swarm optimizer. In *Proceedings of the IEEE Congress on Evolutionary Computation (CEC 1999)*, pages 69–73, Piscataway NJ, 1999.

18. D. H. Wolpert and W. G. Macready. No free lunch theorems for optimization. *IEEE Transactions on Evolutionary Computation*, 1(1):67–82, Apr. 1997.

Inducing Diverse Decision Forests with Genetic Programming

Jan Suchý and Jiří Kubalík

Department of Cybernetics, CTU Prague,
Karlovo náměstí 13, 121 35, Praha 2, Czech Republic
suchyj1@fel.cvut.cz, kubalik@labe.felk.cvut.cz

Abstract. This paper presents an algorithm for induction of ensembles of decision trees, also referred to as decision forests. In order to achieve high expressiveness the trees induced are multivariate, with various, possibly user-defined tests in their internal nodes. Strongly typed genetic programming is utilized to evolve structure of the tests. Special attention is given to the problem of diversity of the forest constructed. An approach is proposed, which explicitly encourages the induction algorithm to produce a different tree each run, which represents an alternative description of the data. It is shown that forests constructed this way have significantly reduced classification error even for small forest size, compared to other ensemble methods. Classification accuracy is also compared to other recent methods on several real-world datasets.

1 Introduction

Classification is a task in which machine learning methods are commonly used. With knowledge of attributes $(x_1, x_2, \ldots, x_n) \in X_1 \times X_2 \times \ldots \times X_n$ of an object the task is to assign a correct class k to it, which is unknown, from a set of possible classes K. A program is sought, called classifier, that correctly describes dependence between the class and the attributes. Decision trees [16] are a popular paradigm for modelling such dependencies. This paper presents an algorithm for decision tree induction from data. Emphasis is put on two important aspects of the problem. First, highly expressive, possibly user-defined tests are allowed in decision tree nodes. This way problem specific knowledge can be incorporated into the algorithm. Strongly typed genetic programming is utilized in the hard task of searching for good such tests. Similar approach has been applied in [13]. Second, the algorithm is designed to induce a whole set of diverse trees that can be grouped together in order to improve classification accuracy. Such formation is often called an *ensemble* or, in the case of trees, a *forest* [10]. Decision of a forest is determined by majority vote of its individual trees. For this method to work, it is important that the forest is diverse [12]. In other words, the individual trees should represent alternative descriptions of the data. This is often achieved by introducing small changes in the data that is input to the induction algorithm [1]. In contrast the method presented in this paper explicitly encourages the

M. Keijzer et al. (Eds.): EuroGP 2005, LNCS 3447, pp. 301–310, 2005.

induction algorithm to produce a different decision tree in each run from the same, unchanged data.

2   Decision Trees

Decision trees divide the process of deciding about object's class into a sequence of tests. The tests are organized into a tree structure. In the tree the tests occupy the internal nodes, the edges determine order in which the tests are applied and the leaves represent final decisions: class labels. Classification of an unknown object starts with test in the root node. The edge that corresponds to the outcome of the test determines which test is applied next. This way the object "falls through" the tree down to a leaf which finally assigns a class label to it.

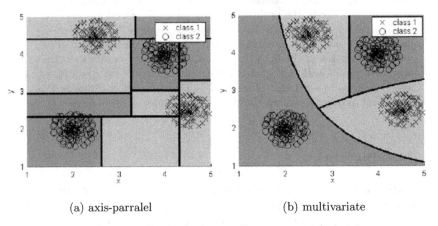

(a) axis-parralel   (b) multivariate

Fig. 1. An example of splitting attribute space with decision trees

Commonly used tests are in the form of conditions $x_i \leq c$ for continuous attributes and $x_i = k$ for discrete attributes. Here c and k are some constants produced by the tree induction algorithm. This kind of tests partition the attribute space with axis-parallel splits, as shown in figure 1(a). It is possible to enrich expressiveness of the trees by allowing more complicated tests in the tree nodes. An example could be $x_1 - \sin x_2 \leq c$. Trees with such tests can be more flexible in partitioning the attribute space, as shown in figure 1(b). They are commonly referred to as *multivariate decision trees* [15]. In this paper the tests are multivariate conditions represented by a tree structure so that they can be straightforwardly searched for with genetic programming.

3   The Tree Induction Algorithm

The algorithm follows the common top-down induction scheme, creating one node at a time. Starting with the root node, the algorithm constructs a test which splits the given set of training examples into two disjoint subsets. The test is constructed so that it maximizes a criterion of optimality, which measures the ability of the test to discriminate examples belonging to different classes. The same is then applied to both subsets obtained with the test. This way the original training set is recursively partitioned until a stopping condition is met.

The individual tests are evolved with genetic programming, using terminal and function sets as specified in section 4. The fitness function coincides with the criterion of test optimality, which is based on measuring the *information gain* [16]. The information gain is the amount of uncertainty (entropy) eliminated by the test from the set it splits. Let M be the training set containing n examples, and s the number of different classes. Let n_i be the number of examples belonging to class i. Then the amount of uncertainty in the set is given by the following equation:

$$H(M) = -\sum_{i=1}^{s} n_i \log_2 \frac{n_i}{n} \ . \tag{1}$$

Further it is defined $H(\emptyset) = 0$ and $0 \log_2 0 = 0$. Every test t splits a set M into M_P and M_N, so that $M = M_P \cup M_N$ and $M_P \cap M_N = \emptyset$. The information gain of test t is computed as:

$$I(t) = H(M) - H(M_P) - H(M_N) \ . \tag{2}$$

While using I as fitness is suitable for single decision trees, it is not appropriate when trees are sequentially induced that are to be combined into a forest. For this scheme a modified criterion is proposed in section 5, which assures that sequentially induced trees differ considerably from each other.

With growing depth of the tree there is an increasing risk of overfitting the data. Therefore at each node a condition is tested which stops the induction if either $\frac{H(M)}{n}$ or the size of M drops below a threshold specified by the user. When this condition is satisfied a leaf node is produced, which assigns to the objects label of the class most frequent in M.

4   The Tests

Higher expressiveness of the trees implies more complex structure of the tests. The tests are expressions in general, which in turn can be easily represented by trees in genetic programming. In this representation the terminal set T consists of attributes and perhaps other entities important for the classification, e.g. random constants. The function set F contains operators, functions and predicates, that express possibly meaningful properties and relations of the attributes. For an expression to be a properly formed test it must evaluate to either **true** or **false**. That is, the function in the root node of the expression must return a boolean value. Now it comes to the serious limitation of standard genetic programming that requires both function and terminal sets to have the closure property. Clearly, one would expect functions that (for example) add or

Table 1. Predefined terminals and functions

Type	Terminals	Functions
Nominal attributes	x, R	$=$
Ordered attributes	x, R	$=$, $>$
Numeric attributes	x, R	$=$, $>$, $+$, $-$, $*$, sin, (> 0)
Logical values		$\wedge$, $\vee$, $\neg$

multiply the attributes in the tests, but this is not possible due to the necessary closure property. In addition one has to often deal with both numeric and nominal attributes in a single classification task, which at last leads to the same problem. To resolve this problem strongly typed genetic programming [14] was used. It introduces *types* of functions and terminals similar to those found in higher programming languages. It also adds a type checking mechanism to the recombination operators so that only valid trees are constructed.

As said above, the function and terminal sets form a sort of language that is used to describe the data. Ideally the user of the algorithm supplies definitions of needed functions and terminals using his or her knowledge of the problem at hand. As this is not always possible a basic set of functions and terminals is predefined within the system. These allow the algorithm to handle nominal attributes, attributes whose domain is an ordered set, and numerical attributes. Table 1 shows a summary of the predefined functions and terminals. For each attribute type (nominal, ordered and numeric) there is a terminal called x in the table, which represents the value of the attribute, and a terminal called R, which represents an ephemeral random constant [11] from the attribute's domain. For each type there is the function $=$, i.e. comparison of values of that type, which evaluates to a logical value **true** or **false**. For ordered and numeric attributes there is also the relational operator $>$. Only for numeric attributes there are also arithmetic operators $+,-,*$ and the sine function sin. For logical (boolean) values there are functions $\wedge$, $\vee$ and $\neg$ (conjunction, disjunction and negation, respectively). With these functions it is possible to form tests that combine attributes of different types, as shown in an example in figure 2.

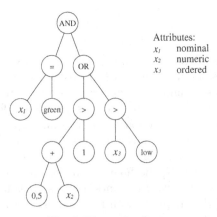

Fig. 2. Example of a test

5   The Forests

Classification accuracy can be often improved by the use of ensemble classifiers. The decision of an ensemble is determined by (weighted) majority vote of the individual classifiers. The underlying concept is that the accuracy of a *diverse* ensemble of classifiers whose accuracy is at least a little better than random guessing is generally better than accuracy of the individual classifiers. The requirement of diversity is important, although precise meaning of the term has not been given clearly yet [12]. A common way to construct different classifiers with an algorithm that outputs a single classifier is to choose different training data for each run of the algorithm. The widely known ensemble methods *bagging* [5] and *AdaBoost* [9] follow this scheme. The algorithm presented here is not deterministic, therefore it produces a different classifier each run even with the same training data. To further improve the resulting ensemble diversity the following approach is proposed.

In the top-down decision tree induction as described in section 3 the root test heavily influences how the other, lower layer tests are formed. Thus by changing the root node test one can substantially alter the whole tree being induced. This is achieved by using a fitness function that causes genetic programming to search for root tests that eliminate uncertainty other than that removed by root tests of previously evolved trees. Tests other than root are evolved with the standard information gain fitness given by equation 2.

It is done in the following way: The root node of the first tree in the forest is evolved with fitness given by equation 2. It should maximize information gain on the training set. The root node of the second tree is evolved with a modified fitness, which favors tests, which eliminate uncertainty not eliminated by the root node of the first tree. Suppose that the first root node test splits the training set M into M_P and M_N, then the fitness used in evolution of the second test is $I(M_P) + I(M_N)$. The root node test of the third tree is evolved so that it eliminates uncertainty not eliminated by either of the root tests of the previously evolved trees. The general formula for fitness of i-th root node test t_i is:

$$J(t_i) = \min_{j \le i-1} \{I(M_{Pj}) + I(M_{Nj})\}, \quad i = 2, \ldots, R, \tag{3}$$

where M_{Pj}, M_{Nj} are sets, in which the root node test of j-th tree splits the original training set. Index j runs over all tests evolved before t_i.

6   Experiments

In all experiments in the following subsections, the algorithm was run in the following setup: Only predefined functions and terminals were used, as specified in section 4. For each rule genetic programming was allowed to run for 100 generations with population of 500 individuals. Maximal number of function and terminal symbols together was limited to 12 in all tests. The induction was

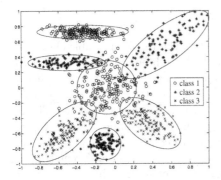

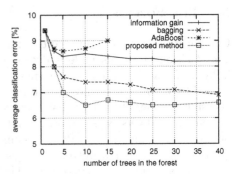

Fig. 3. Synthetic data used for experiment in section 6.1

Fig. 4. Classification accuracy of the compared methods

stopped whenever $\frac{H(M)}{n}$ dropped below $0,65$ bits per example or when the size of M dropped below 5% of its original size.

6.1   Comparison to Other Ensemble Methods

In this section the proposed algorithm is compared to two popular ensemble construction methods bagging and AdaBoost. The comparison was carried out on synthetic data, which allowed to illustrate behavior of the proposed method. At the same time the dataset was meant to emulate an easy but typical classification task with two numerical attributes x, y and three classes c_1, c_2, c_3, sampled from a mixture of Gaussian distributions:

$$p(x, y | c_i) = \sum_j k_{ij} N(\mu_{ij}, \Sigma_{ij}) .$$

The dataset is displayed in figure 3. It is clearly visible that the classes are not perfectly separable. One can achieve classification accuracy approx. 95%.

In the experiment four different methods were used to construct forests of different sizes. For each method and each forest size data consisting of 1000 examples were used for training and testing in 5-fold cross-validation. This procedure was repeated 10 times. All 50 results were then averaged.

First the forests were constructed with pure information gain as fitness, with no changes made in the training data for different trees. This serves as a reference to the other methods. Second, bagging was used as described in [5]. Third, the forests were constructed with AdaBoost.M1 setup exactly as described in [9], section 3. At last, the proposed method was used with fitness given by equation 3 for root node tests and pure information gain for the rest of nodes.

The results are shown in figure 4. The first observation is that when using pure information gain, the induced trees are not as diverse as in the other cases, which leads to lower classification accuracy when compared to the other methods. This is illustrated in figure 5, where trees induced by the proposed method and

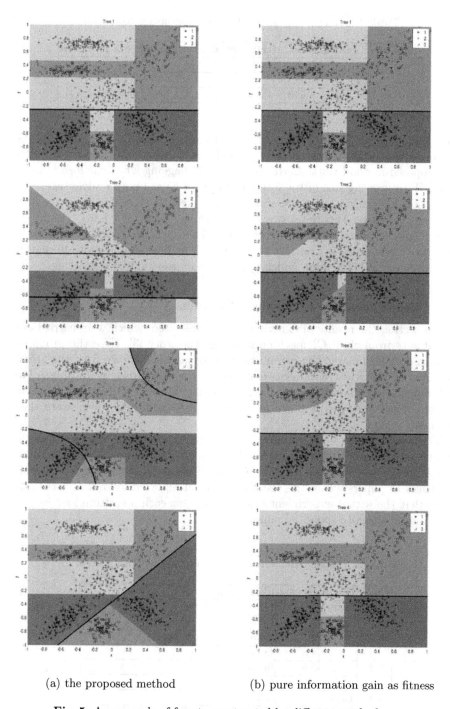

(a) the proposed method   (b) pure information gain as fitness

Fig. 5. An example of forests constructed by different methods

Table 2. An overview of the datasets used in the experiment

dataset	# examples	# attributes	# classes	class distribution
Pima	768	8	2	65% / 35%
Heart	920	13	2	44% / 56%
Breast	699	10	2	65% / 35%

those induced using pure information gain can be visually compared. In the figure the black lines represent the root node splits. Another observation is that the proposed method works better compared to bagging when the forest size is small. With growing size of the forest it ceases to have a major advantage.

Although AdaBoost is more sophisticated method than bagging, it suffers from overfitting on this dataset as it tries to specialize on "hard" examples. These, however, are but noise in this case. The effect is especially noticeable for larger forest sizes. In contrast the proposed method searches for alternative descriptions of *all* examples, which makes it more resistant to overfitting.

6.2   Experiments on Real-World Datasets

In this section the results of experiments are presented, which were carried out on datasets from the UCI machine learning repository [3]. Here the proposed algorithm was compared to several other recent methods known from the literature. A short overview of datasets employed in the comparison is in table 2. For each dataset, the classification accuracy was tested in 10-fold cross-validation, and this procedure was 6 times repeated. The 60 resulting values were then averaged.

The results are summarized in table 3, which contains the values of classification error achieved by each of the algorithms. The results of the algorithms have been reported by their authors. A short description of the algorithms with references to papers where the results have been reported follows. It has to be noted that the actual experiment set-up differs for each algorithm and can be looked up in the corresponding paper.

Table 3. Classification error of the compared algorithms in %

Pima	Heart	Breast	algorithms
25,6	20,0		fuzzy decision trees
28,4	22,2		C4.5
25,7		3,3	C4.5 + Adaboost.M2
26,3		4,8	OC1
26,4		5,9	fuzzy rules
27,1	14,8	4,5	GP rules
24,4	**13,8**	3,3	NN
25,0	21,4	3,8	proposed algorithm – single tree
23,6	18,7	**3,1**	proposed algorithm – forest, 11 trees

Fuzzy decision trees. An evolutionary method for induction of fuzzy decision trees based on clustering [8].

C4.5. Well-known decision tree induction algorithm by J. R. Quinlan [16]. The reported results are taken from [8].

C4.5 + Adaboost.M2. C4.5 combined with the AdaBoost for ensemble construction [9].

OC1. A well-known algorithm for induction of oblique decision trees. In oblique decision trees the tests are in the form of inequalities: $a_1x_1 + a_2x_2 + ... + a_nx_n \leq c$. The reported results are taken from [6].

Fuzzy rules. An expert system based on fuzzy rules constructed with genetic programming. [2].

GP rules. A rule-based system, evolves IF-THEN rules with genetic programming. [7].

NN. A neural network (multi layer perceptron) learned by the back-propagation method. [4].

The results indicate that the proposed algorithm is competitive with the other algorithms. The error reduction achieved by the use of forests is significant even for small sized forests when compared to ensemble sizes used in [5].

7   Conclusions and Future Work

An algorithm was introduced, that induces decision trees with possibly highly expressive tests in their nodes. It allows the user of the algorithm to choose or define "building blocks" of the decision tree tests, appropriate for the problem at hand. As this feature is an advantage theoretically, its practical benefits for real-world problems is still to be investigated in the future. Strongly typed genetic programming is utilized for searching for the tests, which provides a straightforward way to construct tests with different, possibly complex structure.

A special fitness function is proposed, which allows the algorithm to sequentially induce a diverse group of trees, which can be then used as a decision forest. Decision forests constructed this way have considerably higher classification accuracy than the individual trees even for small forest sizes, compared to the general method Bagging. On the other hand they are not as susceptible to overfitting as the AdaBoost algorithm, which is also known to be able to improve classification accuracy even for small ensemble sizes.

Similarly to many other evolutionary algorithms the proposed algorithm has a number of variable parameters. The most important of them are the stopping condition of the tree induction, the maximum allowed size of the evolved tests, and the function and terminal sets used. The parameters used for the experiments conducted in this paper can serve as reasonable default values. As a rule, one should allow only simple tests to be evolved (i.e. tests consisting only of a small number of functions and terminals), when the training sample is small. Otherwise the algorithm is likely to overfit the data. Something similar holds for the stopping condition of the induction algorithm: For small training samples one should stop the induction earlier to avoid overfitting.

The future work will be concentrated on the problem of searching for simple descriptions of data, as they are likely to perform better than complex ones. The authors' observations suggest this, as well as several other studies [15].

References

1. Eric Bauer and Ron Kohavi. An empirical comparison of voting classification algorithms: Bagging, boosting, and variants. *Machine Learning*, 36(1-2):105–139, 1999.
2. P. J. Bentley. Evolving fuzzy detectives: An investigation into the evolution of fuzzy rules. In *Late Breaking Papers at the 1999 Genetic and Evolutionary Computation Conference*, pages 38–47, Orlando, Florida, USA, 1999.
3. C. L. Blake and C. J. Merz. UCI repository of machine learning databases. http://www.ics.uci.edu/~mlearn/MLRepository.html, 1998.
4. M. Brameier and W. Banzhaf. A comparison of linear genetic programming and neural networks in medical data mining. *IEEE Transactions on Evolutionary Computation*, 5(1):17–26, February 2001.
5. L. Breiman. Bagging predictors. *Machine Learning*, 24(2):123–140, 1996.
6. E. Cant-Paz and C. Kamath. Inducing oblique decision trees with evolutionary algorithms. *IEEE Transactions on Evolutionary Computing*, 7(1):56–68, 2003.
7. I. De Falco, A. Della Cioppa, and E. Tarantino. Discovering interesting classification rules with genetic programming. *Applied Soft Computing*, 1(4F):257–269, May 2001.
8. J. Eggermont. Evolving fuzzy decision trees with genetic programming and clustering. In *Proceedings of the 4th European Conference on Genetic Programming, EuroGP 2002*, volume 2278, pages 71–82, Kinsale, Ireland, 3-5 2002. Springer-Verlag.
9. Y. Freund and R. E. Schapire. Experiments with a new boosting algorithm. In *Proc. 13th International Conference on Machine Learning*, pages 148–146. Morgan Kaufmann, 1996.
10. Tin Kam Ho. C4.5 decision forests. In *Proceedings of the 14th International Conference on Pattern Recognition-Volume 1*, page 545. IEEE Computer Society, 1998.
11. J. R. Koza. *Genetic Programming: On the Programming of Computers by Means of Natural Selection*. MIT Press, Cambridge, MA, USA, 1992.
12. L. Kuncheva and C. Whitaker. Measures of diversity in classifier ensembles, 2000.
13. Robert E. Marmelstein and Gary B. Lamont. Pattern classification using a hybrid genetic program decision tree approach. In *Genetic Programming 1998: Proceedings of the Third Annual Conference*, pages 223–231, University of Wisconsin, Madison, Wisconsin, USA, 22-25 July 1998. Morgan Kaufmann.
14. D. J. Montana. Strongly typed genetic programming. BBN Technical Report #7866, Bolt Beranek and Newman, Inc., 10 Moulton Street, Cambridge, MA 02138, USA, 7 May 1993.
15. S. K. Murthy. Automatic construction of decision trees from data: A multidisciplinary survey. *Data Mining and Knowledge Discovery*, 2(4):345–389, 1998.
16. J. R. Quinlan. *C4.5: programs for machine learning*. Morgan Kaufmann Publishers Inc., 1993.

mGGA: The meta-Grammar Genetic Algorithm

Michael O'Neill[1] and Anthony Brabazon[2]

[1] Department of Computer Science and Information Systems,
University of Limerick, Ireland
`Michael.ONeill@ul.ie`
[2] Department of Accountancy, University College Dublin, Ireland
`Anthony.Brabazon@ucd.ie`

Abstract. A novel Grammatical Genetic Algorithm, the meta-Grammar Genetic Algorithm (mGGA) is presented. The mGGA borrows a grammatical representation and the ideas of modularity and reuse from Genetic Programming, and in particular an evolvable grammar representation from Grammatical Evolution by Grammatical Evolution. We demonstrate its application to a number of benchmark problems where significant performance gains are achieved when compared to static grammars.

1   Introduction

The objectives of this study are to investigate the adoption of principles from Genetic Programming [1] such as modularity and reuse (see Chapter 16 in [2]) for application to Genetic Algorithms, and to couple these to an adaptive representation that allows the type and usage of these principles to be evolved through the use of evolvable grammars. The goal being the development of an evolutionary algorithm with good scaling characteristics, and an adaptable representation that will facilitate it's application to dynamic problem environments. To this end a grammar-based Genetic Programming approach is adopted, in which the grammars represent the construction of syntactically correct phenotypes of the Evolutionary Algorithm.

The remainder of the paper is structured as follows. Section's 2 and 3 describes the grammatical approach to Genetic Algorithms, section 4 details the experimental approach adopted and results, and finally section 5 details conclusions and future work.

2   Grammatical Evolution by Grammatical Evolution

The grammar-based Genetic Programming approach upon which this study is based is the Grammatical Evolution by Grammatical Evolution algorithm [3], which is in turn based on the Grammatical Evolution algorithm [4, 5, 6, 7]. This is a meta-Grammar Evolutionary Algorithm in which the input grammar is used to specify the construction of another syntactically correct grammar. The generated grammar is then used in a mapping process to construct a solution.

M. Keijzer et al. (Eds.): EuroGP 2005, LNCS 3447, pp. 311–320, 2005.

In order to allow evolution of a grammar, Grammatical Evolution by Grammatical Evolution $(GE)^2$, we must provide a grammar to specify the form a grammar can take. This is an example of the richness of the expressiveness of grammars that makes the GE approach so powerful. See [4, 8, 9] for further examples of what can be represented with grammars, and [10] for an alternative approach to grammar evolution. By allowing an Evolutionary Algorithm to adapt its representation (in this case through the evolution of the grammar) it provides the population with a mechanism to survive in dynamic environments, in particular, and also to automatically incorporate biases into the search process. In this case we can allow the meta-Grammar Genetic Algorithm to evolve biases towards different building blocks of varying sizes.

In this approach we therefore have two distinct grammars, the *universal grammar* (or grammars' grammar) and the *solution grammar*. The notion of a universal grammar is adopted from linguistics and refers to a universal set of syntactic rules that hold for spoken languages [11]. It has been proposed that during a child's development the universal grammar undergoes modifications through learning that allows the development of communication in their parents native language(s) [12].

In $(GE)^2$, the universal grammar dictates the construction of the solution grammar. In this study two separate, variable-length, genotypic binary chromosomes were used, the first chromosome to generate the solution grammar from the universal grammar and the second chromosome the solution itself. Crossover operates between homologous chromosomes, that is, the solution grammar chromosome from the first parent recombines with the solution grammar chromosome from the second parent, with the same occurring for the solution chromosomes. In order for evolution to be successful it must co-evolve both the meta-Grammar and the structure of solutions based on the evolved meta-Grammar.

3   meta-Grammars for Bitstrings

A simple grammar (referred to as GE) for a fixed-length (example contains 8 bits) binary string individual of a Genetic Algorithm is provided below. In the generative grammar each bit position (denoted as <bit>) can become either of the boolean values. A standard variable-length Grammatical Evolution individual can then be allowed to specify what each bit value will be by selecting the appropriate <bit> production rule for each position in the <bitstring>.

```
<bitstring> ::= <bit><bit><bit><bit><bit><bit><bit><bit>

<bit> ::= 1 | 0
```

The above grammar can be extended to incorporate the reuse of groups of bits (building blocks). In this example all building blocks that are mutliples of two are provided, although it would be possible to create a grammar that adopted more complex arrangements of building blocks. We refer to this grammar as GE+BB.

```
<bitstring> ::= <bbk4><bbk4>
             | <bbk2><bbk2><bbk2><bbk2>
             | <bbk1><bbk1><bbk1t><bbk1><bbk1><bbk1><bbk1><bbk1>

<bbk4> ::= <bit><bit><bit><bit>

<bbk2> ::= <bit><bit>

<bbk1> ::= <bit>

<bit> ::= 1 | 0
```

The above grammars are static, and as such can only allow one building block
of size four and of size two in the second example. It would be nice to allow
evolution to find a number of building blocks of any one size from which a
Grammatical Evolution individual could choose from. This would facilitate the
application of such a Grammatical GA to:

- problems with more than one building block type for each building block
  size,
- to search on one building block while maintaining a *reasonable* temporary
  building block solution,
- and to be able to switch between building blocks in the case of dynamic
  environments.

All of this can be achieved through the adoption of meta-Grammars as were
adopted earlier in [3]. An example of such a grammar (referred to as GEGE+BB)
for an 8-bit individual is given below.

```
<g> ::=
        "<bitstring> ::=" <reps>
            "<bbk4> ::=" <bbk4t>
            "<bbk2> ::=" <bbk2t>
            "<bbk1> ::=" <bbk1t>
             "<bit> ::=" <val>

<bbk4t> ::= <bit><bit><bit><bit>

<bbk2t> ::= <bit><bit>

<bbk1t> ::= <bit>

<reps> ::= <rept>
         | <rept> "|" <reps>

<rept> ::= "<bbk4><bbk4>"
         | "<bbk2><bbk2><bbk2><bbk2>"
         | "<bbk1><bbk1><bbk1><bbk1><bbk1><bbk1><bbk1><bbk1>"

<bit> ::= "<bit>"
        | 1
        | 0

<val> ::= <valt>
        | <valt> "|" <val>

<valt> ::= 1 | 0
```

In this case the grammar specifies the construction of another generative bitstring grammar. The subsequent bitstring grammar that can be produced from

the above meta-grammar is restricted such that it can contain building blocks of size 8. Some of the bits of the building blocks can be fully specified as a boolean value or may be left as unfilled for the second step in the mapping process. An example bitstring grammar produced from the above meta-grammar could be:

```
<bitstring> ::= <bit>11<bit>00<bit><bit>
              | <bbk2><bbk2><bbk2><bbk2>
              | 11011101
              | <bbk4><bbk4>
              | <bbk4><bbk4>

<bbk4> ::= <bit>11<bit>

<bbk2> ::= 11

<bbk1> ::= 1

<bit> ::= 1 | 0 | 0 | 1
```

To allow the creation of multiple building blocks of different sizes the following grammar (referred to as GEGE+KBB) could be adopted (again shown for 8-bit strings).

```
<g> ::=
          "<bitstring> ::=" <reps>
             "<bbk4> ::=" <bbk4>
             "<bbk2> ::=" <bbk2>
             "<bbk1> ::=" <bbk1>
              "<bit> ::=" <val>

<bbk4> ::= <bbk4t>
         | <bbk4t> "|" <bbk4>
<bbk2> ::= <bbk2t>
         | <bbk2t> "|" <bbk2>
<bbk1> ::= <bbk1t>
         | <bbk1t> "|" <bbk1>
<bbk4t> ::= <bit><bit><bit><bit>
<bbk2t> ::= <bit><bit>
<bbk1t> ::= <bit>
<reps> ::= <rept>
         | <rept> "|" <reps>
<rept> ::= "<bbk4><bbk4>"
         | "<bbk2><bbk2><bbk2><bbk2>"
         | "<bbk1><bbk1><bbk1><bbk1><bbk1><bbk1><bbk1><bbk1>"
<bit> ::= "<bit>"
        | 1
        | 0
<val> ::= <valt>
        | <valt> "|" <val>
<valt> ::= 1
         | 0
```

An example bitstring grammar produced by the above meta-grammar is provided below.

```
<bitstring> ::= <bit>11<bit>00<bit><bit>
              | <bbk2><bbk2><bbk2><bbk2>
              | 11011101
              | <bbk4><bbk4>
              | <bbk4><bbk4>

<bbk4> ::= <bit>11<bit>
         | 000<bit>
```

```
<bbk2> ::= 11
        | 00
        | <bit>1

<bbk1> ::= 0
        | 0

<bit> ::= 1 | 0 | 0 | 1
```

Modularity exists above in the ability to specify the size and content (or partial content) of a buiding block through it's incorporation into the solution grammar. This building block can then be repeatedly reused in the generation of the phenotype. An additional mechanism for reuse is through the Wrapping operator of Grammatical Evolution. During the mapping process if we reach the end of the genotype and still have outstanding decisions to make on the construction of our phenotype we can invoke the wrapping operator to move our reading head back to the first codon in the genome. This allows the reuse of rule choices if the codons are used in the same context.

Given that the lengths of binary strings which may need to be represented can grow quite large it is possible to automate the creation of meta-grammars by simply providing the length of the target solution and creating all possible building block structures that can be used to create a bitstring of the target length. In this study the target binary strings are of lengths 60, 90, 120, 180, and 210. The building block sizes incorporated in their corresponding grammars are therefore all integers that divide into the target string lengths (i.e., for a target string of length 60 the building blocks are of sizes 30, 20, 15, 12, 10, 6, 5, 4, 3, 2 and 1).

Meta-grammars are of course not limited to the specification of grammars for binary strings and can be easily extended to the representation of real and integer strings as well as programs, or any structure for which it is possible to represent in a grammatical form.

4   Experimental Setup and Results

The mGGA was applied to three problem types, namely, instances of onemax, instances of a deceptive trap problem, and a dynamic problem instance. Two onemax instances were adopted with target string lengths of 180 and 210.

Similarly, two instances of a Trap5 problem were used having target string lengths of 60 and 90, with these having 12 and 18 subfunctions respectively. The dynamic problem instance has an alternating target every 20 generations between a onemax and zeromax problem with a target string length of 120 bits investigated. In each case the same parameter settings were adopted. These were, a population size of 100, tournament selection, generational replacement, a crossover probability of 0.3 between homologous chromosomes, and a mutation probability of 0.01. Initialisation of the population was performed randomly with individuals having in the range of 1 to 20 codons.

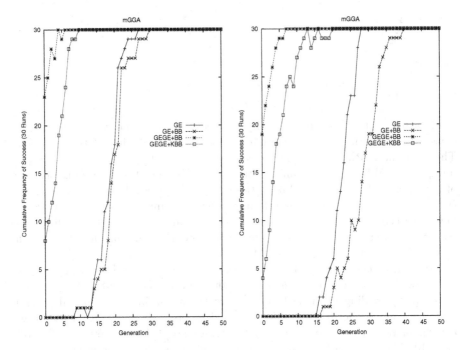

Fig. 1. Plot of the cumulative frequency of success for the OneMax 180 bit problem (left) and the OneMax 210 bit problem (right)

The results for the onemax instances are presented in Fig. 1. It is clear on both instances that the evolvable meta-Grammar's (GEGE+BB and GEGE+KBB) outperform the static grammars (GE and GE+BB) in terms of the speed at which the target solution is reached, although all grammars are capable of finding the perfect solution in every run beyond 50 generations. We would expect, and it is observed, that the performance of the static grammars are close due to their similarity.

Results for the Trap5 instances are presented in Fig. 2. In this case the evolvable meta-Grammar's outperform the static grammars both in terms of their speed at obtaining perfect solutions and in terms of the number of successful runs at the end of 100 generations. The static grammar runs having less than 50% success rate on the 60 bit instance, and less than 33% on the 90 bit instance. This is compared to a 100% success rate for both the GEGE+BB and GEGE+KBB grammars.

The immeadiate success of the GEGE+BB and GEGE+KBB grammars on these static problems can be attributed, in part at least, to the relatively small number of choices that need to be made to generate a perfect solution when compared to making a decision on all 60, 90, 180 or 210 bits individually. In effect if the solution grammar is generated to specify that a solution is comprised of a building block of size 1 bits, and that the building block takes on the value 0, only two codons are required to fully specify a correct solution to the Trap5 problem

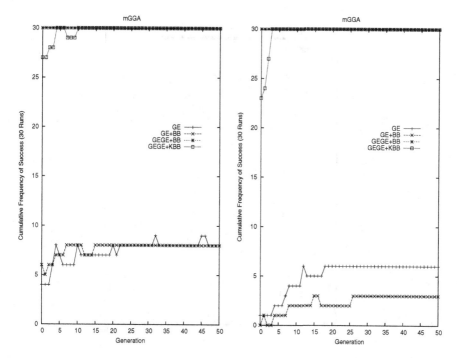

Fig. 2. Plot of the cumulative frequency of success for the 5Trap 60 bit problem (left) and 5Trap 90 bit problem (right)

instances. The evolvable representation adopted contains more redundancy of a form that provides an increased number of avenues by which a perfect solution can be reached relatively quickly within the initial generations, which goes some way to explain the superior performance of the GEGE grammars [13].

Results for the dynamic instance are provided in Fig.'s 3 and 4. We can see that the two static grammars (GE and GE+BB) and the GEGE+BB grammar perform well during the first twenty generations with the majority of runs finding a perfect solution during this time. However, from the first change in target at generation 21 up until generation 40 the performance of the GE and GE+BB grammars degrade significantly in contrast to the two evolvable grammars (GEGE+BB and GEGE+KBB), which have success rates over 50% compared to 0% for their static counterparts. On return to the original target between generations 41 to 60 the static grammars peak at generation 60 with just over a 50% success rate. During this same period the GEGE+KBB grammars success rate is improving steadily towards 66%, while the GEGE+BB grammars performance remains constant just short of a 100% success rate. With the next change in target at generation 61 performance of the static grammars falls off towards a 0% success rate while both the GEGE+BB and GEGE+KBB performance improves. The mean best fitness plot (Fig. 4) supports the trends we observe in the success rate plots. Over the course of the run we see a steady

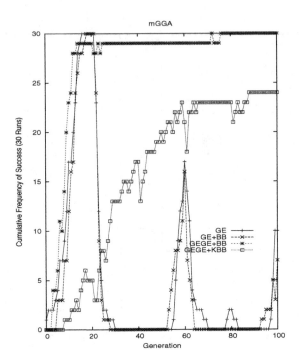

Fig. 3. Plot of the cumulative frequency of success for the Dynamic 120 bit problem

improvement in the mean best fitness for the GEGE+BB and GEGE+KBB grammars, with the performance of the GE and GE+BB grammars fluctuating more distinctly with each change in target solution. Performance for the GE and GE+BB grammars always being better upon return to the original target presented in the first twenty generations.

Two successful, abbreviated, sample solution grammars for the Dynamic 120 problem instance are given below for both of the GEGE+BB and GEGE+KBB meta-grammars. In each solution the same grammar represents the two target solutions that are required and can allow a switch between solutions by changing a single choice made when mapping `<bitstring>`.

```
GEGE+KBB Abbreviated Sample Solution (Dynamic 120 Problem)
<bitstring> ::= <bbk1>..<bbk1>
              | <bbk4>..<bbk4>
              | <bbk1>..<bbk1>
    <bbk4> ::=  00<bit><bit>
    <bbk1> ::=  1
     <bit> ::=  0
```

```
solution: 111111111111111111111111111111111111111111111111111111111111111111111
          111111111111111111111111111111111111111111111111111111111111111111111
```

```
GEGE+BB Abbreviated Sample Solution (Dynamic 120 Problem)
```

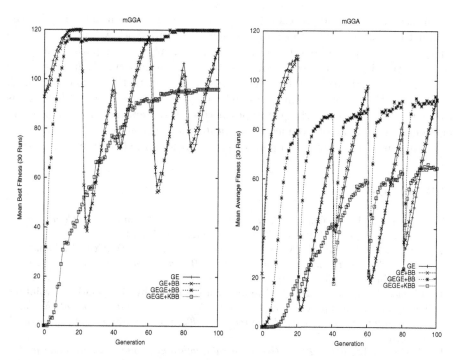

Fig. 4. Plot of the the mean best fitness (left) and mean average fitness (right) for the Dynamic 120 bit problem

```
<bitstring> ::= <bbk8>..<bbk8>
              | <bbk24><bbk24><bbk24><bbk24><bbk24>
              | <bbk5>..<bbk5>
              | <bbk1>..<bbk1>
<bbk24>    ::=  <bit><bit>1110<bit>10110<bit>0010<bit>0<bit><bit>1<bit><bit>
<bbk8>     ::=  0<bit>0<bit><bit>1<bit><bit>
<bbk5>     ::=  0<bit>0<bit><bit>
<bbk1>     ::=  1
<bit>      ::=  0
solution:  000000000000000000000000000000000000000000000000000000000000
           000000000000000000000000000000000000000000000000000000000000
```

5   Conclusions and Future Work

We presented the meta-Grammar Genetic Algorithm (mGGA), illustrating the application of evolvable grammars to Genetic Algorithms. On the three problem domains examined there are clear performance advantages on both the two static problems and the dynamic problem instance for the evolvable grammars GEGE+BB and GEGE+KBB over their static counterparts GE and GE+BB.

In addition to the application to more benchmark problem instances in particular to those belonging to the dynamic class, future work will investigate the effects of the wrapping operator, alternative grammars and comparisons to other Genetic Algorithms from the literature. It would be particularly interesting to analyse the scalability of these algorithms compared to the competent GA's, given that the use of wrapping coupled with the reuse of building blocks has the potential to shorten the genotypes necessary to represent harder problem instances. A number of avenues to facilitate the co-evolution of the grammar and solution such as different operator probabilities will also be investigated.

References

1. Koza, J. R. (1992). *Genetic Programming: On the Programming of Computers by Means of Natural Selection.* MIT Press, Cambridge, MA, USA.
2. Koza, J. R., Keane, M. A., Streeter, M. J., Mydlowec, W., Yu, J., Lanza, G. (2003). *Genetic Programming IV: Routine Human-Competitive Machine Intelligence.* Kluwer Academic Publishers.
3. O'Neill, M., Ryan, C. (2004). Grammatical Evolution by Grammatical Evolution. The Evolution of Grammar and Genetic Code. *LNCS 3003. Proc. of the European Conference on Genetic Programming 2004*, pp. 138-149, Coimbra, Portugal. Springer.
4. O'Neill, M., Ryan, C. (2003). *Grammatical Evolution: Evolutionary Automatic Programming in an Arbitrary Language.* Kluwer Academic Publishers.
5. O'Neill, M. (2001). Automatic Programming in an Arbitrary Language: Evolving Programs in Grammatical Evolution. PhD thesis, University of Limerick.
6. O'Neill, M., Ryan, C. (2001). Grammatical Evolution, *IEEE Trans. Evolutionary Computation.* 2001.
7. Ryan, C., Collins, J.J., O'Neill, M. (1998). Grammatical Evolution: Evolving Programs for an Arbitrary Language.*Proc. of the First European Workshop on GP*, pp. 83-95, Springer-Verlag.
8. Dempsey, I., O'Neill, M., Brabazon, A. (2004). Grammatical Constant Creation. In LNCS 3103, *Proceedings of GECCO 2004*, Part 1, pp. 447-458, Seattle, WA, USA.
9. O'Neill, M., Cleary, R. (2004). Solving Knapsack Problems with Attribute Grammars. In *Proceedings of the Grammatical Evolution Workshop 2004*, GECCO 2004, Seattle, WA, USA.
10. Shan, Y., McKay, R. I., Baxter, R., Abbas, H., Essam, D., Nguyen, H.X. (2004). Grammar Model-based Program Evolution. In *Proceedings of the 2004 Congress on Evolutionary Computation, CEC 2004*, Vol. 1, pp. 478-485, Portland, Oregan, USA.
11. Chomsky, N. (1975). *Reflections on Language.* Pantheon Books. New York.
12. Pinker, S. (1995). *The language instinct: the new science of language and the mind.* Penguin, 1995.
13. Rothlauf, F. (2002). *Representations for Genetic and Evolutionary Algorithms.* Physica-Verlag, 2002.

On Prediction of Epileptic Seizures by Computing Multiple Genetic Programming Artificial Features

Hiram Firpi[1], Erik Goodman[1], and Javier Echauz[2]

[1] Department of Electrical and Computer Engineering, Michigan State University,
East Lansing, Michigan 48824-1226, USA
{firpihir, goodman}@egr.msu.edu
[2] BioQuantix Corp.,
Atlanta, Georgia, USA
echauz@ieee.org

Abstract. In this paper, we present a general-purpose, systematic algorithm, consisting of a genetic programming module and a k-nearest neighbor classifier, to automatically create multiple *artificial features* (*i.e.*, features that are computer-crafted and may not have a known physical meaning) directly from EEG signals, in a process that reveals patterns predictive of epileptic seizures. The algorithm was evaluated in three patients, with prediction defined over a horizon that varies between 1 and 5 minutes before unequivocal electrographic onset of seizure. For one patient, a perfect classification was achieved. For the other two patients, high classification accuracy was reached, predicting three seizures out of four for one, and eleven seizures out of fifteen for the other. For the latter, also, only one normal (non-seizure) signal was misclassified. These results compare favorably with other prediction approaches for patients from the same population.

1 Introduction

Since the invention of the electroencephalograph, great progress has been made in studying many brain disorders. One of the most puzzling disorders is epilepsy, a neurological condition that makes people susceptible to brief electrical disturbance in the brain that produces a change in sensation, awareness, and/or behavior; epilepsy is characterized by recurrent seizures. It affects up to 1% of the world's population, or 60 million people and 25% of patients cannot be fully controlled by current medical or surgical treatment.

Many approaches have been proposed to extract information from EEG signals in order to develop algorithms to predict or detect epileptic seizures [1], [4], [6], [9], [11], [15]. To extract the relevant information that can facilitate such prediction or detection, features are calculated using conventional techniques and methodologies that are time-consuming, trial-and-error processes requiring a great deal of effort from researchers. All of these conventional techniques rely on knowledge of a feature formula or algorithm that may have been obtained from intuition, tradition, the physics of the problem, analogies to problems in other fields, etc. There is no guarantee that any of these conventional features extracts maximally relevant information from the raw data. Nevertheless, seizure prediction studies using

M. Keijzer et al. (Eds.): EuroGP 2005, LNCS 3447, pp. 321–330, 2005.

conventional features increasingly hint at the fact that the information is there, waiting to be fully extracted. Litt *et al.*, in [8], presents evidence that mesial temporal lobe seizures are generated in a series of events that evolve over hours, leading to the clinical seizure onset. This series of events can be recorded by depth intracranial EEG. However, this work was conducted by scoring, manually, many hours of EEGs and the detection was largely limited to what a basic energy feature could reveal about preictal changes. More recently, it has been shown that complicated feature calculations can be realized in miniaturized hardware using a cellular neural network approximation of the feature on a chip [10], [13]. Therefore, the present work seeks to develop an algorithm that can systematically and automatically find features or patterns starting from raw data—in this case, EEG signals. Additionally, this algorithm is intended to address the shortcomings of Genetically, Found Neurally Computed Artificial Features, an algorithm previously proposed in [4].

2 Methodology

We attempt to capture a "pocket" of deterministic dynamics of EEG signals by means of delay-embedding in a stream of sliding windows. First we reconstruct the state-space trajectory of the EEG signal using the standard delay-embedding scheme [14]. Later, this reconstructed trajectory is input into a genetic programming algorithm, which attempts to find a pattern(s) giving the best discrimination between baseline data (nonseizure) and preictal (pre-seizure) data, in the sense of a minimum-error-risk objective function. A universal classifier then performs the categorization task. Fig. 1 is a diagram showing the components of the algorithm.

2.1 State Reconstruction by Means of Delay Embedding

Many authors have applied nonlinear dynamics tools to analyze EEG data. Chaos theory [12] states that within a system displaying apparently disordered random-like data, an underlying order exists. Because of this, one may conjecture that even if precise long-term prediction is impossible, prediction in the short term and with some error allowance may be possible in many systems. Such a property allows us to reconstruct the state-space trajectory of an attractor of the system (in this case, a model of the brain that generates the EEG signals).

We can reconstruct the deterministic component of the state trajectory of the EEG signals by taking previous samples of the observable output and creating an artificial state vector with n_e elements, which we denote λ_k (input to the program in Fig. 1), the embedding vector. This process creates a diffeomorphism (a smoothly distorted copy of the original trajectory) function that preserves dynamic and geometric qualities of the trajectory of the EEG system.

Given such state-space trajectory reconstruction, our approach is to input these pseudo-state vectors (*i.e.*, λ_k), evolving in time, to the genetic programming (GP) module, and by means of the algorithm, to find a transformation, usually nonlinear (although a linear transformation is also possible), that achieves the maximum separability between baseline and preictal data. In other words, the GP algorithm combines the inputs (states) in a (non)linear way and outputs a function that separates

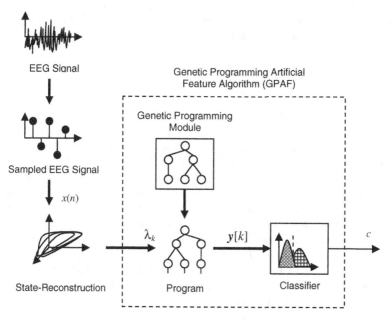

Fig. 1. State-space reconstruction and the components (*dashed box*) of the genetic programming artificial features algorithm

the baseline and preictal classes such that the performance of the classifier is better than or equivalent to categorizing the classes with no transformation or with a benchmark previously defined. To capture the above mathematically, let

$$y[k] = \phi(\lambda),$$ (1)

where ϕ is a transformation function designed by the GP algorithm, and

$$\lambda = \{x(n), x(n-\tau), x(n-2\tau), \cdots, x(n-(n_e-1)\tau)\}$$ (2)

is the set of delayed samples (from a chaos theory perspective, there are the state variables that governs the dynamics of the brain) that the GP will use in any combination to construct the artificial features. The parameter τ is the delay, which will be determined using the autocorrelation function, and n_e is the embedding dimension, i.e., the number of delays (or inputs) in the terminal set. In this work, the embedding dimension is arbitrarily selected to be 6.

As with any conventional feature, such as Fourier transform coefficients, signal energy, etc., in the genetic programming artificial feature (GPAF) algorithm, we need to process the data as viewed through a sliding observation window. The observable window is defined by two parameters: length of the window, denoted L, and displacement, denoted D. The length L defines the number of points that will be evaluated (or analyzed) at one time. On the other hand, D is defined as the number of new points that will be used in the next evaluation or $D = L - O$, where O is the overlap with the previous window. To compress the data, a summation operator is applied in each window position. Recalling (2) and adding to it the sliding window, *i.e.*, a summation operator, the resultant equation is stated in (3).

$$y_i[k] = \sum_{n=1+(k-1)(L-D)}^{k(L-D)+D} \phi\big(x(n), x(n-\tau), ..., x(n-(n_e-1)\tau)\big) \qquad (3)$$

The $y_i[k]$ is the GP artificial feature of the EEG signal $x(t)$, here called the *artificial feature time series*, the subscript i denotes the number of the feature, and k is the discrete time unit of such time-series. The summation operator observes all the points in the interval n delimited by $1 + (k-1)(L-D) \leq n \leq k(L-D)+1$.

The sliding window can be viewed as a reducer of dimensionality of data, *i.e.*, the number of points that an epoch contains. That is, at each window position k, the number of points, L, will be reduced, by means of (3), to a single point at the output. Thus, if we have a signal that contains P_{Tot} points and we set displacement to D, the number of output points after the entire signal is processed by each GP artificial feature is $\kappa = P_{Tot}/D - 1 - L/D$, where the term $1 - L/D$ is an adjustment for the beginning of the signal where the window of length L is not complete and, thus, some points are not calculated. Thus, the discrete time index of the artificial times-series k goes from 0 to $\kappa - 1$ for each feature.

2.2 Genetic Programming

Table 1 shows the function set or building blocks used, which the GP algorithm will use to construct the programs (artificial features), in this work. The maximum tree depth was set to 15 and the population size to 500. The initial population was generated using the ramped half-and-half method. The GP used a crossover operator with a rate of 0.9, fitness proportional selection, and a breeding operator with a rate of 0.1. For this work, the GP algorithm was extended to allow it to evaluate not just one tree per individual, but multiple trees per individual (*i.e.*, a forest), allowing us to generate multiple features. All the trees of the individual were evaluated at the same time; however, there was no crossover among trees (features) of the same individual—crossover was done only among homologous trees.

Table 1. Genetic Programming function set. The mathematical operators used by the GP to build the artificial features

+	−	÷	×
cos	sin	$\log_2$	$\log_{10}$
ln	$\sqrt{|\ |}$	$(\ )^2$	abs

2.3 Objective Function

In this work, we desire that the GP algorithm create a set of artificial features that have the minimum possible error probability. Thus, the metric to be optimized is selected here to be the error risk [5], which assigns risk factors $r > 0$ to the errors, so that their relative costs can be accounted for as

$$R_E = P_{FN} P(S_T) r_{FN} + P_{FP} P(NS_T) r_{FP} \ , \tag{4}$$

where P_{FN} is the probability of false negatives (*i.e.*, when the algorithm misses a seizure), P_{FP} is the probability of false positives, and $P(S_T)$ and $P(NS_T)$ are the prior probabilities of the respective classes (preseizure and baseline). Additionally, r_{FN} is a risk (or cost) factor associated with missing preseizures, and r_{FP} is a risk factor associated with declaring false positives. From experience, we selected $r_{FN} = 0.75$ and $r_{FP} = 0.25$. Thus, the resultant objective function is

$$R_E = (0.75) P_{FN} + (0.25) P_{FP} \ . \tag{5}$$

2.4 Classifier

Although any classifier may be used in the GPAF algorithm, in this work we select the classifier with one primary aspect in mind: applicability in a general-purpose algorithm that automatically creates artificial features from raw data (or conventional features if those are the starting-point data). Therefore, we selected the k-nearest neighbor classifier (k-NN) as a classifier component for the GPAF algorithm (last stage in Fig. 1). This classifier is nonparametric, nonlinear, and capable of producing multiple thresholds or complicated decision boundaries, making it suitable for n-dimensional, multi-modal problems. In addition, the training process is relatively easy, simply involving giving to the classifier all of the training data.

A split-sample method was used in the classification phase in order to train and test using different subsets of data. The experiments were conducted using a point basis, where each point in the artificial feature epochs counts as one example, i.e., each point κ is an example for the classifier. However, statistics reported on a point basis are impractical and not easy to interpret. For instance, if an implantable device (with GPAF features integrated) classifies each t seconds over an incoming EEG signal, the device cannot drug the patient based on the decision that the classifier makes each t seconds. The device needs a longer, fixed-length window, so that it can observe the past evolution of the point-basis classification during a defined period to decide whether a patient will suffer a seizure. Use of this fixed-length window is what we called block-basis classification. Therefore, for experimental purposes, a whole artificial feature epoch on the point basis counts as one example in the block basis. In the experiments carried out in this work, results for testing data will be reported using the block basis. To transform results from the point basis to the block basis, a threshold will be set. If the point-basis classification result (*i.e.*, the classification result of a testing epoch evaluated in the classifier) is equal to or exceeds the threshold, the epoch will be labeled preictal; otherwise, it will be categorized as baseline (nonseizure). In other words, if the threshold is a large number, we are increasing the probability of missing seizures, whereas if we set the threshold to be a small number, we are making the device too sensitive—that is, we would have many false positives.

2.5 EEG Data

The anonymized EEG data used for these experiments were obtained from a tripartite database from Georgia Institute of Technology, Emory University, and the University of Pennsylvania. The EEG data were recorded from epileptic patients undergoing a pre-surgical evaluation. Patients were simultaneously videotaped during their hospital stays, which varied in duration from 4 to 11 days. The EEG signals were recorded at 200Hz, with 12-bit resolution.

3 Results

In the experiments in this section, all baseline and preictal epochs were 10 min long, thus containing 120,000 points each. The length of the sliding window was set to $L =$ 800 (4 s) and $D = 400$ (2 s). Therefore, the number of points κ, after an epoch is evaluated by an artificial feature, is 299 points per epoch. Preictal segments were selected by default to end 5 min before the unequivocal electrographic onset (UEO) of seizure occurred. As stated before, the embedding dimension was set to $m = 6$. Euclidean distance was selected as the metric and number of nearest neighbors for the k-NN classifier was set to $k = 5$, a value commonly used. The training set for each patient was one baseline and one preictal epoch, which were randomly selected from the sets of available data.

The threshold to change from point basis to block basis was set to 65% in order to increase the confidence of the decision beyond the noise floor that a coin flipper at 50% would create. Put in other words, if the point basis classification (*i.e.*, the results from the classifier after evaluation) for an epoch is 65% or greater, the epoch will be labeled preictal. Otherwise, it will be classified baseline. We can see it in another way; if the point basis classification for an epoch is greater than 35%, the epoch will be classified as a baseline epoch. Otherwise, it will be labeled a preictal epoch.

3.1 Patient A

Patient A was diagnosed as having all seizures coming from the left anterior hippocampus. For this patient, we have 7 baseline epochs and 5 preictal epochs, of which one epoch of each set was used for the training stage. The preictal segments were taken in the 10-minute period ending 5 min before the UEO (so that a positive prediction might allow time for intervention). The delay was set to $\tau = 30$ (0.15 s), selected from the first zero-crossing of the autocorrelation plot. Thus, the terminal set (the pseudo state-space vector) for this patient is: $\{x(n), x(n\text{-}30), x(n\text{-}60), x(n\text{-}90), x(n\text{-}120), x(n\text{-}150)\}$.

Equation (8) shows the artificial features found by the GPAF algorithm. Table 2 shows the results obtained from the training data (because the training set was limited to one epoch per class, performance results could only be reported on the point basis).

All baseline epochs were predicted correctly whereas 3 seizures out of 4 were predicted correctly.

$$y_1[k] = \sum_{n=1+400(k-1)}^{400k+400} x(n)(x(n-30))^2$$

$$(8)$$

$$y_2[k] = \sum_{n=1+400(k-1)}^{400k+400} \ln(x(n)) - x(n-90) + (x(n-150))^2$$

3.2 Patient B

Seizures for this patient were found to be multifocal, coming from the left hippocampus and left anterior temporal neocortex. We have 16 baseline epochs and 16 preictal epochs available for the training and validation procedure. The prediction horizon was set to 5 minutes before the UEO occurred. The delay time was computed as $\tau = 49$ (0.245 s). The terminal set for this patient is: $\{x(n), x(n\text{-}49), x(n\text{-}98), x(n\text{-}147), x(n\text{-}196), x(n\text{-}245)\}$. Equation (9) shows the artificial feature found by the GPAF algorithm for this patient. Table 2 depicts the results obtained from the training data. Table 3 shows the results for the validation data. When testing data were evaluated, 14 baseline epochs out of 15 were classified correctly and 11 seizures out of 15 were predicted (*i.e.*, 4 seizures missed).

$$y_1[k] = \sum_{n=1+400(k-1)}^{400k+1} \log_2\left(\frac{x(n-147)}{x(n)}\right)$$

$$(9)$$

$$y_2[k] = \sum_{n=1+400(k-1)}^{400k+1} (\log_{10}(x(n)))^2$$

3.3 Patient C

Seizures of patient C were diagnosed as coming from the left anterior hippocampus region. The delay time obtained for this patient was $\tau = 32$ (0.16 s). Thus, the terminal set was: $\{x(n), x(n\text{-}32), x(n\text{-}64), x(n\text{-}96), x(n\text{-}128), x(n\text{-}160)\}$. This patient had 4 baseline epochs and 4 preictal epochs. The preictal segments were initially selected as ending 5 min before the UEO occurred; however, results obtained were not good. To attempt to get better performance, we moved the prediction horizon to extend until 1 minute before UEO. After this was done, a successful set of features was obtained. Equation (10) displays the three artificial features found by the GPAF algorithm.

$$y_1[k] = \sum_{n=1+400(k-1)}^{400k+1} \log_{10}(\log_{10}(x(n-32)))$$

$$(10)$$

$$y_2[k] = \sum_{n=1+400(k-1)}^{400k+1} \frac{\log_{10}(x(n-160))}{x(n-96)}$$

$$y_3[k] = \sum_{n=1+400(k-1)}^{400k+1} (x(n-128))^4$$

Table 2 shows the results obtained from the training data. Table 3 depicts the results for the validation data. It can be noted that the features provide a transformation such that the classifier achieves a perfect classification (recalling that the threshold to convert from the point- to block-basis was set to 65%).

Table 2. Performance over the training set for each patient. Performance results for training data are presented on the point basis

Patient	A	B	C
Baseline	88.59	85.23	87.87
Preictal	96.98	94.3	67.1
Average	92.79	89.77	77.44

Table 3. Performance over the validation set for each patient. Performance results for validation data are presented on the block basis, with the threshold is set to 65%

Patient	Baseline	Preictal	FPH[1]
A	100.0	75.0	0.0
B	93.33	73.33	0.4
C	100.0	100.0	0.0
Average	97.78	82.78	0.133

4 Discussion

An average of 82.78% seizure prediction was achieved. Overall, 17 of 22 seizures were predicted. On the other hand, 97.78%, or 23 baseline segments out of 24, were correctly classified as baseline (non-seizure) epochs. The overall average rate of false positives obtained was 0.133 per hour. Although we do not present in full a benchmark method to compare the performance obtained from GPAF algorithm, we recall previous work authored by D'Alessandro et al. [3]. In this work, the authors

Table 4. Results for the proposed systems. This table shows the validation results for the different systems proposed

System	Number Patients	Baseline Tested	Preictal Tested	%B$_C$	%PSz$_C$	FPH
D'Alessandro	4	276	46	90.47	62.50	0.278
GPAF	3	24	22	97.98	82.72	0.133

used a genetic algorithm to find the "best" set of features for each patient. The set of features was composed by handcrafted features as energy of the signal, nonlinear energy, curvelength, spectral entropy, and energy of wavelet packets. Another set of more elemental features was also used. The GA selected a combination of features such that when EEG data were processed by those features, the classifier, a probabilistic neural network, could discriminate correctly between baseline and preictal epochs. For

---

[1] FPH = False positives per hour is the number of false alarms per hour, *i.e.*, when the algorithm warns an attack will occur but it does not.

that work, the authors used the same EEG database that we did. Table 4 shows details and results of D'Alessandro's work and the results of this work for comparative purposes. The table contains number of patients, number of baseline tested epochs (10 min duration epochs), number of preictal tested epochs, rate of baseline epochs classified correctly (%B$_C$), rate of preictal epochs classified correctly (%PSz$_C$), and average of false positive per hour. Even though is not a comparison on completely identical data, because training and validation sets are not the same, it gives us an idea how we are doing with the current methods. It does not demonstrate that our algorithm is outperforming D'Alessandro's approach, but simply indicates that the GPAF algorithm produced promising results that are worthy of further research.

5 Conclusion

It is clear that the artificial feature formulas for each patient are relatively simple, but are far from trivial and are not likely to come from intuition or knowledge of the physics of the problem. The artificial features designed by the GPAF algorithm are "optimized" for each particular patient based only on their available raw EEG recordings. By construction, these features match or exceed the performance of traditional conventional features and are thus a viable subject for further improvement and research. For example, in future work, we intend to perform leave-one-out training and validation, which is expected to improve significantly on the performance obtained.

This work also presents evidence that a unique prediction horizon for all patients may not be practical to determine. Good artificial features seem to vary from patient to patient, as do also the prediction horizons. EEG signals in some patients present more obvious abnormalities before the seizure than those in other patients. Additionally, we need to consider that the feature search was limited by the number of generations, in this case between 100 and 150, because of the time/computational expense required to execute the algorithm.

References

1. Chen, H., Zhong, S., Yao, D.: Detection Singularity Value of Character Wave in Epileptic EEG by Wavelet. IEEE International Conference on Communications, Circuits and Systems and West Sino Expositions 2002, Vol. 2. (2002) 1094-1097
2. D'Alessandro, M.: The Utility of Intracranial EEG Feature and Channel Synergy for Evaluating the Spatial and Temporal Behavior of Seizure Precursors. Ph.D. Dissertation. Georgia Institute of Technology, Atlanta (2001)
3. M. D'Alessandro, R. Esteller, G. Vachtsevanos, A. Hinson, J. Echauz, and B. Litt.: Epileptic seizures prediction using hybrid feature selection over multiple intracranial EEG electrode contacts: a report of four patients. IEEE Transactions on Biomedical Engineering, Vol. 50, No. 5. (2003) 603-615
4. Esteller, R., Vachtsevanos, G., Echauz, J., D'Alessandro, M., Bowen, C., Shor, R., Litt, B.: Fractal Dimension Detects Seizures Onset in Mesial Temporal Lobe Epilepsy. Proceedings of the First Joint BMES/EMBS Conference: Serving Humanity, Advancing Technology. (1999) 442

5. Firpi, H.: Genetically Found, Neurally Computed Artificial Features with Applications to Epileptic Seizure Detection and Prediction. Master's Thesis. University of Puerto Rico-Mayagüez, Mayagüez, PR. (2001)
6. Iasemidis, L., Savit, R., Sackellares, J.: The Use of Dynamical Analysis of EEG Frequency Content in Seizure Prediction. Proceeding American Electroencephalographic Annual Conference. New Orleans, LA. (1993) abstract
7. Iasemidis, L., Shiau, D.-S., Chaovalitwongse, W., Sackellares, J.C., Pardalos, P.M., Principe, J.C., Carney, P.R., Prasad, A.P., Veeramani, B., Tsakalis, K. : Adaptive Epileptic Seizure Prediction System. IEEE Transactions on Biomedical Engineering, Vol. 50, No. 5, (2003)
8. Litt, B., Esteller, R., Echauz, J., D'Alessandro, M., Shor, R., Henry, T., Pennell, P., Epstein, C., Bakay, R., Dichter, M., Vachtsevanos, G.: Epileptic Seizures May Begin Hours in Advance of Clinical Seizures: A report of five patients. Neuron. Vol. 29, No. 4. (2001)
9. Mirbagheri, M.M., Badie, K., Hashemi, R.M., Golpayegani, Amir Ahmadi, M.: A Neural Network Approach to EEG Classification for the Propose of Differential Diagnosis Between Epilepsy and Normal EEG States. Proceedings of the Annual International Conference of the IEEE Engineering in Medicine and Biology Society, Vol. 14. (1992) 2649-2650
10. Niederhoefer, C., Gollas, F., Chernihovskyi, A., Lehnertz, K., Tetzlaff, R.: Detection of Seizure Precursors in the EEG with Cellular Neural Networks. Epilepsia, Vol. 45, supplement 7. (2004) 245 (abstract)
11. Szilágyi, L., Benyó, Z., Szilágyi, S.M.: A New Method for Epileptic Waveform Recognition Using Wavelets Decomposition and Artificial Neural Networks. Proceedings of the Second Joint EMBS/BMES Conference. (2002) 2025-2026
12. Sprott, J.C.: Chaos and Time-Series Analysis. Oxford University Press, New York (2003)
13. Sowa, R., Mormann, F., Chernihovskyi, A., Niederhoefer, C., Tetzlaff, R., Elger, C., Lehnertz, K.: Seizure Prediction: Measuring EEG Phase Synchronization with Cellular Neural Networks. Epilepsia, Vol. 45, supplementary 7. (2004) 244 (abstract)
14. Takens, F.: Detecting Strange Attractors in Turbulence. Dynamical Systems and Turbulence, Warwick 1980. Lecture Notes in Mathematics 898. Springer-Verlag Berlin, Germany (1981) 336-381
15. Tetzlaff, R., Niederhofer, C., Fischer, P.: Feature Extraction in Epilepsy Using a Cellular Neural Network Based Device: First Results. Proceedings of the International Symposium on Circuit and Systems 2003, Vol. 3. (2003) 850-853

Relative Fitness and Absolute Fitness for Co-evolutionary Systems

Nanlin Jin and Edward Tsang

Department of Computer Science, University of Essex,
Colchester CO4 3SQ, United Kingdom
{njin, edward}@essex.ac.uk

Abstract. The commonly adopted fitness which evaluates the performance of individuals in co-evolutionary systems is the relative fitness. The relative fitness measure is a dynamic assessment subject to co-evolving population(s). Researchers apparently pay little attention to the use of absolute fitness functions in studying co-evolutionary algorithms. The first aim of this work is to define both the relative fitness and the absolute fitness for co-evolving systems. Another aim is to demonstrate the usage of the absolute and relative fitness through two case studies. One is for the Iterated Prisoners' Dilemma. Another case is for solving the Basic Alternating-Offers Bargaining Problem, for which a co-evolutionary system has been developed by means of Genetic Programming. Experiments using the relative fitness function have discovered co-adapted strategies that converge to nearly game-theoretic solutions. This finding suggests that the relative fitness essentially drives co-evolution to perfect equilibrium. On the other hand, the absolute fitness measuring co-evolving populations monitors the development of co-adaptation. Having analyzed the micro-behavior of the players' strategies based on their absolute fitness, we can explain how co-evolving populations stabilize at the perfect equilibrium.

1 Introduction

The objective of this study is to analyze co-evolutionary systems through using two types of fitness functions: a relative fitness function and the chosen absolute fitness functions.

The original concept of co-evolution comes from nature. Biologists observe that, in nature, one species modifies itself to adapt to the changes from its co-existing species in their shared physical surrounding. Such modifications, in turn cause its co-evolving species to change themselves accordingly. This sort of reciprocal evolutionary changes in interacting species is known as *co-evolution* in biology.

Computer scientists, inspired by this natural phenomenon, create co-evolutionary algorithms that have achieved considerable success in solving a wide range of problems. Holland [6] initiates co-evolution for artificial ecology systems; Miller [9] uses co-evolution to study Iterated Prisoners' Dilemma; Koza

M. Keijzer et al. (Eds.): EuroGP 2005, LNCS 3447, pp. 331–340, 2005.

[7] introduces the term of *relative fitness* and employs co-evolution to disclose minimax strategies for a two-player finite extensive-form game. Co-evolution is also successful in other fields: factory organization, robotics, predator-prey systems, sorting networks and social sciences.

Most applications of evolutionary algorithms (EA) have clear objectives, which can be mathematically expressed, without interference with other evolving objects. In contrast, to most applications of the co-evolution, such absolute fitness functions usually are unavailable. In the case of two-player games, for example, the performance of a player's strategy depends on the strategy his opponent uses. The opponents strategy set is generally huge in size. Therefore, any strategy cannot be taken as the best response without knowing the opponent's behavior. Because of the difficulty of discovering underlying objectives for some problems, researchers turn to use co-evolution to solve these problems for which reliable and precise objectives are unknown, whereas a kind of "reciprocal" relationship between (among) species (players in games), can be assumed. Applications have shown that co-evolution using the relative evaluation makes it possible to solve a class of problems.

Recent researches disclose that under certain conditions, the absolute fitness may provide the same information as the relative fitness. de Jong and Pollack [5] have proved that by identifying a complete evaluation set, an algorithm named DELPHI is able to generate tests that probably have the same results as testing against co-evolving population(s). One purpose of introducing this ideal evaluation is to avoid inaccuracy in co-evolutionary algorithms. Luke and Wiegand [8] argue the possibility of existing an objective measure that may make EA exhibit similar dynamics and generates similar results to a single-population co-evolution. Our work tries to analyze the absolute fitness and relative fitness, and furthermore to understand the behaviors of co-evolving strategies and their incremental learning.

In the following sections we first present formal definitions of the relative fitness and the absolute fitness. Section 3, reviews literature of two studies on Iterated Prisoner's dilemma and compares the different results generated by the absolute fitness and by the relative fitness. In section 4, we introduce an Alternating-Offers Bargaining Problem, describe the design of a co-evolutionary system for solving this bargaining problem and present experimental results by using a relative fitness function. Section 5 summarizes observations from experiments using two absolute fitness functions. Conclusions will be drawn in section 6.

2   Relative and Absolute Fitness in Co-evolution

In terms of co-evolutionary algorithms, the relative fitness [7] (also called subjective fitness), measures how fit co-evolving species (players in games) are to each other, and is usually the straightforward option for applications of the co-evolutionary algorithms. Relative evaluation functions are dynamic, updating over evolving time. In other words, individuals in different generations are evaluated by probably different functions. In contrast, the absolute fitness [7] (objective fitness) evaluates the fitness upon static targets, similar to the objective fitness in evolutionary algorithms. Using the absolute fitness is implicit in applications of conventional evolutionary algorithms where the objective function is assumed to be identical over all generations. In co-evolution, the absolute fitness can behave as a monitor rather than a part of co-evolving process. It provides information concerning the co-adaptive process.

To these two concepts, "the classic analogy is the co-evolutionary arms race (in nature): a plant has chemical defenses and an insect evolves the biochemistry to detoxify these compounds. The plant in turn evolves new defenses that the insect in turn 'needs to further detoxify" [12]. The (relative) fitness of the insect depends on the evolutionary state of the plant, so the relative fitness directs the insect to adjust its behaviors to detoxify the plant's *current* chemical defenses. From the insect's relative fitness, we know the situation of co-adaptation of these two species at certain time, in other words, how fit the plant to the insect. To discover how the insect adapts to the plant progressively, it is necessary to investigate the insect' biochemistry (the insect's absolute fitness) at every stages of the co-evolution.

We formalize the above two concepts. Suppose in a simple co-evolving system, two species exist in a stable physical environment. The two populations (P, P') are the sets of individuals of these two species which are simultaneously co-evolving over time in terms of the generation g, a positive integer. Assume these two species start evolving at the same time and spend exactly same amount of time per generation. The pair of populations at the generation g is $(P(g), P'(g))$ which means co-evolving populations are functions of the generation.

In this co-evolutionary system the Relative Fitness Function r of an individual $x \in P$ is a function of x and of the state of the other population P': $r(x, P')$. If the generation is specified, we get: $r(x, P'(g))$. It simply tells that the individual x, has a relative fitness implicitly depending on the evolving time g because r is a function of P' that is changing over time. Unlike the relative fitness function r, the absolute fitness function f evaluates an individual x in P not in connection with the co-evolving P' but upon a static objective $f(x)$ independent of the time g.

Note that in general it is relative fitness that motivates the co-evolutionary improvement, (to detoxify the plant's present chemical defenses is what the insect tries to do now), but the relative one records the co-evolving history (at different evolving times, the insect has different biochemistry materials). It is impossible to transform a relative fitness into an absolute one or vice versa, because the relative fitness is always a function of time and generation-dependent, but the absolute fitness is not. Only if the time is frozen at a certain moment g, can the relative fitness at that time g be also interpreted the absolute fitness at g.

3   Studies of Iterated Prisoners' Dilemma

Two studies on a well-known controversy of the Iterated Prisoners' Dilemma (IPD) are good expositions of how an absolute fitness function and a relative fitness function can generate different results to the same problem. Axelrod's GA experiments [1] evolves player's strategies against a fix environment, "eight representatives". The set of these eight chosen representatives is an absolute fitness evaluator that is independent of the evolving time. His GA experimental results support the claim that TIT FOR TAT and its variants are the best responses to IPD problem.

However, this claim is questioned by the results from co-evolutionary experiments where a relative fitness function is in use. Darwen and Yao [3] follow the strategy representation by Axelrod's [1] , but a strategy fitness is the scores it achieves from playing against all the other strategies in its population. This setting is a typical one-population co-evolution with the use of a relative fitness measure. In the end of their experiments [3], "only cooperative strategies survive", however some behavioral patterns claimed to be parts of TIT FOR TAT and its variants in [1] have not been reported: "Be provocable" and "accept a rut" which play defections.

4   A Co-evolutionary System for the Bargaining Problem

We thereafter exemplify the uses of the absolute fitness and relative fitness in another case study on a bargaining problem.

4.1   Alternating-Offers Bargaining Problem

A classic bargaining problem modeled and solved by Rubinstein [13] is named as the *Basic Alternating-Offers Bargaining Problem*. Its bargaining scenario describes as two players making proposals on dividing a cake in an alternating manner. At any given time when one makes an offer, the other one can either accept thus the game ends with agreement, or reject then the game continues and the player who rejected the previous offer makes a new proposal. The player i's bargaining cost is expressed by a discount factor δ_i, which means his partition from a cake with the size of $\pi = 1$, shrinking to δ_i^t at the time t, a non-negative integer.

In terms of game theory, this bargaining problem is an infinite extensive-form game with complete and perfect information. All combinations of players' behaviors are infinite. Game theorists intelligently reduce the solution space by imposing strict assumptions on players' rationality, which players have all relevant knowledge, well-defined and stable utility functions and full computational capacity [11]. Technical treatments and proofs are available in [13], and [2]. The unique Equilibrium taken as the formula solution of this game is *Perfect Equilibrium Partition* (P.E.P) in which the first player obtains:

$$\pi_1^* = \frac{1 - \delta_2}{1 - \delta_1 \delta_2} \tag{1}$$

4.2   A Co-evolutionary System and Experimental Outcomes

We have applied co-evolution to solving this bargaining problem, which has been described by [4] in great detail [1].

We hypothesize there are reciprocal interactions between players' behaviors, and players learn through trial-and-error experiences. We implement a co-evolution system for this problem by Genetic Programming (GP). There are two populations, one for the player A and another for the player B. Each population consists of 100 candidate solutions: strategies. A strategy decides its player's behaviors on all possible contingencies. The function-oriented representation of a time-dependent strategy of the player i is $s_i = g_i \times (1 - r_i)^t$ where t is the bargaining time and r_i is the i's discount rate, $\delta_i \equiv e^{-r_i}$. A genetic program, a function g_i is the part of a strategy to be co-evolving, consisting of all or some elements in the primitive set of $\{1, -1, \delta_i, \delta_j, +, -, \times$ and $\div$ (protected) $\}$.

The set (population) S_x have m candidate strategies, of the player x. The payoff of i ($i \in S_x$) from an agreement with $j \in S_y$ is denoted as $p_{i \rightarrow j}$. S_y has n strategies. In this case, $n = m$. The fitness function of i is:

$$f_i = \frac{\sum_{j \in S_y} p_{i \rightarrow j}}{n} \tag{2}$$

f_i returns the average payoff a strategy received from agreements with either the co-evolving opponent or the monitor. When S_y is a co-evolving set, f_i calculates the *relative fitness* of i at the co-evolving time when S_y emerges. If S_y is static, f_i returns i's *absolute fitness*.

This system performs relatively stable while the crossover rate is within 0 to 0.1 and the mutation rate ranges from 0.01 to 0.5. In a typical run shown in Fig. 1, the behaviors of two players tend stable before the 200th generation. Fig. 1 illustrates the GP programs' values, shares π_A from the agreements and payoffs of the best-of-generation strategies of the first player A. To terminate runs at the time of the 300th generation is practically efficient to stabilize fitness.

The table 1 displays the experimental outcomes from using the relative fitness function. A strategy's fitness at the generation g is determined by its actual payoff from bargaining against his opponent's strategies at the corresponding time. In the Table 1, the average shares of π_As from agreements are shown, as the π_Bs are that of complement. The π_As in final agreements converge to the neighborhood of the theoretical prediction P.E.P. From a regression analysis on the results in table 1, the R-square value is 0.9928, which is very close to 1. Its coefficient variable is 0.9588 and coefficient intercept is 0.0257. These tell how much of P.E.P is approximated by experimental results. In our case, the

---

[1] [4] includes the motivations, alternative assumptions of the EA method to the game-theoretic assumptions, together with the experimental setting-up, outcomes and conclusions. To make this paper self-contained, we outline critical information to permit replication and present major results.

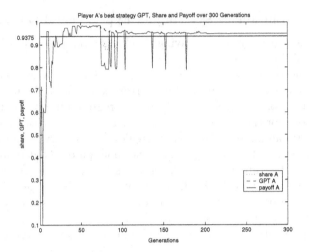

Fig. 1. A typical run: best-of-generation strategies for the first player A. The pair of discount factors is (0.9, 0.4). The line $y = 0.9375$ is the P.E.P of the player A. The overlaps of π_A, p_A, and g_A imply that agreements are settled down at time $t = 0$. No bargaining cost incurs [4]

Table 1. Shares π_A obtained by the best-of-generation individual in the player A's population at the 300th generation. The mean of 100 runs for each game setting

Discount Factors	P.E.P π_A^*	Experimental Mean π_A	π_A's Deviation
(0.1 , 0.1)	0.9091	0.9226	0.0308
(0.4 , 0.1)	0.9375	0.9987	0.0064
(0.4 , 0.6)	0.5263	0.5092	0.0102
(0.4 , 0.9)	0.1563	0.1444	0.1155
(0.5 , 0.5)	0.6667	0.6754	0.0271
(0.9 , 0.1)	0.9890	0.9989	0.0030
(0.9 , 0.4)	0.9375	0.9104	0.0091
(0.6 , 0.9)	0.2174	0.1551	0.0458
(0.9 , 0.9)	0.5263	0.5141	0.1194
(0.9 , 0.99)	0.0917	0.1167	0.0585

regression result is $\pi*_A = 0.9588 \times \pi_A + 0.0257$, so P.E.P is nicely approximated by the experimental results. This demonstrates that the co-evolution experiment is a convincing approximator of the game-theoretic method. These experimental results indicate that the relative fitness is a satisfied assessment that successful guides the co-evolving players to perform approximate Perfect Equilibrium strategies.

5   Absolute Fitness Evaluation

In the aforementioned experiments, the members of the populations at the end of co-evolving time are co-adapted strategies for a particular bargaining setting. These outcomes are inadequate to answer relevant questions that are important in observing the co-adapting process and evaluating the co-adaptive strategies' performance. How a co-evolving population develops in order to find the best responses to the opponent's population? Can co-adapted strategies out-perform the theoretical solution? Another question is during the evolving time, whether co-evolutionary learning helps players to adapt to more diverse environments, or only perform well to its co-adapted population S_y in (2), a dynamic but known environment.

5.1   Experimental Design

Taking into account the property of the absolute fitness, and the knowledge of game theoretic solutions, we design two external indicators to continually assess the co-adapting individuals, without replacing the relative fitness measure for the co-evolving. At every generation, each co-evolving population is evaluated against the set of P.E.P strategies and against another static set of randomly generated strategies.

First player A's co-evolving population A_C starts from a randomly generated initial population A_R. A_P is his P.E.P strategy. Similarly, player B' B_C, B_R, and B_P refer to B's co-evolving population, random population and P.E.P strategy, respectively. All random strategies make proposals uniformly distributed in the range of $[0, 1]$. The following sets of experiments are to be executed, in which the symbol "∘" means "to bargain against": (i) $A_C \circ B_R$ and $A_R \circ B_C$: random strategies against co-evolving strategies; (ii) $A_R \circ B_P$ and $A_P \circ B_R$: random strategies against P.E.P strategies. A_R and A_R could utilize the initial populations which are also generated randomly; (iii) $A_C \circ B_P$ and $A_P \circ B_C$: co-evolving strategies against P.E.P strategies.

5.2   Experimental Results

We choose an absolute fitness function that S_y in (2) is the A's P.E.P, to study the co-evolving population B. In a typical run shown in Fig. 2, it is observed that in a very short period of time immediately after the beginning, the highest A_C performs better than A_P because that P.E.P is unable to exploit some inexperienced strategies of B_C who propose or accept a partition of cake less than the P.E.P. But some strategies in the randomly generated A_C can take advantages of these weakly performed strategies in B_C by asking lager shares than P.E.P.

In the following of evolving time, the average absolute fitness of A_P declines after first twenty of generations and stabilizes after around $150th$ generation. This suggests that his opponent B_C is learning to gain payoff more than he did at the beginning when he had no experience. But B_C' improvement does not reflect on his relative fitness against his co-evolving opponent A_C. A_Cs relative fitness remains rather stable after the $20th$ generations. Therefore, it is the ab-

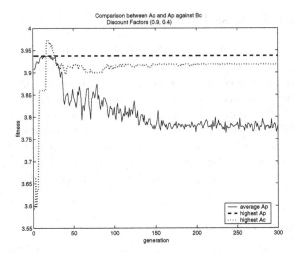

Fig. 2. The highest and the average of relative fitness of A_C, and that of absolute fitness of A_P, both against the co-evolving B_C. The average A_C does not display as it is much smaller than the other three values

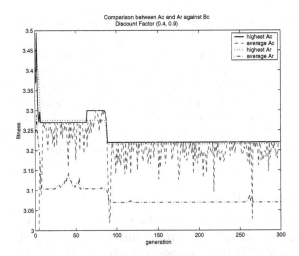

Fig. 3. the highest and average absolute fitness of random A_R together with the relative fitness of co-evolving A_C, against co-evolving B_C

solute fitness from the fix opponent population (A_P) that provides information concerning a co-evolving population's (B_C's) adaptation and improvement.

Another absolute fitness function for studying co-evolving population B_C is using a fix randomly generated population A_R. In Fig. 3, the highest absolute fitness of A_R almost overlaps the highest relative fitness of A_C. Both values quickly decline from 3.5 to 3.23. This implies that B_C quickly discovers the approximation to his P.E.P and also improves its competitive strength against both its co-evolving population A_C and the random A_R. Due to the diversity property of the random population A_R, the absolute fitness of A_R does not show as strong indication of B_C's adaptation as the above example.

6   Conclusions

This work formalizes two important concepts about the fitness evaluations in artificial co-evolution: the absolute and relative fitness. These two fitness evaluations have resulted in inconsistent outcomes to the IPD problem, reported in literature. Taking an application of the co-evolutionary algorithms for solving the bargaining problem as another example, we analyze the co-adapting process of two distinct populations of a co-evolutionary system through observing development of individuals' relative and absolute fitness. We gain insights into not only the empirical justification to the game-theoretic P.E.P, but also the importance of the adoption of absolute evaluations to co-evolutionary adaptive systems. On the ground of experimental observations, the relative fitness continues pushing individuals to co-adapt. The absolute evaluation, on the other hand, provides information on the co-evolving process, by measuring against fixed objectives. We have noticed that the absolute fitness functions chosen in this example, limit on overspecialization and problem-specific knowledge. Future studies on the absolute evaluations for more general applications will be examined.

Acknowledgment

We would like to thank Alberto Moraglio for his important comments on the draft. Anonymous reviewers comments are very helpful for the revision. Edwin de Jong and Hui Li helped to improve the quality of the final submission. This research is partly sponsored by University of Essex Research Fund.

References

1. Axelrod, R.: The evolution of strategies in the iterated prisoner's dilemma. In Lawrence Davis (Ed.), Genetic Algorithms and Simulated Annealing. Morgan Kaufmann (1987)
2. Bierman, H. and Fernandes, L.: Game Theory with Economic Applications. Addison-Wesley, New York (1998)
3. Darwen, P. and Yao. X.: On evolving robust strategies for iterated prisoner's dilemma. In Progress in Evolutionary Computation, Lecture Notes in Artificial Intelligence, Vol. 956, Springer-Verlag, Heidelberg (1995)
4. Jin, N., Tsang, E.: Using Evolutionary Algorithms to study Bargaining Problems. IEEE Symposium on Computational Intelligence and Games (2005) to appear.
5. de Jong, E. and Pollack, J.: Ideal Evaluation from Coevolution. Evolutionary Computation 12(2), (2004)
6. Holland, J.: Adaptation in Natural and Artificial Systems. Second edition. MIT Press. (1992)

7. Koza, J.: Genetic Programming: On the Programming of Computers by Means of Natural Selection. The MIT Press (1992)

8. Luke, S., and Wiegand, P.R.: Guaranteeing Coevolutionary Objective Measures. In Foundations of Genetic Algorithms VII, Morgan Kaufman (2002)

9. Miller, J.: The Co-evolution of Automata in the Repeated Prisoner's Dilemma. Journal of Economic Behavior and Organization, 29(1), (1996)

10. Schmitt, L.M.: Classification with Scaled Genetic Algorithms in a Coevolutionary Setting. GECCO (2004)

11. Simon, H.: A Behavioral Model of Rational Choice, The Quarterly Journal of Economics, Vol. 69, NO.1 (1955)

12. Rand, D.: The course note of BIO48, Brown University (2004)

13. Rubinstein, A.: Perfect Equilibrium in a Bargaining Model. Econometrica, 50: 97-110 (1982)

Teams of Genetic Predictors for Inverse Problem Solving

Michael Defoin Platel[1], Malik Chami[2], Manuel Clergue[1], and Philippe Collard[1]

[1] Laboratoire I3S, UNSA-CNRS, Sophia Antipolis, France
[2] Laboratoire d'Oceanographie de Villefranche sur Mer, France

Abstract. Genetic Programming (GP) has been shown to be a good method of predicting functions that solve inverse problems. In this context, a solution given by GP generally consists of a sole predictor. In contrast, Stack-based GP systems manipulate structures containing several predictors, which can be considered as *teams of predictors*. Work in Machine Learning reports that combining predictors gives good results in terms of both quality and robustness. In this paper, we use Stack-based GP to study different cooperations between predictors. First, preliminary tests and parameter tuning are performed on two GP benchmarks. Then, the system is applied to a real-world inverse problem. A comparative study with standard methods has shown limits and advantages of teams prediction, leading to encourage the use of combinations taking into account the response quality of each team member.

1 Introduction

A direct problem describes a Cause-Effect relationship, while an inverse problem consists in trying to recover the causes from a measure of effects. Inverse problems are often far more difficult to solve than direct problems. Indeed, the amount and the quality of the measurements are generally insufficient to describe all the effects and so to retrieve the causes. Moreover, since different causes may produce the same effects, the solution of the problem may be not unique. In many scientific domains, solving an inverse problem is a major issue and a wide range of methods, either analytic or stochastic, are used. In particular, recent work [6][4] has demonstrated that Genetic Programming (GP) is a good candidate.

GP applies the Darwinian principle of survival of the fittest to the automatic discovery of programs. With few hypothesis on the instructions set used to build programs, GP is an universal approximator [16] that can learn an arbitrary function given a set of training examples. For GP, as for Evolutionary Computation in general, the way individuals are represented is crucial. The representation induces choices about operators and may strongly influence the performance of the algorithm. The emergence of GP in the scientific community arose with the use, *inter alia*, of a tree-based representation, in particular with the use of the Lisp language in the work of Koza [8]. However, there are GP systems manipulating linear structures, which have shown experimental performances equivalent to Tree GP (TGP) [1]. In contrast to TGP, Linear GP (LGP) programs are sequences of instructions of an imperative language (C, machine code, ...). The evaluation of a program can not be performed in a recursive way and so needs to use extra memory mechanisms to store partial computations. There are at least two kind of

M. Keijzer et al. (Eds.): EuroGP 2005, LNCS 3447, pp. 341–350, 2005.

LGP implementation. In the first one [1], a finite number of registers are used to store the partial computations and a particular register is chosen to store the final result. In the other one, the stack-based implementation [12],[15] and [3], the intermediate computations are pushed into an operand stack and the top of stack gives the final result. It is important to note that in TGP, an individual corresponds to a sole program and its evaluation produces a unique output. In LGP, an individual may be composed of many independent sub-programs and its evaluation may produce several outputs and some of them may be ignored in the final result. In this paper, we propose to combine those sub-programs into a team of predictors.

The goal of Machine Learning (ML) is to find a predictor trained on set of examples that can approximate the function that generated the examples. Several ML methods are known to be universal approximators, that is they can approximate a function arbitrarily well. Nevertheless, the search of optimal predictors might be problematic due to the choice of inadequate architecture but also to over-fitting on training cases. Over-fitting occurs when a predictor reflects randomness in the data rather than underlying function properties, and so it often leads to poor generalization abilities of predictors. Several methods have been proposed to avoid over-fitting, such as model selection, to stop training or combining predictors, see [14] for a complete discussion. In this study, we mainly focus on combining predictor methods, also called ensemble, or committee methods. In a committee machine, a team of predictors is generated by means of a learning process and the overall predictions of the committee machine is the combination of the predictions of the individual team members. The idea here, is that a team may exhibit performances unobtainable by a single individual, because the errors of the members might cancel out when their outputs are combined. Several schemes for combining predictors exist, such as simple averaging, weighted combination, mixtures of experts or boosting. In practice, ANN ensembles have already been strongly investigated by several authors, see for instance [9]. In the GP field, there has been some work on the combination of predictors, see for examples [17][11][2].

In this paper, our goal is initially to improve the performance of stack-based GP systems by evolving teams of predictors. The originality of this study is that teams have a dynamic number of members that can be managed by the system. In Section 2, we describe different ways to combine predictors and how to implement them using a stack. In Section 3, evolutionary parameters are tuned and the performances of several combinations are tested on two GP benchmarks and on a real-world inverse problem. Finally, in Section 4, we investigate the relationship between uncontrolled growth of program size in GP and dynamical size of teams.

2   Teams of Genetic Predictors

2.1   Stack-Based GP

In stack-based GP, numerical calculations are performed in *Reverse Polish Notation*. According to the implementation proposed by Perkis[12], an additional type of closure

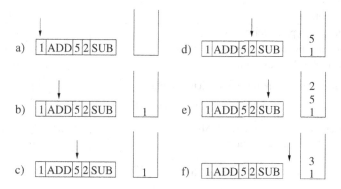

Fig. 1. Evaluation of a program in Stack-based GP

constraint is imposed on functions: they are defined to do nothing when arity is unsatisfied by the current state of the operand stack. In Figure 1, a basic example of program execution is presented. We can see the processing of the program "1,ADD,5,2,SUB". Initialization phase corresponds to step a where the stack is cleared. During steps b, d and e, numerical constants 1, 5 and 2, respectively, are pushed onto the operand stack. During step f, since the operand stack stores enough data (at least two), the "SUB" instruction is computed and the result ($3 = 5 - 2$) is pushed onto the stack. During step c the stack contains only one value, so computation of the "ADD" instruction is impossible: the instruction is simply skipped with no effect on the operand stack. In most cases, problems addressed using GP are defined with multiple fitness cases and a compilation phase can easily removed those unexpected instructions to speed up evaluation.

Let us note that after step f, the operand stack contains the values 1 and 3. This means that there are two independent sub-programs, "1" and "5,2,SUB", in the sequence "1,ADD,5,2,SUB". We propose to make the whole set of sub-programs involved in the fitness evaluation. The idea is to combine the results of each sub-programs, i.e. all the elements of the final stack, according to the nature of the problem addressed. For example, in the case of a *Symbolic Regression Problem*, any linear combination of the sub-programs outputs may be investigated. Using this kind of evaluation, the evolutionary process should be able to tune the effect of the genetic operators by changing the number and the nature of sub-programs and so modifying the contribution of each of them to the final fitness. Moreover the amount of useless code usually found in stack-based representation programs is significantly decreased leading to improved performance. This approach may be viewed as the evolution of teams of predictors corresponding to the combination of sub-programs.

2.2 Different Combinations

Let us consider a program with m instructions giving a team T having $k \in [1, m]$ predictors p_i. The team output $O(T)$ consists of a combination of the k predictors outputs $o(p_i)$. Several combinations have been tested :

Team combination	$O(T)$
Sum : Sum of each predictor output	$\sum_{i=1}^{k} o(p_i)$
AMean : Arithmetic mean of predictors outputs	$\frac{1}{k}\sum_{i=1}^{k} o(p_i)$
Pi : Product of each predictor output	$\prod_{i=1}^{k} o(p_i)$
GMean : Geometric mean of each predictor output	$\sqrt[k]{\prod_{i=1}^{k} o(p_i)}$
EMean : Mean of each predictor output weighted by their error	$\frac{1}{\sum_{i=1}^{k} w_i}\sum_{i=1}^{k} w_i o(p_i)$
WTA : Winner predictor output Takes All	$o(argmin_{p_i, i \in [1,k]}(E(p_i)))$
Top : Top of stack predictor output	$o(p_k)$

with $E(p_i)$ the training error of p_i and with $w_i = e^{-\beta E(p_i)}$, β a positive scaling factor.

3   Experimental Results

3.1   Symbolic Regression

In this section, the evolutionary parameters' tuning is extensively investigated on a Symbolic Regression Problem. We choose the Poly 10 problem [13], where the target function is the 10-variate cubic polynomial $x_1x_2 + x_3x_4 + x_5x_6 + x_1x_7x_9 + x_3x_6x_{10}$. In this study, the fitness is the classical Root Mean-Square Error. The dataset contains 50 test points and is generated by randomly assigning values to the variables x_i in the range $[-1, 1]$. We perform 50 independent runs with various mutation and crossover rates. Populations of 500 individuals are randomly created according to a maximum creation size of 50. The instructions set contains: the four arithmetic instructions ADD, SUB, MUL, DIV, the ten variables $X_1 \ldots X_{10}$ and one stack-based GP specific instruction DUP which duplicates the top of the operand stack. The evolution, with elitism, maximum program size of 500, 16-tournament selection, and steady-state replacement, takes place over 100 generations [1]. We use a statistical unpaired, two-tailed t-test with 95% confidence to determine if results are significantly different.

In Table 1, the best performances on the Poly 10 problem with different combinations of predictors are presented, using the best settings of evolutionary parameters found, crossover rate varying from 0 to 1.0 and mutation rate from 0 to 2.0. Let us notice that a mutation rate of 1.0 means that each program involved in reproduction will undergo, on average, one insertion, one deletion and one substitution. In the first row, results obtained using a Tree GP implementation[2] are reported in order to give an absolute reference. We see that the classical Top of stack and WTA methods work badly, which is to be expected, since in this case, teams' outputs correspond to single predictor outputs. The Sum and AMean methods report good results compared to the Tree method. In contrast, the Pi and GMean methods are not suitable for this problem, perhaps because the Poly 10 problem is easiest to decompose as a sum. Finally, EMean undoubtedly outperforms

---

[1] In a steady state system, the generation concept is somewhat artificial and is used only for comparison with generational systems. Here, a generation corresponds to a number of replacement equal to the number of individual in the population, i.e. 500.

[2] We note that an optimization of parameters has been also performed for Tree GP.

Table 1. Best Results on Poly 10

Team	Train			
	Mean	Std Dev	Best	Worst
Tree	0.250	0.072	0.085	0.407
Top	0.353	0.105	0.172	0.520
WTA	0.443	0.048	0.347	0.541
Sum	0.173	0.031	0.120	0.258
AMean	0.165	0.034	0.109	0.251
Pi	0.361	0.063	0.220	0.456
GMean	0.222	0.027	0.151	0.301
EMean	**0.066**	**0.017**	**0.029**	**0.141**

Table 2. Results on Mackey-Glass (10^{-2})

Team	Train		Valid		Test	
	Mean	Std Dev	Mean	Std Dev	Mean	Std Dev
Tree	**0.61**	**0.14**	1.06	0.43	1.21	0.71
Top	1.09	0.24	0.99	0.35	0.95	0.33
WTA	1.87	1.04	1.97	1.06	1.95	0.36
Sum	0.64	0.05	0.83	0.21	1.38	0.39
AMean	0.63	0.06	0.81	0.20	0.83	0.40
Pi	0.71	0.15	0.84	0.27	1.10	0.44
GMean	0.67	0.04	0.78	0.14	0.76	0.15
EMean	**0.59**	**0.03**	**0.74**	**0.07**	**0.71**	**0.06**

other methods. We note that results presented here correspond to a scaling factor β of 10. In Figure 2, the average training error is plotted as a function of β. The training error is clearly related to β and gives a minimum for $\beta = 10$.

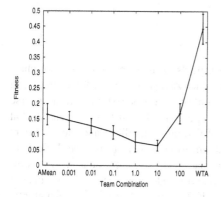

Fig. 2. Average train error for AMean, WTA and EMean with different β on Poly 10 problem

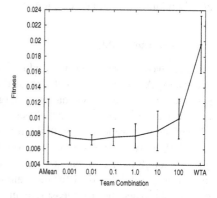

Fig. 3. Average test error for AMean, WTA and EMean with different β on Mackey-Glass problem

3.2   Chaotic Time Series

In this Section, we choose the IEEE benchmark Mackey-Glass chaotic time series (see http://neural.cs.nthu.edu.tw/jang/benchmark/, $\tau=17$, 1201 data points, sampled every 0.1) to examine the performances of different teams in the context of over-fitting. This problem has already been tested in LGP, see [10] for example, and seems to have sufficient difficulty to allow appearance of over-fitting behaviors.

We have discarded the first 900 points of the dataset to remove the initial transients and we have decomposed the last 300 in three sets of same size, the Train, Validation and Test sets. The goal for GP is, given 8 historical values, to predict the value at time $t + 1$. Thus, each set contains 100 vectors with values at times $t - 128, t - 64, t - 32,$

$t - 16, t - 8, t - 4, t - 2$ and t. The evolutionary parameters are identical to those used to solve the Poly 10 problem. The instruction set is limited to handle only the 8 inputs and an Ephemeral Random Constant (cf. [8]) in the range [-1, 1] has been added onto the instructions set.

In Table 2, the performances on the Mackey-Glass problem with different combinations of predictors are presented, using the best settings of evolutionary parameters found on Poly 10 problem. The train errors reported here tend to confirm the previous results on Poly 10. Indeed, the Top and WTA combinations give the worst results while, with other combinations, we have obtained performances equivalent to Tree GP. We see important variations between Train and Test errors, in particular for Tree and Sum combination. However we have found only 2 or 3 programs with very bad test errors (among 50 per each combination) that are responsible of the main part of these variations. The EMean team has obtained the best results, for both train and test errors, and seems to over-fit less than the others. It is important to note that the programs discovered during our different experiments on the Mackey-Glass problem with Stack GP have approximatively the same number of instructions than programs obtained with the Tree GP implementation.

The results of EMean combination presented here correspond to a scaling factor β of 0.01. In Figure 3, the average test error is plotted as a function of β. The test error is clearly related to β and gives a minimum for β=0.01.

3.3   Inverse Problem

The real-world inverse problem addressed here deals with atmospheric aerosol characteristics. An accurate knowledge of these characteristics is central, for example, in satellite remote sensing validation. A large amount of data is necessary to train inverse models properly. The teaching phase should be performed with truthful data so that inverse models are learned with the influence of natural variabilities and measurement errors. However, according to the inverse problem to deal with, it may be difficult to gather enough measurements to cover the data space with sufficient density. It is common to use simulated data for the training step and to add some geophysical noise in the data set to account for measurement errors [7][4]. Therefore, in this paper, synthetic data obtained with a radiative transfer model were used to perform the learning step.

We want to inverse a radiative transfer model, called *Ordres Successifs Ocean Atmosphere* (OSOA) model, which is described in detail by Chami et al [5]. The direct model solves a radiative transfer equation (RTE) by the successive orders of scattering method for the ocean-atmosphere system. It takes into account the multiple scattering events and the polarization state of light for both atmospheric and oceanic media. The atmosphere is a mixture of molecules and aerosols. Aerosols are supposed to be homogeneous spheres. The aerosol model is defined (refractive index and size distribution) and its optical properties (phase function, single scattering albedo) are computed using Mie theory. The air-water interface is modeled as a planar mirror. Consequently, the reflection by a rough surface is not taken into account in the radiative transfer code. The OSOA outputs the angular distribution of the radiance field and its degree of polarization at any desired level (any depth, the surface or the top of atmosphere).

In this study, the solar zenith angle is fixed to 70 degrees. Each row of the dataset corresponds to sky radiances at 440 nm, 675 nm and 870 nm for 10 scattering angles

Table 3. Results on Inverse Problem

Team	Train Mean	Std Dev	Valid Mean	Std Dev	Test Mean	Std Dev
Tree	0.018	0.004	0.018	0.004	0.021	0.006
Top	0.035	0.006	0.032	0.006	0.035	0.006
WTA	0.040	0.013	0.037	0.012	0.041	0.014
Sum	0.019	0.004	0.018	0.004	0.020	0.005
AMean	0.021	0.004	0.019	0.004	0.022	0.005
Pi	0.022	0.004	0.021	0.004	0.023	0.005
GMean	0.017	0.002	0.019	0.006	0.022	0.009
EMean	**0.013**	**0.002**	**0.012**	**0.001**	**0.014**	**0.003**

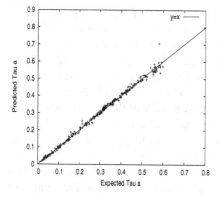

Fig. 4. Scatter plot of Expected vs Predicted τ_a on test data for the best team found.

Fig. 5. Average test error for AMean, WTA and EMean with different β on inverse problem.

ranging from 3 to 150 degrees. From these 30 inputs, the inverse model has to retrieve the aerosol optical depth τ_a at 675 nm. To avoid over fitting, we have decomposed the dataset into three parts, 500 samples are devoted to the learning step, 250 to validation and test steps. The evolutionary parameters are identical to those used to solve the Poly 10 problem. The instruction set is extended to handle the 30 inputs. Finally, a LOG instruction ($ln|x|$) and an Ephemeral Random Constant (cf. [8]) in the range [-10, 10] have been also added onto the instructions set.

In Table 3, the performances on the inverse problem with different combinations of predictors are presented. As previously, for similar reasons, we see that the classical Top of stack and WTA methods works badly. Contrary to results reported for Poly 10 problem, EMean is the only method that significantly outperforms Tree. We note that results presented here correspond to a scaling factor β fixed to 0.1. Figure 4 shows performances on test data of the best team found with EMean and $\beta = 0.1$. We see that a good agreement is obtained between expected and predicted τ_a giving a nearly perfect inversion.

In Figure 5, the average test error is plotted as a function of β. The test error is clearly related to β and gives a minimum for $\beta = 0.1$. We have also trained two ANNs on this inverse problem : one ANN with feed-forward topology and no hidden layer (to handle linear relationships only) and one ANN with feed-forward topology and one hidden layer (optimal number of nodes found experimentally). The corresponding test errors are also plotted in Figure 5. We see that the use of teams of predictors improves the performances of the GP system in such a way that it outperforms a linear ANN. Nevertheless, the accuracy of a non-linear ANN can not be obtained with the parameters used in this study. Let us notice that they are many sophisticated techniques, such as Automatic Define Functions, Demes, ..., well known in the GP community, that could significantly improve the results presented here. Our aim was not to compare ANN and GP, but rather to quantify the benefit of optimizing the β parameter.

4   Discussion

In this paper, a team of predictors corresponds to a combination of the sub-programs of a stack-based GP individual. Contrary to previous studies addressing teams of predictors in GP, here, the number of members of a team is not apriori fixed but can change during the evolutionary process. We had hoped that the dynamic of the algorithm would optimize the number of members. Unfortunately, the team size tends to increase quickly as of the early generations and is strongly correlated to the size of individuals. Moreover, it is well known that GP suffers from an uncontrolled growth of the size of individuals, a phenomenon called bloat. Thus, the team size can not be directly managed by the system. In Figure 6, the average number of predictors on the inverse problem is reported for EMean with $\beta = 0.1$. The limit, around 200 predictors, is probably due to the 'maximum allowed size of programs' parameter (500 instructions). We have also reported the number of predictors having a weight w_i (almost) equal to zero and the number of predictors whose weight is positive. We see that around a third of the predictors are all but eliminated from teams. So, the use of combinations of predictors that take into account the response quality of each team member gives a way to control team size.

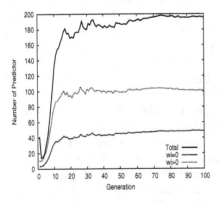

Fig. 6. Evolution of average number of predictors with $\beta = 0.1$ on inverse problem.

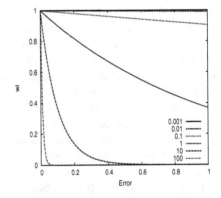

Fig. 7. Weights w_i as a function of the predictors error with $\beta = 0.001, 0.01, 0.1, 1, 10$ and 100.

However, we have seen that the best tuning of β depends on the problem addressed. Figure 7 shows the weights w_i as a function of the predictor errors for different β. We have limited the error to the range $[0, 1]$. We see that for small values of β, weights are almost equal to 1, as for the AMean combination, whereas when β is equal to 100, the majority of weights are null, as for WTA combination. From Figures 2, 3 and 5, we see that EMean is able to cover the entire performance spectrum between AMean and WTA.

5   Conclusion

In this paper, our aim is not to compare our work with other learning methods, such as boosting for example, but rather to improve the performance of stack-based GP systems. Thus, we start with one of their main drawbacks : they provide many outputs, that is sub-programs, instead of only one. A naive way to overcome this is to keep only one arbitrary output as the output of the program. This method exhibits very poor results, notwithstanding the waste of resources needed to evolve part of programs that are never used. Keeping only the best sub-program leads to even worse results, as the system undergoes premature convergence. Taking the arithmetic mean or the sum of sub-programs outputs gives our system the same performances as a tree-based GP system, which evolves individual predictors.

The best results we obtain are when the output of the program is a weighted sum of the sub-programs, where each sub-program receives a weight depending on its individual performance. This way programs are not penalized by bad sub-programs, since they do not contribute, or only lightly, to the final output. Moreover, this supplementary degree of freedom promotes the emergence of dynamically sized teams. Indeed, even if the total number of predictors in a team is strongly correlated to the maximum number of instructions in programs, only some of them contribute to the program output and so really participate in the team. The number of such sub-programs is free. Genetic operations may cause it to vary, with the introduction of a new "good" sub-program or the destruction of a former "good" one.

This paper represents the first step of our work and empirical results should be confirmed by experiments on other problems and theoretical studies. Moreover, some choices we have made are arbitrary, such as the use of a linear combination of sub-programs. Other types of combinations, e.g. logarithmic ones may improve performances. Also, we need to better understand how the value of β influences performance, and its effect on team composition.

References

1. M. Brameier and W. Banzhaf. A comparison of linear genetic programming and neural networks in medical data mining. *IEEE Transactions on Evolutionary Computation*, 5(1):17–26, 2001.
2. M. Brameier and W. Banzhaf. Evolving teams of predictors with linear genetic programming. *Genetic Programming and Evolvable Machines*, 2(4):381–407, 2001.

3. W. S. Bruce. The lawnmower problem revisited: Stack-based genetic programming and automatically defined functions. In *Genetic Programming 1997: Proceedings of the Second Annual Conference*, pages 52–57, Stanford University, CA, USA, 13-16 1997. Morgan Kaufmann.

4. M. Chami and D. Robilliard. Inversion of oceanic constituents in case i and case ii waters with genetic programming algorithms. *Applied Optics*, 40(30):6260–6275, 2002.

5. M. Chami, R. Santer, and E. Dilligeard. Radiative transfer model for the computation of radiance and polarization in an ocean-atmosphere system: polarization properties of suspended matter for remote sensing. *Applied Optics*, 40(15):2398–2416, 2001.

6. P. Collet, E. Lutton, F. Raynal, and M. Schoenauer. Polar IFS+parisian genetic programming=efficient IFS inverse problem solving. *Genetic Programming and Evolvable Machines*, 1(4):339–361, 2000.

7. L. Gross, S. Thiria, R. Frouin, and B. G. Mitchell. Artificial neural networks for modeling the transfer function between marine reflectances and phytoplankton pigment concentration. *J. Geophys. Res.*, C2(105):3483–3495, 2000.

8. J. R. Koza. *Genetic Programming: On the Programming of Computers by Means of Natural Selection*. MIT Press, 1992.

9. A. Krogh and J. Vedelsby. Neural network ensembles, cross validation, and active learning. *NIPS*, 7:231–238, 1995.

10. W. B. Langdon and W. Banzhaf. Repeated sequences in linear gp genomes. In *Late breaking paper at GECCO'2004*, Seattle, USA, June 2004.

11. G. Paris, D. Robilliard, and C. Fonlupt. Applying boosting techniques to genetic programming. In *Artificial Evolution 5th International Conference, Evolution Artificielle, EA 2001*, volume 2310 of *LNCS*, pages 267–278, Creusot, France, October 29-31 2001. Springer Verlag.

12. T. Perkis. Stack-based genetic programming. In *Proceedings of the 1994 IEEE World Congress on Computational Intelligence*, volume 1, pages 148–153, Orlando, Florida, USA, 27-29 1994. IEEE Press.

13. R. Poli. A simple but theoretically-motivated method to control bloat in genetic programming. In *Genetic Programming, Proceedings of EuroGP'2003*, volume 2610 of *LNCS*, pages 200–210, Essex, 14-16 April 2003. Springer-Verlag.

14. W. Sarle. Stopped training and other remedies for overfitting. In *Proceedings of the 27th Symposium on Interface*, 1995.

15. K. Stoffel and L. Spector. High-performance, parallel, stack-based genetic programming. In *Genetic Programming 1996: Proceedings of the First Annual Conference*, pages 224–229, Stanford University, CA, USA, 28–31 1996. MIT Press.

16. Xin Yao. Universal approximation by genetic programming. In *Foundations of Genetic Programming*, Orlando, Florida, USA, 13 1999.

17. Byoung-Tak Zhang and Je-Gun Joung. Time series prediction using committee machines of evolutionary neural trees. In *Proceedings of the Congress of Evolutionary Computation*, volume 1, pages 281–286. IEEE Press, 1999.

Understanding Evolved Genetic Programs for a Real World Object Detection Problem

Victor Ciesielski[1], Andrew Innes[1], Sabu John[2], and John Mamutil[3]

[1] School of Computer Science and Information Technology,
RMIT University, GPO Box 2476V, Melbourne, Vic Australia 3001
{vc, ainnes}@cs.rmit.edu.au
[2] School of Aerospace, Mechanical and Manufacturing Engineering,
RMIT University
[3] Braces Pty Ltd, 404 Windsor Road, NSW 2153, Australia

Abstract. We describe an approach to understanding evolved programs for a real world object detection problem, that of finding orthodontic landmarks in cranio-facial X-Rays. The approach involves modifying the fitness function to encourage the evolution of small programs, limiting the function set to a minimal number of operators and limiting the number of terminals (features). When this was done for two landmarks, an easy one and a difficult one, the evolved programs implemented a linear function of the features. Analysis of these linear functions revealed that underlying regularities were being captured and that successful evolutionary runs usually terminated with the best programs implementing one of a small number of underlying algorithms. Analysis of these algorithms revealed that they are a realistic solution to the object detection problem, given the features and operators available.

1 Introduction

Genetic programming has been used with considerable success for a wide range of problems, both toy and real world. In the area of image processing there have been a number of successful applications to difficult problems. One of the drawbacks of genetic programming solutions is that the evolved programs are very hard to understand and can be very big. This makes it hard to sell genetic programming solutions as many engineers and other professionals, particularly in the medical area, are very unhappy with black box solutions. They want to understand the operation of any solution to better appreciate its limitations and generality.

There has been very little prior work on the problem of understanding evolved program trees. Most work is focused on finding acceptable solutions to problems without any major concern for their understandability. However, [1] describes 'trait mining', a method of finding sub-trees, or traits, in the evolved programs. Traits are associated with high fitness and provide some insight into the evolved solutions. The basic idea is that even if the whole program cannot be understood,

M. Keijzer et al. (Eds.): EuroGP 2005, LNCS 3447, pp. 351–360, 2005.

being able to understand some of the subtrees means that the evolved program is not a black box. [2] describes the evolution of classification rules with genetic programming. The rules are intended to be understandable to humans. This is achieved by evolving the rules in a predetermined if-then format, rather than as full genetic programs. [3] describes an approach to finding the underlying regularities in evolved programs which use pixel inputs for texture discrimination. A strategy of limiting the function set and penalising large programs was used. Analysis of the resulting programs revealed that a number of masks of pixel positions were being used to discriminate textures. Systematic regularities were consistently being discovered in the different runs, arbitrary positions of pixels were not being used.

We have recently had considerable success in using genetic programming to evolve object detection programs for finding orthodontic landmarks in cranio-facial X-Rays [4]. Orthodontists are not comfortable with genetic programming and there is a need to explain the evolved programs in ways that are credible to them. The aim of the work presented in this paper is to determine whether we can explain how evolved programs work for the object detection problem. In particular we are interested in:

1. Can we develop a methodology for analysing evolved programs?
2. Are regularities being discovered? If so what are they?
3. Can we express, in some reasonable way, the underlying algorithms in the evolved programs.

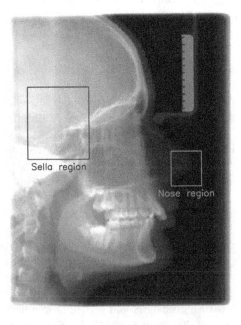

Fig. 1. A digital cephalogram depicting two regions containing the mid nose and sella landmarks

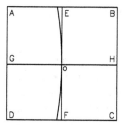

square size=14

Features		Local regions
μ	σ	
M_1	S_1	full square A-B-C-D
M_2	S_2	left half A-E-F-D
M_3	S_3	right half E-D-C-F
M_4	S_4	2 centre columns E-F
M_5	S_5	2 centre rows G-H

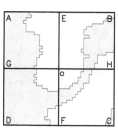

square size=40

Features		Local regions
μ	σ	
M_1	S_1	shaded region
M_2	S_2	unshaded region
M_3	S_3	full square A-B-C-D
M_4	S_4	top left A-E-o-G
M_5	S_5	top right E-B-H-o
M_6	S_6	bottom left G-o-F-D
M_7	S_7	bottom right o-H-C-F

Fig. 2. The diagrams in the left column depict the feature maps used for the nose and sella landmarks. The features consist of the mean and standard deviation calculated for each shape from grey level intensities. The corresponding pictures in the middle column depict the size of the feature map (shown as the white square) relative to the image

2   The Object Detection Problem

The object detection problem is to locate, to within 2mm, a number of key landmarks used by orthodontists in treatment planning. Full details of this problem can be found in [4, 5, 6]. For this paper we focus on just two landmarks, an easy one, the tip of the nose and a very difficult one, the sella. These landmarks are shown in figures 1 and 2. We consider the finding of each landmark as an independent object detection problem. Using prior knowledge of facial geometry we identify regions in the picture where the landmark should be with respect to the edge of the ruler which is easy to find by classical image processing methods. Each of these individual problems is solved independently using the method described in the next section. It might appear that the problem of finding the nose tip would be very straight forward, but this is not the case due to the wide variation between humans in nose shapes and various noise effects on the X-Rays.

3   Finding Object Detection Programs by Genetic Programming

The basic method is to take an input window of pixels and slide the window over all positions in the region to find the location of the landmark [4]. This is similar to the template matching approach in classical vision except that the template is implemented as a genetic program. The output of the program is interpreted as the presence or absence of the landmark at the pixel position at the centre of the window. Since we know that there can be only one landmark in the region and that it must occur, we use the position with the largest output as the predicted position of the object.

The evolved programs use a feature set of pixel statistics of regions, or shapes, customised for each landmark. The features are obtained from a square input window centred on the landmark and large enough to contain key landmark characteristics, but not large enough to introduce unnecessary clutter, as shown in figure 2. For the nose point the input window has been partitioned manually into the shapes shown and the means and standard deviations of the pixels in each shape are the features. For the sella point the input window has been segmented into shapes by the use of pulse coded neural networks as described in [7]. The feature values are used as the terminals and the function set is $\{+, -, \times, \%\}$ (% denotes protected division).

A data base of images marked up by an orthodontist is used in training. To evaluate fitness during training, each program is applied, in moving window fashion, across each of the training images and the detection rate is computed. The detection rate is used as the fitness measure.

The method was tested on a number of landmark points, ranging from relatively easy to very difficult. Detection performance on the easier points was excellent and the performance on the difficult points was quite good [4].

4   Methodology for Analysing Evolved Programs

The evolved programs from the system described in the previous section are large and difficult to understand. To evolve smaller programs which have a higher likelihood of being understood we repeat the evolutionary process with:

1. A size penalty for large programs. The fitness function is

$$fitness = (1 - Dr) \times 100 + \frac{Program\ Size}{511} \times \frac{1}{10}, \qquad (1)$$

   where Dr is the detection rate.

   Program Size is the number of nodes in the program and there are 511 nodes in a full tree of depth 9, which is the max depth used in the runs. The size component represents constant parsimony pressure. The weightings of the two terms ensures that accuracy is the primary objective and size is secondary.

2. A reduced function set. For the nose point we use the two function sets $\{+\}$ and $\{+, -\}$. By keeping the operators to plus and minus we ensure that an evolved program will be a linear function of the inputs, possibly making it less accurate, but hopefully, easy to analyse.

3. A reduced terminal set. For the sella point we first use a terminal set which contains only the most obviously useful features together with the terminal set $\{+, -\}$. We then extend the analysis to the full terminal set.

5   Analysis of Evolved Programs

5.1   Easy Landmark: Nose

Function Set is $\{+\}$*:* The fittest individual, which was evolved in all 80 evolutionary runs, was:

$$Output = S_5 \tag{2}$$

Recalling that the position of the highest output is the predicted position of the object, and noting from figure 2 that S_5 is the standard deviation of the centre 2 rows of pixels, this program implements a reasonable algorithm for detecting the nose point. Inspection of figure 2 confirms that S_5 will be at a maximum when the input window is positioned over the nose tip. This program has a detection accuracy of 65.9%. It fails on a number of images, like the lower image in figure 3, in which there are some edge effects near the nose. However, it appears that the methodology is producing the best program possible given the restricted function set.

Function Set is $\{+, -\}$*:* In this situation all of the evolved programs can be simplified to the form given in equation 3. All α_i and β_i will be integers.

$$Output = \alpha_1 M_1 + \beta_1 S_1 + \alpha_2 M_2 + \beta_2 S_2 + \cdots + \alpha_5 M_5 + \beta_5 S_5 \tag{3}$$

Analysis of the best programs evolved in 80 runs revealed the following underlying algorithms:

$$Output = M_1 - S_2 - S_3 - 2M_4 + M_5 \tag{4}$$
$$Output = M_2 - 2S_2 - 2M_4 + M_5 \tag{5}$$
$$Output = 2M_1 - S_1 - 2M_4 - S_4 \tag{6}$$
$$Output = M_2 - 2S_2 - 2M_4 + M_5 \tag{7}$$

Analysis of an Individual Program: The following is an analysis of how the detection program described in equation 4 predicts the position of the landmark based on the values of features at six carefully chosen positions on an image. Each position used in figure 3 is indicative of regular patterns that occur in the training images. The window is located on soft tissue(1), the background(2), soft-tissue/background edge(3) and centred on the known position of the nose

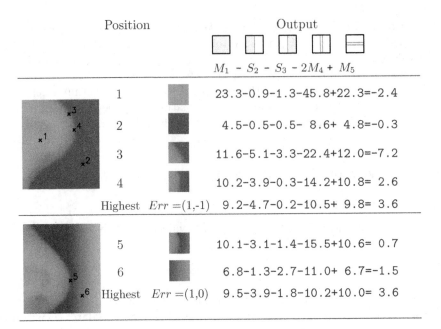

Position   Output

$$M_1 - S_2 - S_3 - 2M_4 + M_5$$

	Position		Output
	1		23.3-0.9-1.3-45.8+22.3=-2.4
	2		4.5-0.5-0.5- 8.6+ 4.8=-0.3
	3		11.6-5.1-3.3-22.4+12.0=-7.2
	4		10.2-3.9-0.3-14.2+10.8= 2.6
	Highest $Err =(1,-1)$		9.2-4.7-0.2-10.5+ 9.8= 3.6
	5		10.1-3.1-1.4-15.5+10.6= 0.7
	6		6.8-1.3-2.7-11.0+ 6.7=-1.5
	Highest $Err =(1,0)$		9.5-3.9-1.8-10.2+10.0= 3.6

Fig. 3. Sample output using detection program, $M_1 - S_2 - S_3 - 2M_4 + M_5$, applied to six different positions. Output is based on features calculated using a greyscale image

landmark(4). Positions 5 and 6 are on an image which has some edge effects near the nose.

If the program is evaluated when the input window is located on an area of constant brightness, such as positions 1 and 2, the output of the program is $\approx$ 0. The mean pixel intensities of each of the shapes, M_i, will be approximately the same and the standard deviations, S_i, will be $\approx$ 0. On edge position (3) the program output will be negative because S_1 and S_2 will be large. When the input window is centred on the true position (4), the program produces a high output in comparison to the previous positions. If we compare the outputs at positions 3 and 4 we observe that when the input window is located on a diagonal edge the most significant component for varying the output is terminal M_4. If the input window moves either side of the soft-tissue/boundary, the value of either S_2, the standard deviation of the left half, or S_3, the standard deviation of the right half, will decrease the output because of the negative coefficients. Figure 3 also shows the value of the highest output of the evolved program and the error in pixels in the x and y directions from the true position.

The analysis is consistent with the graph shown in Figure 4 which is a visualisation of the output of the program superimposed on a grey level image. From this analysis we can conclude that the program implements a reasonable algorithm for finding the nose point. A similar analysis performed for equations 5-7

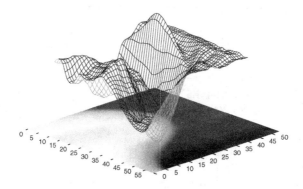

Fig. 4. Output from individual, $M_1 - S_2 - S_3 - 2M_4 + M_5$

reveals that they also implement reasonable algorithms. The accuracy achieved by these programs is no different from the accuracy achieved by programs using the full function set $\{+, -, \times, \%\}$ which indicates a linear problem.

Analysis of regularly occurring patterns across programs: It is clear that the function set used contains some redundancy, M_1, M_2 and M_3, for example, are obviously related. If we remove some of the redundancy by using the relationships $M_1 = \frac{1}{2}(M_2 + M_3)$ and $M_1 \approx M_5$, we obtain the programs shown in the central column of table 1. The program of equation 4 reduces to the fourth one in this table. Simplifying the best programs from 80 runs in the same way revealed that a number of underlying algorithms were repeatedly being evolved. For example, 13 runs produced the program shown in the first line of the table. Also, in all evolved programs that achieved a 100% detection rate, the sum of the coefficients of the M_i features used was 0 and the coefficients of the S_i features were always negative.

Figure 5 shows that there is a very large variation at the LISP level in programs that implement the same underlying algorithm.

5.2   Difficult Landmark: Sella

Due to the complexity of this landmark and the large number of features, we have carried out the analysis of the evolved programs in two stages. In the first stage we analyse runs using a reduced terminal set and in the second stage we

Table 1. Frequency of occurrence of detection program in 80 runs

Frequency	Program	Detection Rate
13/80	$1.5M_2 - 2S_2 + .5M_3 - 2M_4$	100% (82/82)
13/80	$1.5M_2 - S_2 + .5M_3 - 2M_4 - S_4$	100% (82/82)
7/80	$0.5M_2 - S_2 - .5M_3 - S_3 + S_5$	98.7% (81/82)
6/80	$M_2 - S_2 + M_3 - S_3 - 2M_4$	100% (82/82)
4/80	$-S_1 + M_2 + M_3 - 2M_4 - S_4$	100% (82/82)

(- (- M5 S2) (+ M4 (+ S2 (- M4 M2))))

(- (+ (- M5 (+ M4 (+ M4 (- S2 M2)))) S5) (+ M1 S2))

(+ (+ (+ (- (- S5 M4) S2) (- (- (- S2 (+ S2 M3)) M4) (+ S1 S5)))
  (+ M4 M3))(- (- (+ (+ M4 M5) (+ (+ M5 S1) (+ M3 M1))) (+ M3 M1))
  (+ (+ S5 (+ (+ M4 M3)M4)) (- M2 (+ (- M2 S2) (+ M1 M2))))))

Fig. 5. Programs equivalent to $1.5M_2 - 2S_2 + 0.5M_3 - 2M_4$

use the full terminal set. In both stages we use the function set $\{+, -\}$. Runs with just $\{+\}$ for the function set did not result in programs that were accurate enough to be worth analysing.

Reduced Terminal Set. The full feature set used for this landmark are shown in the lower half of figure 2. Features M_1, M_2, S_1 and S_2 were obtained from a segmentation of training images using pulse coupled neural networks [7] and are clearly the most discriminating features. Due to the large number of features we first analyse runs which use just these four features. After performing 80 runs and simplifying the evolved programs, we found that the best programs were all variants of two underlying algorithms:

$$Output = 5M_1 - 5S_1 - 5M_2 - 2S_2 \tag{8}$$
$$= M_1 - S_1 - M_2 - 0.4S_2$$
$$Output = 3M_1 - 3S_1 - 3M_2 - S_2 \tag{9}$$
$$\approx M_1 - S_1 - M_2 - 0.3S_2$$

Programs implementing equation 8 achieved a test accuracy of 70.7% while programs implementing equation 9 achieved a test accuracy of 69.5%.

Analysis of an Individual Program: We performed similar analyses to that shown in figure 3 for equations 8 and 9. As before, the analyses revealed that the algorithms are reasonable ones in the context of the window sweeping through the image with the position of highest output chosen as the predicted position of the sella landmark. In summary, features M_1 and M_2 assist with differentiating between the true position and a scene of constant brightness by the positive output of $M_1 - M_2$, while S_1 and S_2 are used for differentiating between cluttered scenes and the known position.

Analysis of Regularly Occurring Patterns Across Programs: As for the nose point, in all of the best evolved programs, the sum of the coefficients of the M_i features used was 0 and the coefficients of the S_i features were always negative.

Full Terminal Set. Eighty evolutionary runs using the full set of fourteen features shown in figure 2 were carried out. The linear functions shown in Equations 10-13 are the fittest programs from four randomly chosen evolutionary runs.

$$Output = 5M_1 - 2M_2 - S_2 - 2M_3 - 3S_3 - 2S_4 + S_5 - M_6 + S_7 \tag{10}$$

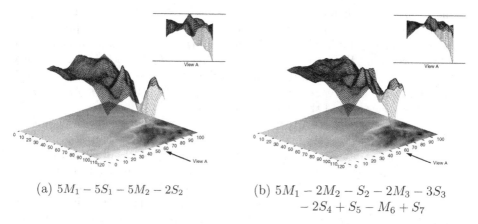

(a) $5M_1 - 5S_1 - 5M_2 - 2S_2$

(b) $5M_1 - 2M_2 - S_2 - 2M_3 - 3S_3$
$- 2S_4 + S_5 - M_6 + S_7$

Fig. 6. Visualisation of output landscape, sella point, reduced function set (a), full function set (b)

$$Output = 6M_1 - 2S_1 - 5M_2 - 3S_2 - M_3 - 3S_4 + M_5 + S_5 \qquad (11)$$
$$Output = 3M_1 - 3S_1 - M_2 - 4S_2 - M_3 + 4S_3 - S_4 - M_6 + S_7 \qquad (12)$$
$$Output = 6M_1 - S_1 - 5M_3 - 3S_3 - 2S_4 + S_5 - 2M_6 + M_7 + S_7 \qquad (13)$$

Analysis of an Individual Program: Comparison of these functions with equations 8 and 9 reveals that equation 11 could be considered as a refinement of equation 8. However, other similarities are difficult to find. Placing the input window in a number of positions and computing the outputs as before, shows that the output is highest around the true position of the landmark, but, unlike the previous two cases, does not reveal any intuition about why the program works.

Analysis of regularly occurring patterns across programs: After performing similar simplifications and substitutions as for the nose point, we found that the sum of the coefficients of the M_i terms in an equation was 0, as in the previous two situations. We have not been able to determine the significance of this, however the regularity is striking. However, the coefficients of the S_i were no longer all negative, as before. We have examined three dimensional views, such as figure 4, of program output. Many of these views are minor variations of the landscapes shown in figure 6. This provides additional evidence that the same underlying algorithms are being evolved.

6   Conclusions

Our aim was to determine whether we could develop explanations of how the evolved programs work for an object detection problem. We have succeeded in this to a large extent. The programs no longer need to be presented as some kind of black box that works by magic. We have shown that a methodology of simplifying function and terminal sets and simplifying the resulting programs to linear functions yields insight into the underlying algorithms implemented in the evolved programs. We found that underlying regularities were being consistently discovered in many repeated runs and that the regularities could be expressed as linear functions which were realistic in the context of the input window sweeping across the image to find the point of interest.

In the context of this particular object detection problem, the methodology works well for up to 4 terminals. For this number of terminals it is possible to comprehend the inter-relationships between the components of the linear function in the context of the sweeping process. For more than 4 terminals this is very difficult. However, the fact we have been able to identify the underlying algorithms for simplified problem instances gives us confidence that, in situations where there are too many terminals and functions to permit understanding, that the evolved programs are still capturing regularities of the domain.

We were surprised that the accuracy of the best programs did not drop with the reduced function set. It appears that genetic programming is very good at finding linear models for these kinds of problems.

References

1. Walter Alden Tackett. Mining the genetic program. *IEEE Expert*, 10(3):28–38, June 1995.
2. I. De Falco, A. Della Cioppa, and E. Tarantino. Discovering interesting classification rules with genetic programming. *Applied Soft Computing*, 1(4):257–269, May 2001.
3. Andy Song and Vic Ciesielski. Texture analysis by genetic programming. In Garrison Greenwood, editor, *Proceedings of the 2004 Congress on Evolutionary Computation (CEC2004)*, volume 2, pages 2092–2099. IEEE, June 2004.
4. Vic Ciesielski, Andrew Innes, John Mamutil, and Sabu John. Landmark detection for cephalometric radiology images using genetic programming. *International Journal of Knowledge Based Intelligent Engineering Systems*, 7(3):164–171, July 2003.
5. J. Cardillo and M.A. Sid-Ahmed. An image processing system for locating craniofacial landmarks. *IEEE Transactions on Medical Imaging*, 13(2):275–289, Jun 1994.
6. Y. Chen, K. Cheng, and J. Liu. Improving cephalogram analysis through feature subimage extraction. *IEEE, Engineering in Medicine and Biology Magazine*, 18(1):25–31, Jan–Feb 1999.
7. Andrew Innes, Vic Ciesielski, John Mamutil, and Sabu John. Landmark detection for cephalometric radiology images using pulse coupled neural networks. In Hamid Arabnia and Youngsong Mun, editors, *Proceedings of the International Conference on Artificial Intelligence (IC-AI'02)*, volume 2, pages 511–517, Las Vegas, June 2002. CSREA Press.

Undirected Training of
Run Transferable Libraries

Maarten Keijzer[1], Conor Ryan[2], Gearoid Murphy[2], and Mike Cattolico[3]

[1] Prognosys, Netherlands
mkeijzer@xs4all.nl
[2] University of Limerick, Ireland
{Conor.Ryan,Gearoid.Murphy}@ul.ie
[3] Tiger Mountain Scientific, Seattle, USA
mike@tigerscience.com

Abstract. This paper investigates the robustness of Run Transferable Libraries(RTLs) on scaled problems. RTLs provide GP with a library of functions which replace the usual primitive functions provided when approaching a problem. The RTL evolves from run to run using feedback based on function usage, and has been shown to outperform GP by an order of magnitude on a variety of scalable problems.

RTLs can, however, also be applied across a *domain* of related problems, as well as across a range of scaled instances of a single problem. To do this successfully, it will need to balance a range of functions. We introduce a problem that can deceive the system into converging to a sub-optimal set of functions, and demonstrate that this is a consequence of the greediness of the library update algorithm.

We demonstrate that a much simpler, truly evolutionary, update strategy doesn't suffer from this problem, and exhibits far better optimization properties than the original strategy.

1 Introduction

Run Transferable Libraries (RTLs) [7] have recently been introduced as a method for scaling evolutionary computation methods by performing learning across many runs. An example of an RTL implementation applied to a general Genetic Programming system [2] has shown that the system can acquire knowledge about a problem or problem domain as it is iteratively applied to different problem instances.

By implementing an evolutionary outer loop around a GP system, RTLs allow information gathered in specific runs to influence future runs via the manipulation of the library contents, which are made available to GP in place of the usual functions set. In effect, potentially useful motifs, patterns or functions are evolved. RTL systems search for a library from which an evolutionary system can produce an answer with less effort than a system using a standard function set used by GP. In this view, RTLs are similar to programming libraries used by human programmers. A library of functional primitives represents the API

M. Keijzer et al. (Eds.): EuroGP 2005, LNCS 3447, pp. 361–370, 2005.
© Springer-Verlag Berlin Heidelberg 2005

through which a task can be accomplished. A well designed library allows for "better"/brief/less-complex programming to accomplish the tasks targeted by the library.

A key component within an RTL system is the library maintenance/evolution. Each time an RTL system is applied to a problem instance the library may be updated. An RTL update algorithm, based on the usage counts of the library members, performs reproduction and variation operators on the members, increasing the number (and hence, initial bias towards) of good performers, removing those that didn't enjoy a lot of use and creating new functions based on the good performers using variation operators.

Previous work, using a basic library update algorithm, demonstrated that RTL enabled GP could solve the Parity and Multiplexer problems with orders of magnitude less effort than a standard GP system. Each of these classic ADF friendly test problems was employed to test the ability of the library to acquire the functional knowelege of the problem. However, the problems can be made much easier by favorably biasing the selection of the *single* correct function; XOR in the case of Parity, and IF-THEN-ELSE in the case of Multiplexer, which RTL facilitates. Availablity of useful functions alone does not necessarily make the problems trivial, rather it is a combination of first *discovering* of the function, and then the update algorithm's subsequent *biasing* of the library to that function so that it is over represented in the initial function set.

In this paper we test the RTL update algorithm on a significantly more difficult problem, a combination of a Parity problem and a Majority problem. With this we can check the capability of the algorithm to maintain a diverse set of functions. Interestingly, however, it turns out that a phenomenon of destructive feedback can occur, which forces the library to a suboptimal point in the library configuration, a local optimum.

It is hypothesized that this destructive feedback is caused for the most part by the use of *structural* information from a run, that is, the counts of function usage. In order to examine this, a very simple alternative algorithm is developed that looks at the performance of the entire library as a whole, rather than trying to optimize individual parts of it.

2   Background

Programming libraries are an important tool for any programmer. They provide collections of related functions that can be helpful when decomposing tasks. Analogies for GP include ADFs,[3] [4] GLib [1], ARL [6], amongst others. Typically, these systems generate modules in parallel with the population that use them, although *Subtree Encapsulation* [5] system attempts to use the effort from each run to bootstrap the following one.

See [7] and [2] for comprehensive review of module acquisition methods. As described in section 3, the libraries in these cases turned out to be seething with activity, from the time that the most useful function was initially discovered until the library was suitably biased towards it. This is because, as the library converged towards an optimal bias, other, sub-optimal but still useful, functions fought for control of the library, and these struggles often continued long after the library reached its maximal capabilities.

The concept of Run Transferable Libraries (RTLs) overturns the notion of independent runs for GP. The basic insight that inspired the approach is that in order for a GP to scale from simple to more difficult problems, some method needs to be in place that can transfer knowlege, accrued in previous runs, to further runs on the same problem, as well as on different problems from the same domain. In the RTL approach, this knowledge takes the form of a library, a set of concise functionality that provides *functional bias* to individual runs. Particularly, it is hypothesized that gathering knowledge on simpler problems in a problem domain can be useful to tackle harder, more interesting problems. Previous work made use of a sequential update mechanism that harvests usage information from a particular run and transmits it to further runs. A complete description of this mechanism can be found in [7].

A generic RTL implementation provides algorithms for library member referening as well as library contents maintainance. These provide for the execution of the library member code and the timing and style of updates to the library contents. An RTL system has the ability to cause evolution of the library at a time scale which is separate from the the evolutionary time scale of the solving a problem instance. The purpose of evolving/changing the library is to change the landscape of primitives available to the next run, enhancing the system's ability to reach a solution to a problem instance with less evaluations. This work continues to use the basic framework from the the previous system.

The library is initially made up of randomly generated functions of various arities (typically 0 to 3), each of which can be addressed by a tag, usually, but not necessarily, represented by a real-valued number. These functions are constructed using the function set typically associated with the problem at hand, but do not contain any problem specific terminals, instead using parameters where appropriate. The individuals are constructed from the terminal set and *Dynamic Linking Nodes*, which are effectively pointers into the library. These *DLNodes* take from 0 to 3 arguments as appropriate. When an individual is evaluated, each DLNode is replaced with the corresponding library member.

In contrast with the system used in [2], the system used here is an RTL system that is limited to the domain of induction of boolean functions. It was designed to remove the need to synthesize functions, in order to study *selection* of functions. This system implements all possible arity 1,2 and 3 boolean functions in the GP system in the form of lookup-tables. This leads to a constant function set consisting of $4 + 16 + 256 = 276$ possible functions. An RTL consists simply of 276 floating point numbers, representing proportions of occurance of the 276 functions. In the individual runs, these proportions are used during initialization and mutation, thereby mimicking a library of multiply occuring functions. The update routines manipulate these proportions to model the changes in the library over many runs.

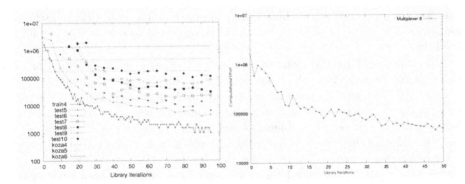

Fig. 1. (Left) Number of individuals needed to be processed to obtain a solution with 99% probability on the parity problems using a library trained on parity 4, and applied to higher parity problems. As a benchmark, the results presented by Koza with ADFs are drawn as straight lines. (Right) number of individuals needed to be processed to obtain a solution with 99% probability on the 6-multiplexer problem using a population of 500. Koza reported 4,478,500 using a similar population. All results are averaged across fifty independent libraries

The update routine described in detail in [2], is implemented in this system as a simple manipulation of proportions: during the run, the usage of the functions are tracked and proportions are increaased or decreased according to their usage; using an exponential moving average to wash out initial bias. Variation is simulated by maintaining a low background proportion of each possible functions.

3   Library Convergence

RTL was initially tested on a number of classic ADF problems. Figure 1 shows how the system performed on parity and multiplexer

For the parity problem, an initially random library was trained on parity-4 for 50 iterations (serial GP runs), and then subsequently tested on a range of problems from parity-5 to parity-10, although there was *no* learning permitted during testing. That is, each of the higher parity problems was tackled using the same library. This set of experiments was to test how well a library could scale across different problem instances when trained on a simple version.

Even as the difficulty of the problem was scaled up, the RTL enabled method was usually able to solve the problem an order of magnitude more quickly than standard GP with ADFs. Particularly when using problems of smaller scale to *prime* the library, the performance on larger problems was enhanced.

The key to the success of RTL on these problems was the identification of useful functions (XOR in the case of Parity, IF-THEN-ELSE for the Multiplexer) early on, and their subsequent application. As the library was trained, it became more biased to these functions so that they were more common in the initial generations of individual runs, and so the performance kept improving.

If we wish to apply RTL to larger scale problems, it is reasonable to assume that it will need to maintain a balance of functions in the library. To this end, we introduce a new, synthetic problem, which we refer to as the *Mix* problem. A Mix problem requires individuals to solve two (or more) related but distinct problems, and is easily scalable in terms of the number of problems required to solve, and the size of the problems.

Much previous work has looked at the *n-parity* problem, as this can easily be scaled simply by increasing the number of inputs. Another popular test problem for module acquisition strategies is the multiplexer problem, which can also be easily scaled, but not with the same kind of fine grained control as parity, because not all numbers of inputs are legal test cases for the multiplexer, i.e. three, six and eleven inputs can be used for multiplexers, but not four, five, seven etc.

As well as parity, we use a boolean problem that can be scaled in almost as fine a grain manner as parity, the *Majority* problem. Majority takes an odd number of inputs and returns whichever value is in the majority. Thus, a Mix-2-X problem is a combination of two problems, each of which have X inputs. All test cases will have $X + 1$ inputs, with the extra bit indicating to which problem the test case belongs, effectively forcing the individuals to perform a multiplexing operation also.

The optimal function for the Parity problem is known to be XOR. Once it has been discovered, the problem can be solved relatively easily through the judicious application of this function. Similarly, an optimal Majority function also exists, which we term the **Majority-3** function. This arity three function takes three arguments and returns whichever is in the majority. Typically majority solutions are larger than parity solutions, because the inputs need to be used multiple times to solve the problem. For the Mix problem, it is not quite clear what the optimal solution needs: even though the problem is constructed in such a way that it is solvable with a mix of XOR and Majority functions, it is quite possible that other strategies are viable.

3.1   Running the Update Routine on the Mix Problem

1500 full library runs of 50 iterations were performed on the Mix-2-5 problem. The GP system using the library used a population size of 500 individuals for 50 generations. This amounts to 1,250,000 individuals evaluated per library run, and 1,875,000,000 individuals evaluated in total. At each iteration the best of run was recorded. As a control run, 1000 runs of the GP system with an initial uniform bias were performed, which lead to an estimated mean performance of 53.1 +/- 0.02 for a 'standard' GP system. The average performance of the library is displayed in figure 2. Although the library improves upon the baseline performance, the improvements quickly level off, and actually decline when the run is prolonged.

Closer inspection reveals that a local optimum is found in the search space, the solution to the Majority-5 problem, which only uses Majority-3 primitives.

This function only uses 13 nodes, and scores 54 hits (84% accuracy), as it also scores 22 points on the Odd-Parity-5 part of the problem. The Majority-5 solution functions as an attractor. When it is found in a run, it reinforces the use of the Majority-3 primitive maximally, to the detriment of other functions. The reinforced library in its turn makes it more likely to find the Majority-5 solution, which leads to a further round of reinforcement of the Majority-3 primitive, ultimately leading to a library containing such a high proportion of Majority-3 functions that no other solution will be found. Libraries that have surpassed the performance level of the Majority-5 solution are also likely to get trapped, because in order to solve the problem they need a fair proportion of Majority-3 primitives. This also makes it more likely to induce the Majority-5 solution leading to an ever-present risk of becoming trapped in the attractor.

3.2   Reducing the Feedback Strength

We hypothesize that the prime cause of the strength of the attractor lies in the tight coupling between usage of the primitives and reinforcement in the library. This direct feedback loop between primitive usage and library update without regards to the performance of the particular run seems to lead to a situation where a library can easily get worse at solving a single problem, let alone a problem class over time. Altering the update routine so that it takes into account the performance of the individual runs is a possible route to take, but this does lead to a further set of parameters and issues when applying the library on new problems. It will also not solve the issue of getting trapped in the local optimum at the onset. To investigate the feasibility of guiding the library using performance metrics only, we implemented a simple ES-type of outer loop, where the content and proportions of the library functions is evolved without looking at the usage of the elements in the individual runs at all. It is thus a pure evolutionary meta-algorithm for optimizing the library, where an individual fitness case is a single run of a GP system using that library.

The ES-style algorithm forms a very simple wrapper approach around the main run of the system. It implements an (1+5) strategy, where the current library (a set of numbers designating proportions of functions), is mutated 5 times using:

$$p \leftarrow \exp(\log(p) + N(0,1))$$

All six libraries, the parent and its offspring, are evaluated a single time on the problem at hand, and the library that has best performance is chosen for the next iteration. There is therefore no direct interaction with the run through module harvesting and/or updating, which makes this approach significantly different from our previous approach, but also from Module Acquisition, ADFs and GLib. Any and all feedback from the individual runs is transferred through the competition between six libraries that are allowed to evolve a solution a single time. For performance reasons this very simple strategy is used, but with availability of computational resources it is clear that the method can be extended with a significantly larger population of competing libraries.

Although true synthesis of new functions is not implemented in the system due to the presence of all possible functions up to arity 3 in the library, it seems straightforward to implement such a system undergoing mutation and crossover on the level of library functions. Future work will address this; the idealised situation of having all possible functions is used here to test if the deceptive nature of the mix problem can be overcome by decoupling the evolving library and the problem-solving runs even further than previously.

4   Experiments

Our first set of experiments were designed to test the new library ES-style update strategy. For this we used a Mix-2-5 problem, with a population of 500 running for 50 generations. A meta-run of the ES algorithm consists of 50 generations, leading to 300 evaluations (GP runs) per library run.

Figure 2 (left) shows the performance of the system using the greedy update strategy, while figure 2 (right) describes the performance using the new, ES-style strategy. The greedy updating strategy is quickly lead into the deceptive local optima and, after a peak in performance after around 15 iterations it starts to deteriorate. The error bars show that this is not accidental. Clearly, this is something that should never happen with RTL, as this means that even useful functionality can be washed out of a library by inapropriate feedback. The ES approach, on the other hand, does not suffer from this problem, and the performance of the system is still increasing even after 300 evaluations. In sequential terms, this is equivalent to 15,000 generations, which makes it all the more impressive that RTL is still improving even after all that time. Even though the ES run uses 300 evaluations, it is already abundantly clear that the system improves upon the greedy routine even after 50 evaluations (appr. 8 meta-generations). The undirected variations from the ES-algorithm steer the libraries away from the local optimum.

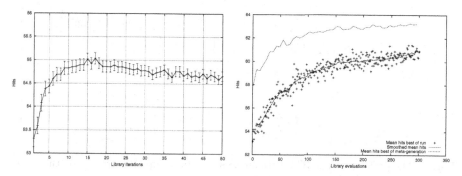

Fig. 2. A comparison of the performance of RTL on the Mix-2-5 problem using the original greedy scheme (left) and an ES style library update strategy (right). A meta-generation is the set of libraries created on each library generation. Notice the difference in scales

Table 1. Comparison on mean hits after 50 library evaluations for the ES and update routines. The difference in mean hits is tested for significance using a two-sided t-test

Experiment	Best	P value
Mix-2-5	ES	10^{-12}
Parity 9	greedy	0.49
Multiplexer 6	greedy	0.20

A second set of experiments was performed to test how well the ES strategy performed on the original problems (that is, Parity and Multiplexer) in order to see how it would handle situations that so clearly suited the greedy strategy. Also here the ES system performs adequately and inspection of the libraries reveals that the contents show a much greater diversity of functions than with the update routine. This diversity comes at a cost however, the ES algorithm being slower than the update mechanism to focus on the appropriate functionality. To check the differences in performance between the two methods, 500 runs of 50 iterations of the update routine are tested against the ES-style algorithm. The results of the test can be found in Table 1. Even though the standard update routine performed slightly better on the Multiplexer and Parity problems than the ES algorithm if we concentrate on the performance after around 50 library evaluations, the performance improvement is statistically inconclusive. The performance difference on the Mix problem however is however highly significant.

The fact that the greedy update scheme performs better (although not statistically significantly so) on the original problems is simply a consequence of a greedy algorithm performing better on a problem with a clear path to the optimum. However, the ES style algorithm is considerably more robust, performing extremely well on the difficult problem, yet still capable of attaining a comparable result even on these, more direct problems.

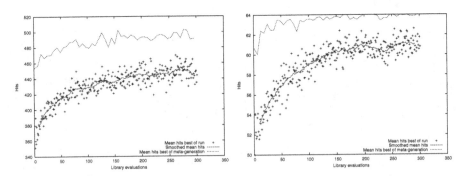

Fig. 3. The performance of RTL with an ES style library update strategy on the parity 9 (left) and 6-multiplexer (right) problems

5   Conclusions and Future Work

The paper has described an alternative method for updating Run Transferable Libraries, a method for scaling GP by transferring knowledge across runs. RTLs operate with an inner loop (normal GP) and an outer loop (library maintenance) which give the system a combination of characteristics from both Darwinian and Lamarckian evolution. This is possible because the libraries endeavour to capture some of the salient characteristics of successful individuals from earlier runs, effectively Lamarckian evolution, with a view to seeding standard runs that evolve in the normal way (Darwinian evolution).

This paper presented a boolean problem, a mix of Parity and Majority, to test the ability of a RTL to maintain a diverse set of different functionality in a library. It was found that for this particular problem, the greedy update routine used in previous work was susceptible to begin trapped in a local optimum: the solution to the Majority problem. It will even get trapped when, on the surface, it has already surpassed the level of performance of this local optimum. Closer inspection reveals that the feedback loop between library usage and library re-inforcement is the most likely cause of this sub-optimal behaviour. To tackle this issue, a true evolutionary meta-algorithm was developed that receives feedback over individual runs only through the performance of the runs: no updating according to usage and also no module-harvesting is performed. It is experimentally verified that this method does not get trapped in the local optimum, and also that it performs adequately on problems where a greedy strategy is more optimal.

We demonstrate that, on this difficult problem, direct feedback to the library about the structures in use from individual runs can be detrimental to the search efficiency. It is shown that the performance of the library on individual runs is a critical component in improving the search. However, the performance of individual runs is a stochastic quantity, as it is subject to all stochastic effects present in standard GP systems. It was shown that a very simple strategy, where a library is tested using a single run is adequate to solve a range of problems in the boolean domain.

It was further demonstrated that, on simpler problems, where the best strategy for the library *is* to converge on a single function, there is no statistically significant difference in the performance of the original and new library update systems.

5.1   Future Work

The investigations in this paper give some indications that the intuition that using detailed information from a run of a GP to configure a library useful for other runs may be misguided. GP is a system that is not 100% reliable and using the results of a GP run as a starting point for another run might not be the best approach. The alternative to this approach is to view the process of creating a library as an optimization problem in its own right, where the library evolves and varies in an independent manner, only using feedback in the form of the library's capability in solving the problem at hand.

Initial work [7] has been carried out on identifying the quality of the online performance of the libraries, in which it was shown that the size of the best of run individuals produced by a library, relative to others produced by a similar library, can be used to roughly gauge the generalisation properties of the library. Although it was demonstrated in this paper that there was no statistically significant different between the two update methods on the simpler problems, it could be the case that there may be situations where the original, greedy method may be advantageous. The ability to determine the generalisation qualities of a library would facilitate the construction of a hybrid system.

All our results to date have indicated that RTL has the potential to scale in ways normally unassociated with Evolutionary Computation. Our next priorities will be to examine increasingly larger Mix problems and to examine real world domains for their applicability to RTL.

References

1. P. J. Angeline and J. B. Pollack. The evolutionary induction of subroutines. In *Proceedings of the Fourteenth Annual Conference of the Cognitive Science Society*, Bloomington, Indiana, USA, 1992. Lawrence Erlbaum.

2. Maarten Keijzer, Conor Ryan, and Mike Cattolico. Run transferable libraries — learning functional bias in problem domains. In Wolfgang Banzhaf, Jason Daida, Agoston E. Eiben, Max H. Garzon, Vasant Honavar, Mark Jakiela, and Robert E. Smith, editors, *Proceedings of the Genetic and Evolutionary Computation Conference*. Springer Verlag, 13-17 July 2004.

3. John R. Koza. Genetic programming: A paradigm for genetically breeding populations of computer programs to solve problems. Technical Report STAN-CS-90-1314, Computer Science Department, Stanford University, 1990.

4. John R. Koza. *Genetic Programming II: Automatic Discovery of Reusable Programs*. MIT Press, Cambridge Massachusetts, May 1994.

5. Simon C. Roberts, Daniel Howard, and John R. Koza. Evolving modules in genetic programming by subtree encapsulation. In Julian F. Miller, Marco Tomassini, Pier Luca Lanzi, Conor Ryan, Andrea G. B. Tettamanzi, and William B. Langdon, editors, *Genetic Programming, Proceedings of EuroGP'2001*, volume 2038 of *LNCS*, pages 160–175, Lake Como, Italy, 18-20 April 2001. Springer-Verlag.

6. Justinian P. Rosca and Dana H. Ballard. Hierarchical self-organization in genetic programming. In *Proceedings of the Eleventh International Conference on Machine Learning*. Morgan Kaufmann, 1994.

7. Conor Ryan, Maarten Keijzer, and Mike Cattolico. Favourable biasing of function sets. In Rick Riolo, Una-May O'Reilly, and Tina Yu, editors, *Proceedings of the second Genetic Programming Theory and Practice Workshop*. MIT Press, To Appear 2004.

Zero Is Not a Four Letter Word: Studies in the Evolution of Language

Chris Stephens[1], Miguel Nicolau[2], and Conor Ryan[2]

[1] Instituto de Ciencias Nucleares, Universidad Nacional Autonoma de Mexico,
stephens@nuclecu.unam.mx
School of Theoretical Physics, Dublin Institute for Advanced Studies
stephens@stp.dias.ie
[2] Department Of Computer Science And Information Systems,
University of Limerick, Ireland
{Conor.Ryan, Miguel.Nicolau}@ul.ie

Abstract. We examine a model genetic system that has features of both genetic programming and genetic regulatory networks, to show how various forms of degeneracy in the genotype-phenotype map can induce complex and subtle behaviour in the dynamics that lead to enhanced evolutionary robustness and can be fruitfully described in terms of an elementary algorithmic "language".

1 Introduction

Evolutionary algorithms (EAs) have been applied with a remarkable degree of success to a large variety of problems. However, this is often done with little or no understanding of the dynamics of the system, and practitioners often find themselves unable to explain why a particular tweak has apparently improved their system. Features such as degeneracy and neutral evolution are generally accepted to aid evolution, but little detailed work, particularly in Genetic Programming (GP), has been carried out to investigate how GP exploits these features (with a few notable exceptions [1]). This paper presents a study in evolutionary dynamics, demonstrating how a detailed analysis detects complex and subtle structure formation. Phenomena such as competing conventions, robustness and redundancy are examined and we demonstrate how it is natural to describe these phenomena in the framework of natural language.

In section 2 we describe the representation of our system, showing how it has all the salient features of GP systems, as well as a Genotype-Phenotype map (GPM) inspired by genetic regulatory networks. We also describe the different types of degeneracy inherent in the system. Section 3 takes an in depth look at a representative run, illustrating some surprising strategies that are adopted by the system. The paper concludes with a summary and discusses areas in which our system can be used to develop a deeper understanding of bloat, degeneracy and neutrality, all of which are crucial to the development of more robust EAs.

M. Keijzer et al. (Eds.): EuroGP 2005, LNCS 3447, pp. 371–380, 2005.

2   Representation

The representation we use is based on two existing systems; Grammatical Evolution (GE) [2] and that of [3], where a GPM inspired by gene expression models and the phenomenon of cellular division was used. Both exhibit genetic redundancy: GE via a *degenerate* mapping scheme, where many *codons* map to the same item, while in [3] the redundancy was at the gene expression level, in that only a certain proportion of genes from an individual needed to be expressed to produce a fully specified phenotype.

The underlying representation is similar to GE in that we employ a binary string representation, but in this case, a fixed length string is used. The genotype consists of N_g genes, with each gene consisting of N_c codons, where each codon takes a symbolic value taken from an alphabet of size N_a. In our experiments we consider $N_c = 4$ and $N_g = N_a = 8$, so each codon is described by three bits.

The first step is to *transcribe* the genome from the binary representation to eight genes of four codons each. Once that is done, the first gene, known as the *switchboard* gene, establishes which genes are inhibited or promoted, by using each of its four codons as indexes to the promoted genes. Then, in a manner not entirely dissimilar to cellular growth, the initial structure is replaced with a new one, consisting of four genes. Fig. 1 illustrates this process.

Once the activated genes have been identified, the system reverts to a standard GE type mapping, with each codon being used to make a choice in a grammar. Consider the grammar below, with each production rule numbered. As the codons are represented by three bits, each decodes to a value from 0 to 7. This is referred to as a *closed* grammar [2], that is, there is only one non-terminal and, hence, just a single context.

```
<e> :: = tanh(<e>)      (0)    add(<e>,<e>)   (4)
           tanh(<e>)      (1)    add(<e>,<e>)   (5)
           tanh(<e>)      (2)    X              (6)
           tanh(<e>)      (3)    0              (7)
```

Consider the following individual, already reduced to its activated genes:

4567 1623 0021 4401

Each codon will *always* make the same choice, e.g. codon 7 will always perform the mapping e -> 0. The mapping steps are as follows:

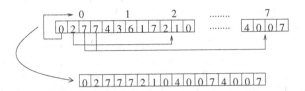

Fig. 1. An example of the switchboard gene in operation. Each codon in the switchboard gene (in the current version, always located in the first position) acts as an index into the *entire* genome. Notice how this particular switchboard gene indexes itself

```
4 -> add(e,e)
5 -> add(add(e,e),e)
6 -> add(add(X,e),e)
7 -> add(add(X,0),e)
1 -> add(add(X,0),tanh(e))
6 -> add(add(X,0),tanh(X))
```

Notice that only six codons are used for the mapping; in the case where all codons have been used and still non-terminals remain, a fitness of zero is given. This straightforward mapping scheme permits us to use a more convenient notational representation to facilitate human interpretation of codons. That is, a shorthand for each codon value can be inserted as follows: 0, 1, 2, 3 → h; 4, 5 → +; 6 → X and 7 → 0. Thus, the individual above can be rewritten as:

++X0 hXhh hhhh ++hh

We refer to genes described in this way as *words*. Because only six codons were used, we could, using schema notation, describe the above "sentence" as

++X0 hX ✱✱✱✱ ✱✱✱✱**

Notice that the second word is not made up of four distinct codons/letters. More generally, the last word *used* can vary in length from one to four letters.

2.1   Neutrality

Degeneracy leads to the existence of neutral networks [4], where individuals from different areas in the search space have the same fitness.In this work degeneracy exists at several levels; that is, there are several ways of describing the same functionality, e.g. $add(X, 0)$ and $add(0, X)$. Operators are neutral when the individual they produce is genetically different to the individual that they operated on, but phenotypically the same. Point mutation can be neutral at several of the levels above, e.g. changing a codon value such that it still selects the same rule as before, changing a value on the switchboard gene so that it generates the same word as before, but from a different gene, etc. It is also possible to perform neutral crossover.For example, crossover might only effect non-activated genes, both parents might have a copy of a required gene, etc.

2.2   Gene Expression as a Language

The kinds of words that one would expect to be produced by this grammar depend on the fitness function. Consider the function $f(X) = 4X$; this can be described by using a number of different sentences.One possible solution is:

+++X XXX✱

This particular sentence maps to a minimal solution of $(+(+XX)(+XX))$. Another solution, which maps to the same phenotype is :

+X+X +XX✱

Notice how these sentences are fundamentally different because their first words differ. This does not suggest a lack of robustness, however, as one would not expect a single evolving population to balance two such different solutions at the same time for very long, although, as described below, it is possible for a number of distinct optimal solutions to appear throughout a run, often competing with each other for dominance of the population, until one becomes extinct.

The role of modularity in most complex problem solving systems, including nature, cannot be over estimated. Difficult problems are often best solved by decomposing them into a set of smaller ones, each of which can be solved more easily than the whole. Similarly, simple modules or strategies which can be reused several times, either on different problems or while solving a single problem are likely to be preferred over more complex ones. Consider the individual:

+++X +++X X000 X000

This maps to $(+(+X(+(XX))(+00))(+0X))$, and can be reduced to $4X$.

3   Results

We applied the system to the problem of performing symbolic regression on the function $f(X) = 4X$. By normal GP standards, this is a trivial problem which one would expect to appear in a reasonably sized population with a good initialisation scheme. However, we are concerned with making a detailed analysis of the dynamics, and a simple function like this keeps the analysis tractable.

In total 30 runs were conducted, all of which discovered an optimal solution. Typically, the solutions initially consisted of three or four expressed genes, but shorter solutions almost always appeared, reducing the length to usually two or sometimes three genes. Repeated activation of the same gene was ubiquitous. Typically, each run discovered several ways to represent an optimal solution. A population of 100 individuals was used with a mutation rate, implemented at the bit level, of 0.01. One-point recombination was used with probability 0.9 and restricted to occur only at the boundaries between genes. For selection, a rank-based method was used, where the ranking was applied only to individuals that successfully mapped onto syntactically correct expressions.

3.1   Description of Algorithmic Language

In this section we consider a detailed description of a particular run in order to show the complexity of the dynamics associated with the GPM, even in the case of our very simple search problem. With the production rules specified in Section 2, starting off with a random population one finds, as expected, that the initial codon frequencies are approximately: $h = 50\%$, $+ = 25\%$, $X = 12.5\%$ and $0 = 12.5\%$. Later on, however, one sees structure begins to emerge - for instance, usage of codon h is significantly less over the majority of the run, while usage of codon X is significantly greater, as it plays a useful role in fit solutions. With codon 0 there is a greatly increased usage over the middle part of the run, for a reason that will become apparent shortly.

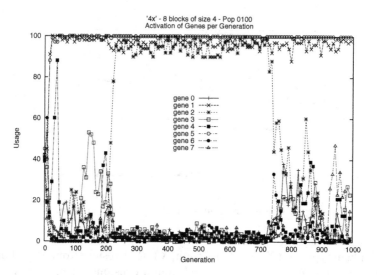

Fig. 2. Gene activity per generation. The graph plots the percentage of individuals per generation that have a particular gene activated

As on the switchboard gene the distribution of codons is random, one finds initially a random distribution of activated genes, i.e. that any gene is active on average in 50% of the population. Later on, however, gene activity patterns emerge with more structure, as can be seen in Fig. 2. Very early on in the evolution an ordering convention is arrived at, whereby the content of the switchboard gene is largely fixed, with more than 90% of the individuals activating genes 5 and 1. Interestingly, there is a redundant usage, whereby these codons are very frequently repeated in the switchboard. As we shall see this leads to enhanced robustness. At generation 25 more than 95% of switchboard genes in the population are of the form $5 * **$ while over 60% are of the form $55 * *$. Similarly, 92% possess $* * 1*$ and 83% possess $* * 11$.

In Fig 3 we show the different codon frequencies in gene 5 as a function of time. Before an optimal solution is found gene 5 consists of almost 50% of codons of type $+$, i.e. producers. This is about double what would be expected with a random distribution. Similarly, block 1 contained nearly 90% more codons of type X than would be expected by chance. Gene 5 is clearly the most important, or "core", gene as its codon content is so stable that there must be strong selection pressure in order to maintain this stability. Gene 1 by contrast showed more variation, though the important role played by codons X and 0 was evident. Note that when expressed, gene 5 precedes gene 1, hence, as expected, the system puts more producers to the left and more terminals to the right.It is interesting to note that optimal solutions are detected from time to time but it is not until after generation 200 that they become established in the population. Also, fitter solutions tend to use only two different active genes - 5 and 1 - and three expressed ones - $551*$ - the final non-coding tail not being expressed. The most

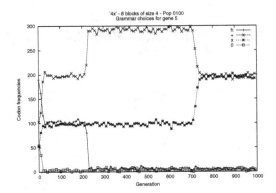

Fig. 3. Grammar choices for gene 5 per generation

common phenotype is $f(X) = 2X + 2\tanh(X)$, that results from a combination of a repeated producing gene $5 = ++hX$ and a consuming gene $1 = XX00$.

At generation 213 optimal solutions finally begin to successfully propagate through the population. At this point the number of $+$ codons in gene 5 increases by 50% while the number of 0 codons in gene 1 doubles. The first solution found is shown below. Notice that the switchboard block appears twice, once in decimal form to facilitate reading, and once in the normal form, $++hh$ in this case.

```
5533.++hh.XX00.0+h+.X000.++hX.+++X.++Xh.+Xhh
```

```
(+(+(+X(+(+(+X X ) 0) 0)) 0)) X) = 4X
```

Note that this founder does not activate gene 1 as the majority of the population. It is based on the same switchboard template $55**$ of previous suboptimal solutions, but is achieved through a single mutation of the 5 gene $++hX \rightarrow +++X$ which, repeated, combines with gene $3 = X000$ to give $4X$. 5533 however, is not the dominant switchboard gene, that role being kept by 5512, and therefore the dominant optimal solution had 5512 as a founder switchboard. Moreover, with this switchboard, in general all four genes 5, 5, 1 and 2 are expressed, as can be seen in the huge increase in use of gene 2 after generation 200 in Fig. 2.

Given that all these solutions are optimal one might expect that, on average, evolution preserves their structure, apart from the effect of neutral drift. However, this is not the case. For the next 500 generations this solution spreads and evolves. Early on, among the optimal solutions, in gene 1 over 60% of the individuals have a $+$ codon. Later on this percentage has dropped to zero!There is no direct selection pressure for this, as we are talking about the structure of this gene only for optimal individuals. In terms of effective fitness [5] however, there is a clear explanation: any $+$ in gene 1 means that in order to maintain an optimal solution more codons from gene 2 must be expressed. As these are subject to mutational damage there is an effective selection pressure to make the solution more robust within the context of a repeated "core" gene, $+++X$, that needs a minimum of 5 more terminal codons. The elimination of the $+$ codon reduces the total number of codons that are expressed in the optimal solutions and therefore increases robustness against mutational damage.

At generation 704 a new optimal solution appears - still based on the switchboard gene 5512 but with a mutated core gene 5 of the form $+X + X$. An immediate advantage of this solution is that it uses fewer expressed codons and therefore will be more mutationally robust. However, another complementary and more subtle effect appears: $55 * *$ in the switchboard gene requires a final terminating 0 codon in the first position of the following gene (or some other codon combination, such as $hh0*$, that evaluates to zero) in order to provide an optimal solution. In the initial population, when $+X + X$ solution is first found, this is provided by gene 1, associated with the switchboard $551*$, as more than 50% of the individuals that used the $+ + +X$ core gene already had a 0 codon in the first position there. However, of the 67 optimal individuals at generation 704 only 5 had a 0 first position codon. 300 generations later, an examination of genes 4, 6 and 7, which are neither activated by the switchboard nor expressed, show that, from the 246 such genes associated with the 82 optimal individuals, 101, i.e. 41% have a 0 in the first position. This is almost three times the percentage (15%) expected if the distribution were random! What is the reason for this extraordinary self-organization - the origin of our somewhat tongue-in-cheek title? The answer is that from $551*$ a mutation on the switchboard of the third codon to activate any gene that has a 0 codon in first position would result in an optimal string. In this sense many of the non-activated genes are acting as a genetic "reserve" to protect againstmutations of the switchboard gene.

As in [5] one can summarize the algorithmic "language" that has emerged. This is an evolving language, in the sense that the system continually finds "fitter" genes (the "words" of the language) and "fitter" ways of expressing them through the switchboard (the "syntax" of the language), where "fitter" means an effective fitness that also measures evolutionary robustness. In Table 1 we give a description of the algorithmic language that emerged after 1000 generations.

One may be surprised that the system based itself on $(+X(+X(+X(+X\ 0))))$, which actually uses more expressed codons - 9 - than the minimal solution $(+X(+X(+X\ X)))$, which uses only 7 and therefore might be surmised to be more evolutionarily robust. The answer is, of course, that because the first solution uses a repeated building block - $+X + X$ - this block may be expressed twice. In this case the effective number of codons that are subject to mutational damage of $(+X(+X(+X(+X\ 0))))$ is only 5 compared to the 7 used by $(+X(+X(+X\ X)))$.

We have discussed the evolution of robustness and believe that the above description of the results of the experiment offers unequivocal evidence for it. However, one can determine more rigorous and quantitative measures. If one can think of one optimal solution as being more robust than another then this should imply that the more robust state's "neighbours" are on average fitter, where the notion of neighbour depends on the operator we are thinking of robustness with respect to; for mutation, it is natural to use Hamming distance as a measure of

Table 1. Description of the algorithmic language that has emerged by generation 1000

Words of the Language		
Gene	Codon content	Logic
Gene 5	$+X+X$	Gives possibility of at least 2X when expressed once and 4x when expressed twice
Gene 1	$0***$	Terminates the twice expressed $00**5$ gene with 0 codons that do not affect the 4X expression
Genes 4,6,7	$0***$	Genetic Reserve - predominantly not expressed but backup in case of mutation in third switchboard codon
Genes 2,3	$****$	"Non-coding" regions - either do not form part of the language or have unknown functionality
Syntax of the Language		
Switchboard	$551*$	Puts producing gene 5, expressed twice, before consuming gene 1 to given optimal ordering

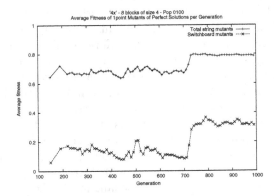

Fig. 4. Average fitness of 1-point mutants of perfect individuals, and of 1-point mutants of each perfect individual's switchboard

neighbourhood. To this end, for every optimal solution we consider the average fitness of its 96 one-mutant neighbours. In Fig. 4 we see a graph of this quantity as a function of time. The key observation is: upon discovery of the solution $(+X(+X(+X(+X\ 0))))$ at generation 704, the system now has a solution that uses in the genotype only five distinct codons, other than the switchboard gene, whereas the previous solution, based on $+++X$, used 9 distinct codons. This implies that, as the $+++X$ solution has $(32-13)=19$ non-expressed codons, the $+X+X$-based solution, which uses 4 less should have $4/19 \sim 20\%$ more optimal one-mutant neighbours which is roughly the increase seen in Fig 4.

The evolution of robustness is even more pronounced if we examine the fitness of the one-mutant neigbours of optimal solutions where we consider only the switchboard gene, i.e. only 12 neighbours. Fig 4 also shows the temporal evolution of the average fitness of the one-mutant neighbours of the switchboard. We have also examined the evolution of the ratio of one-point mutants of both the entire genotype and the switchboard that generate optimal and valid solutions respectively. The behaviour is similar to that for fitness, showing marked increases when a more robust solution is found. Naturally, in the case of the switchboard, due to the important role this gene plays in generating the syntax, and the lack of neutral mutations due to the fact that all four codons are used to activate other genes, one notes that the average fitness of the one-mutant neighbours is small. This is a sign that the switchboard is more brittle than many of the other genes. However, even in the case where the optimal solution uses 4 expressed genes, there is still some degree of robustness. As the average number of optimal mutants can be as high as 13% this means that up to 2 switchboard codons could be mutated and still leave an optimal individual. After generation 704 the new class of optimal solution uses only three activated genes hence the fourth codon on the switchboard loses its importance. Therefore a minimum of 25% of the one-mutant neighbours should be optimal. However, it was observed that $27 - 28\%$ is the norm and hence, the system had evolved robustness above and beyond just finding a solution that uses fewer expressed codons.

While our conclusions thus far have been gleaned from an examination of a single (though representative) run, all our experiences with other runs suggested that the phenomena we observed are common across a large number of runs. It is however legitimate to ask what happens over those runs. The problem with this is that many of the observed phenomena are contingent: the switchboard structure, and subtleties such as genetic reserve, will look quite different in different runs. Two basic related phenomena associated with robustness that can be seen over many runs are: the tendency to activate more than once the same gene - especially the core gene - and a tendency to use fewer expressed genes. In fact, solutions that use two expressed genes occured considerably more frequently than three-gene solutions, (typically 10-15 times more often) while four-gene solutions are rare indeed. This tendency is the equivalent of the more familiar phenomena of bloat in standard GP. In both case there is a tendency for the system to reach a state where the ratio of coding material to non-coding material is minimized. In GP this is achieved mainly by increasing the amount of non-coding material while here is it by minimizing the amount of coding material. Obviously, the payoff is enhanced evolutionary robustness via resistance to mutational damage.

4   Conclusions

In this paper we investigated how the existence of a degenerate GPM can lead to the evolution of robustness and the emergence of an algorithmic language as a result of the self-organization of the GPM.

In distinction to previous work, we concentrated on an in depth analysis of a single run, to give an idea of the tremendous subtlety and complexity of the phenomena that can occur, even in this simple scenario. We saw that the manner in which the system can build robustness can be very varied, from simply developing solutions that require fewer expressed genes, to influencing the content of non-coding parts of the genome and the pattern of gene expression, such as the creation of genetic reserves. We saw and quantified a tendency to reduce the size of the effective coding region - a phenomena analogous to bloat in GP. We saw that robustness can evolve both continuously and in a more punctuated manner, as when passing between solutions with different numbers of expressed genes.

Our study was motivated by a desire to offer a phenomenological predictive framework and description of the evolution of robustness in the context of a genetic model with some language-like features. There exists a formal mathematical framework in which to describe these phenomena - induced symmetry breaking of the genotype-phenotype map and effective fitness as a quantitative measure of this fitness [5]. We will return to a description within this framework at a later date. We believe that further studies of our model and framework will lead to a much deeper understanding of the phenomena of bloat, as well as help in the design of better genetic operators and therefore more competent EAs. A further motivation is that of [3] - to understand the origins and evolution of language.

References

1. Banzhaf, W.: Genotype-Phenotype Mapping and Neutral Variation - A case study in Genetic Programming. In: Davidor et al., (Eds.): Proceedings of the third conference on Parallel Problem Solving from Nature. Lecture Notes in Computer Science, Vol. 866. Springer-Verlag. (1994) 322–332
2. O'Neill, M. and Ryan, C.: Grammatical Evolution - Evolving programs in an arbitrary language. Kluwer Academic Publishers. (2003)
3. Angeles, O., Stephens, C.R., Waelbroeck, H.: Emergence of Algorithmic Language in Genetic Systems. BioSystems **47**. (1998) 129-147
4. Van Nimwegen, E., Crutchfield, J.P., and Huynen, M.: Neutral Evolution of Mutational Robustness. Proc. Natl. Acad. Sci. USA 96. (1996) 9716-9720
5. Stephens, C.R. and Mora, J.: Effective Fitness as an Alternative Paradigm for Evolutionary Computation. Gen. Prog. Evol. Hardware 2. (2000) 7-32.
6. Keller, R. and Banzhaf, W. : Genetic Programming using Genotype-Phenotype Mapping from Linear Genomes into Linear Phenotypes. In: Genetic Programming 1996: Proceedings of the First Annual Conference. MIT Press. (1996)
7. Wong, M. and Leung, K. Inductive Logic Programming Using Genetic Algorithms. I.I.A.S. (1994)

Author Index

Lecture Notes in Computer Science

For information about Vols. 1–3340

please contact your bookseller or Springer

Vol. 3389: P. Van Roy (Ed.), Multiparadigm Programming in Mozart/OZ. XV, 329 pages. 2005.

Vol. 3388: J. Lagergren (Ed.), Comparative Genomics. VII, 133 pages. 2005. (Subseries LNBI).

Vol. 3387: J. Cardoso, A. Sheth (Eds.), Semantic Web Services and Web Process Composition. VIII, 147 pages. 2005.

Vol. 3386: S. Vaudenay (Ed.), Public Key Cryptography - PKC 2005. IX, 436 pages. 2005.

Vol. 3385: R. Cousot (Ed.), Verification, Model Checking, and Abstract Interpretation. XII, 483 pages. 2005.

Vol. 3383: J. Pach (Ed.), Graph Drawing. XII, 536 pages. 2005.

Vol. 3382: J. Odell, P. Giorgini, J.P. Müller (Eds.), Agent-Oriented Software Engineering V. X, 239 pages. 2005.

Vol. 3381: P. Vojtáš, M. Bieliková, B. Charron-Bost, O. Sýkora (Eds.), SOFSEM 2005: Theory and Practice of Computer Science. XV, 448 pages. 2005.

Vol. 3380: C. Priami, Transactions on Computational Systems Biology I. IX, 111 pages. 2005. (Subseries LNBI).

Vol. 3379: M. Hemmje, C. Niederee, T. Risse (Eds.), From Integrated Publication and Information Systems to Information and Knowledge Environments. XXIV, 321 pages. 2005.

Vol. 3378: J. Kilian (Ed.), Theory of Cryptography. XII, 621 pages. 2005.

Vol. 3377: B. Goethals, A. Siebes (Eds.), Knowledge Discovery in Inductive Databases. VII, 190 pages. 2005.

Vol. 3376: A. Menezes (Ed.), Topics in Cryptology – CT-RSA 2005. X, 385 pages. 2005.

Vol. 3375: M.A. Marsan, G. Bianchi, M. Listanti, M. Meo (Eds.), Quality of Service in Multiservice IP Networks. XIII, 656 pages. 2005.

Vol. 3374: D. Weyns, H.V.D. Parunak, F. Michel (Eds.), Environments for Multi-Agent Systems. X, 279 pages. 2005. (Subseries LNAI).

Vol. 3372: C. Bussler, V. Tannen, I. Fundulaki (Eds.), Semantic Web and Databases. X, 227 pages. 2005.

Vol. 3371: M.W. Barley, N. Kasabov (Eds.), Intelligent Agents and Multi-Agent Systems. X, 329 pages. 2005. (Subseries LNAI).

Vol. 3370: A. Konagaya, K. Satou (Eds.), Grid Computing in Life Science. X, 188 pages. 2005. (Subseries LNBI).

Vol. 3369: V.R. Benjamins, P. Casanovas, J. Breuker, A. Gangemi (Eds.), Law and the Semantic Web. XII, 249 pages. 2005. (Subseries LNAI).

Vol. 3368: L. Paletta, J.K. Tsotsos, E. Rome, G.W. Humphreys (Eds.), Attention and Performance in Computational Vision. VIII, 231 pages. 2005.

Vol. 3367: W.S. Ng, B.C. Ooi, A. Ouksel, C. Sartori (Eds.), Databases, Information Systems, and Peer-to-Peer Computing. X, 231 pages. 2005.

Vol. 3366: I. Rahwan, P. Moraitis, C. Reed (Eds.), Argumentation in Multi-Agent Systems. XII, 263 pages. 2005. (Subseries LNAI).

Vol. 3365: G. Mauri, G. Păun, M.J. Pérez-Jiménez, G. Rozenberg, A. Salomaa (Eds.), Membrane Computing. IX, 415 pages. 2005.

Vol. 3363: T. Eiter, L. Libkin (Eds.), Database Theory - ICDT 2005. XI, 413 pages. 2004.

Vol. 3362: G. Barthe, L. Burdy, M. Huisman, J.-L. Lanet, T. Muntean (Eds.), Construction and Analysis of Safe, Secure, and Interoperable Smart Devices. IX, 257 pages. 2005.

Vol. 3361: S. Bengio, H. Bourlard (Eds.), Machine Learning for Multimodal Interaction. XII, 362 pages. 2005.

Vol. 3360: S. Spaccapietra, E. Bertino, S. Jajodia, R. King, D. McLeod, M.E. Orlowska, L. Strous (Eds.), Journal on Data Semantics II. XI, 223 pages. 2005.

Vol. 3359: G. Grieser, Y. Tanaka (Eds.), Intuitive Human Interfaces for Organizing and Accessing Intellectual Assets. XIV, 257 pages. 2005. (Subseries LNAI).

Vol. 3358: J. Cao, L.T. Yang, M. Guo, F. Lau (Eds.), Parallel and Distributed Processing and Applications. XXIV, 1058 pages. 2004.

Vol. 3357: H. Handschuh, M.A. Hasan (Eds.), Selected Areas in Cryptography. XI, 354 pages. 2004.

Vol. 3356: G. Das, V.P. Gulati (Eds.), Intelligent Information Technology. XII, 428 pages. 2004.

Vol. 3355: R. Murray-Smith, R. Shorten (Eds.), Switching and Learning in Feedback Systems. X, 343 pages. 2005.

Vol. 3354: M. Margenstern (Ed.), Machines, Computations, and Universality. VIII, 329 pages. 2005.

Vol. 3353: J. Hromkovič, M. Nagl, B. Westfechtel (Eds.), Graph-Theoretic Concepts in Computer Science. XI, 404 pages. 2004.

Vol. 3352: C. Blundo, S. Cimato (Eds.), Security in Communication Networks. XI, 381 pages. 2005.

Vol. 3351: G. Persiano, R. Solis-Oba (Eds.), Approximation and Online Algorithms. VIII, 295 pages. 2005.

Vol. 3350: M. Hermenegildo, D. Cabeza (Eds.), Practical Aspects of Declarative Languages. VIII, 269 pages. 2005.

Vol. 3349: B.M. Chapman (Ed.), Shared Memory Parallel Programming with Open MP. X, 149 pages. 2005.

Vol. 3348: A. Canteaut, K. Viswanathan (Eds.), Progress in Cryptology - INDOCRYPT 2004. XIV, 431 pages. 2004.

Vol. 3347: R.K. Ghosh, H. Mohanty (Eds.), Distributed Computing and Internet Technology. XX, 472 pages. 2004.

Vol. 3346: R.H. Bordini, M. Dastani, J. Dix, A.E.F. Seghrouchni (Eds.), Programming Multi-Agent Systems. XIV, 249 pages. 2005. (Subseries LNAI).

Vol. 3345: Y. Cai (Ed.), Ambient Intelligence for Scientific Discovery. XII, 311 pages. 2005. (Subseries LNAI).

Vol. 3344: J. Malenfant, B.M. Østvold (Eds.), Object-Oriented Technology. ECOOP 2004 Workshop Reader. VIII, 215 pages. 2005.

Vol. 3343: C. Freksa, M. Knauff, B. Krieg-Brückner, B. Nebel, T. Barkowsky (Eds.), Spatial Cognition IV. Reasoning, Action, and Interaction. XIII, 519 pages. 2005. (Subseries LNAI).

Vol. 3342: E. Şahin, W.M. Spears (Eds.), Swarm Robotics. IX, 175 pages. 2005.

Vol. 3341: R. Fleischer, G. Trippen (Eds.), Algorithms and Computation. XVII, 935 pages. 2004.